O9-AIC-653

THE WORLD'S MISSILE AND SPACE PROGRAMS, 1920–1995

	Germany and Europe	United States Military and CIA	United States Civilian and NASA	Soviet Union
1920	—	—	— Goddard—*A Method of Reaching Extreme Altitudes*	—
	Oberth—*Die Rakete*			
1925	—	—	—	—
			Goddard—first liquid-fueled rocket	
	Film—*Frau im Mond*		Guggenheim funds	Glushko—liquid engi
1930	— Nebel—Raketenflugplatz	—	— Goddard	—
	First liquid rockets Army supports von Braun			Red Army funds MosGIRD
			Pendray's liquid rockets American Rocket Society	First liquid-rocket te flights
	First Army rockets			
1935	—	—	—	—
	First plans for A-4 Peenemunde		Von Karman—rocket research at Caltech	Tukhachevsky execut Arrest of Korolev
		Air Corps funds Caltech		
1940	—	—	—	—
	First flight of V-2	Reaction Motors, Inc. Aerojet Jet Propulsion Laboratory		
	V-2 operational			
1945	— Defeat of Germany	—	—	— Ustinov's missile program
		Bollay at North American Aviation V-2 at White Sands Proving Ground		V-2 at Kapustin Yar
		Santa Susana engine tests		Atomic bomb
1950	—	— Cape Canaveral	—	—
		Hydrogen bomb Navaho engines for ICBM Tea Pot report; Atlas	*Collier's* on spaceflight	Glushko's ICBM engi Stalin approves ICBM Korolev's R-7 progra
1955	—	— ICBM program	— Vanguard program	— Hydrogen bomb

	France and Europe	United States Military and CIA	United States Civilian and NASA	Soviet Union and Russia
55	—	— Thor, Titan, Jupiter, WS-117L, Polaris programs	—	—
				First ICBM; first sputniks
		Minuteman, Corona programs	NASA organized	
			First plans for Apollo	Luna 3 to lunar farside
60	—	— Discoverer satellites	— Tiros, Echo satellites	— Nedelin's disaster
		"Missile gap" disproved	JFK commits to the moon	Yuri Gagarin in orbit
	ELDO and ESRO formed	Cuban missile crisis		
		First flight of Gambit close-look satellite	Project Mercury flights; Telstar satellite	Valentina Tereshkova
				First three-man craft
			Ranger lunar impact	
65	— France launches satellite	— First flight of Titan III	— Gemini two-man flights; Early Bird satellite	— First spacewalk; first flight of Proton
			Three die in Apollo fire	Komarov dies in Soyuz 1
	Kourou, French Guiana launch center		First flight of Saturn V	Zond lunar missions
			Astronauts land on moon	N-1 moon rocket blows up
70	—	— Manned Orbiting Laboratory canceled	— Intelsat system is global	—
			Mariner 9 orbits Mars	
		First flight of KH-9 (Big Bird)	Nixon approves space shuttle	Last launch of N-1
			Skylab space station	Salyut 3 military station
75	— European Space Agency; Ariane program	—	—	— Salyut 4 civilian station
		First flight of KH-11	Viking craft on Mars	
				Salyut 6 for long-duration flights
			Voyager probes reach Jupiter and Saturn	
80	— First flight of Ariane	—	—	—
			First flight of space shuttle	
			Reagan approves space station	
85	—	— Air Force orders Titan IV	—	—
			Challenger disaster	Mir, permanently inhabited
		Air Force orders Delta and Atlas		
				Flight of Energiya
				Flight of Buran
90	— Ariane dominates commercial market	—	— Magellan orbits Venus	—
				Fall of Communism
			Bush-Yeltsin agreement	
			Gore-Chernomyrdin pact for a joint space station	Lockheed markets Proton
		Global Positioning System is operational		
95	—	—	—	— Russian engines for Atlas

Countdown

Other Books by T. A. Heppenheimer

Colonies in Space
Toward Distant Suns
The Real Future
The Man-Made Sun
The Coming Quake
Turbulent Skies

Countdown

A History of Space Flight

T. A. Heppenheimer

John Wiley & Sons, Inc.
New York • Chichester • Weinheim • Brisbane • Singapore • Toronto

This text is printed on acid-free paper.

Published by John Wiley & Sons, Inc.

Library of Congress Cataloging-in-Publication Data:

Heppenheimer, T. A.
Countdown: A history of space flight / T. A. Heppenheimer.
p. cm.
Includes bibliographical references and index.
ISBN 0-471-14439-8 (cloth)
1. Space race—History. 2. Astronautics and state—United States—History. 3. Astronautics and state—Soviet Union—History. 4. Cold War—History. I. Title.
TL789.8.U5H49 1997
387.8'09—dc20 96-28245

Printed in United States of America

10 9 8 7 6 5 4 3 2 1

To Phyllis LaVietes, mother of my children

Contents

Acknowledgments

Of the people who have helped make this book possible, the historian Peter Gorin stands out. He steered me to source material and prepared translations from the Russian. He generously shared his considerable knowledge of the Soviet and Russian space programs and read drafts of this book's manuscript with care, noting errors. The historian Asif Siddiqi merits similar note, for he too reviewed this manuscript while sending me his own material. In addition, Theodore Rockwell, a specialist in naval nuclear power and the author of *The Rickover Effect,* has reviewed portions of chapter 7. Another specialist, James Harford, has shared the manuscript of his forthcoming book, *Korolev.*

Other people gave me books and papers, some of them unpublished: Evgeniia Albats, J. Leland Atwood, Anita Gale, Aleksandr Gurshtein, John Logsdon, John R. Moore, Dale Myers, Jim Oberg, Marcia Smith, George Sutton, and Charles Vick. I received similar support from archivists: Lee Saegesser at NASA, Michael Baker and Nancy Stilson at Redstone Arsenal, and Ann Warren at the Central Intelligence Agency.

Public-affairs people at several corporations sent me material and set up interviews. Joyce Lincoln, now retired from Rocketdyne, stands out for her good help over a number of years. Others include Michelle Lyle of Arianespace, Evelyn Smith at McDonnell Douglas, and Montye Male and Dan McClain at TRW.

Others have helped with artwork and photos: Don Dixon, a colleague of long standing; Christine Kaske of the National Air and Space Museum; Hugh Morgan at the Air Force Museum; Dan Gauthier; and Charles Vick. For editorial and production work, there are Robert Tabian, literary agent; Hana Lane and John Cook, editors at John Wiley; and Phyllis LaVietes, secretary.

This book draws on several decades of personal interest and on articles I have written for magazines. It is a pleasure to acknowledge my editors: George Larson, Linda Shiner, and Diane Tedeschi at *Air & Space;*

Richard Snow and Fred Allen at *American Heritage;* and Scott DeGarmo at *Science Digest.*

While researching those articles, as well as the present book, a number of people have been kind enough to grant interviews:

Air Force: Colonel Edward Hall, Jack Neufeld, General Bernard Schriever, Charles Terhune

Douglas Aircraft: C. J. Dorrenbacher, Jim Gunkel, Bob Johnson

Jet Propulsion Laboratory: Al Hibbs, Dan Schneiderman

Lockheed: Max Hunter, Ben Rich

MIT: Richard Battin, Hal Laning

North American Aviation: J. Leland Atwood, Scott Crossfield, John R. Moore, Dale Myers, Ed Redding, Harrison Storms

NASA: Joseph Allen, Max Faget, Robert Freitag, Hans Mark, Jesco von Puttkamer, David Scott

Reaction Motors: Bob Holder, Peter Palen, Bob Seaman

Redstone Arsenal: Konrad Dannenberg, Karl Heimburg, Lieutenant Colonel Lee James, William Lucas, Ernst Stuhlinger, General John Zierdt

Rocketdyne: David Aldrich, Jim Broadston, Paul Castenholz, Tom Dixon, Bill Ezell, Sam Hoffman, Dom Sanchini

TRW: Richard Booton, Allen Donovan, George Gleghorn, Frank Lehan, Ruben Mettler, George Mueller, George Solomon, Dolph Thiel

White House staff: George Keyworth, Jim Muncy, Doug Pewitt, Victor Reis, Colonel Gilbert Rye

Others: Harold Agnew, Eugene Bollay, Mrs. Jeanne Bollay, Wolf Demisch, Philip Klass, Admiral Kinnaird McKee, Jonathan McDowell, John Pike, Robert Truax

My thanks also go to longtime friends, notably Don Dixon, Anita Gale, Jim Oberg, and Dave Ross, for useful and timely discussions.

T. A. Heppenheimer

Fountain Valley, California
September 18, 1996

Introduction

AMID THE FLAT TIDELANDS southeast of Houston, at the Johnson Space Center of the National Aeronautics and Space Administration, one sees a powerful and melancholy sight: a full-size moon rocket, set on its side for public display. It is not a mockup or a replica. It once was capable of flight, with five first-stage engines each nineteen feet tall, with a cavernous second stage that would reach for orbit, and a third stage ready to carry a lunar expedition. Today owls nest in it.

In the National Air and Space Museum in Washington, D.C., one finds a similar monument. It is Skylab 2, a duplicate of a full-size space station that reached orbit and housed nine astronauts during 1973 and 1974. This too is no mockup, but NASA could not find reason to justify launching it. It could have flown atop the very rocket that stands in Houston as a lawn ornament, but instead, NASA donated it to the Smithsonian. Today children wander through it.

Why did these things happen? Why, at the outset, did we go to the moon? Why did NASA build these magnificent items of equipment, only to abandon them? *Countdown* addresses these issues and then goes further. It presents an overview of the major features of the world's other principal space programs, including those of the Soviet Union, Europe, and America's Central Intelligence Agency.

The time is right for such an overview, for the end of the Cold War and the opening-up of Russia now present the aerospace historian with a trove of new information. These disclosures make it possible to present the CIA as a leader that stood in the forefront of America's space program. In addition, one can now discuss Soviet activities on the same basis as those of NASA, and to the same degree of completeness.

This book presents the rise of the Soviet space program by tracing the experiences of its founder and leader, Sergei Korolev. His name is little known in the United States; as recently as 1979, Tom Wolfe, in *The Right Stuff,* referred to him only as "the Chief Designer." Yet among the space leaders of history, Korolev stands alongside Wernher von Braun. He did more than shape his own country's program: When President Kennedy approved the Apollo program, with its goal of landing astronauts on the moon, he was responding to Korolev's challenge.

Countdown also breaks new ground by emphasizing the significance of satellite reconnaissance. America's first true space projects addressed this goal, as early as 1955. They received overriding priority; the early civilian satellite project, Vanguard, was subordinated to serve the interests of the Air Force and the CIA. Vanguard's low priority then allowed Moscow to be first into space.

But reconnaissance satellites repaid their high priority as early as 1961, by giving a clear view of Russia's missile program. The need for powerful rockets, to launch such spacecraft, has driven the pace of America's development of space boosters from the 1950s to the present day. And in the Soviet Union, an early reconnaissance-satellite program provided the Vostok spacecraft that carried Yuri Gagarin to orbit as the first man in space.

This book also presents the Soviet manned moon effort that paralleled this country's effort with Apollo. Many people now appreciate that this race grew out of a cynical ploy by Premier Nikita Khrushchev, in which he funneled his limited resources into the narrow field of space flight, hoping thereby to create an illusion of great technical strength. We see that this ploy nearly succeeded, as the Russians came close to beating America to the moon. The Soviets flew a successful lunar mission in 1968, leaving out only the cosmonauts. It took a last-minute surge by NASA to win this race.

This book also seeks historical perspective. In addressing the question of why we went to the moon, it views Apollo not as a drive toward a new human future, but as a high-water mark of a particular time. This era featured far-reaching federal activism: We would make war on poverty and carry out sweeping social change. Significantly, that era also looked to Washington as a source of the most promising new technologies. During the 1960s, these included interstate highways, jet airliners, and nuclear power plants.

But there was a darker side to Kennedy's commitment, for his Democratic Party carried the burden of having held power when China fell to the Communists in 1949. He knew that further losses in the Third World were unacceptable, and he saw Soviet spaceflights as powerful propaganda that might sway shaky regimes. A major rationale for Apollo then lay in its use as counter-propaganda that might discourage Third World leaders from looking to Moscow. A similar rationale lay behind the Vietnam War. It is not surprising that America's twin commitments, to Vietnam and to Apollo, rose and fell in tandem.

Another major theme presents the increasing significance of electronics. Modern circuitry makes it possible to build spacecraft so capable that modest numbers of them suffice to provide complete worldwide services in a number of important fields: communications, cable TV, weather observation, navigation, and air traffic control. Electronics has also fulfilled the classic hope of exploring the planets.

By contrast, NASA has continued to emphasize manned flight. This book argues that while the leaders of NASA view it as a high-tech agency, in this matter it has been serving goals that date back to a much earlier time. This is particularly true in the matter of space stations.

The concept of such an orbiting station originated in 1923 and was developed at length by 1929, in an era when the only serious electronics was radio. This concept proposed that astronauts in space would perform a host of useful activities: operating telephone switchboards, observing military bases, watching weather patterns. It then made sense that all these people should live in a single space facility, an orbiting station that would provide for their comfort.

But as early as the 1950s it was clear that through electronics, these tasks would fall to unmanned spacecraft. Even so, the space-station concept could beguile; one remembers the great rotating wheel of Stanley Kubrick's *2001: A Space Odyssey.* However, its advocates have never found much for it to do. This raises interesting questions as to whether NASA has been reinventing the past and calling it the future. For since 1984 this agency has pursued, as its next major project . . . a space station.

This book argues that orbiting stations have a major flaw: They are far too costly for the services they can provide. This criticism holds for the space shuttle as well. Hence, manned flight cannot expect to follow unmanned spacecraft into the commercial world, where willing buyers turn to space missions because they provide valuable returns. Instead, the shuttle and space station will remain a sport of governments, pursued for political advantage.

Yet this advantage may unfold against an important recent trend, wherein the space programs of the United States and Russia have been merging into a single global enterprise. Already the firm of Lockheed Martin is marketing a Russian launch vehicle, the Proton, while a new version of its Atlas rocket features Russian engines. In addition, the two nations are pursuing the space station as a joint effort. This venture can dramatize a central fact of our time: The Cold War has given way to a new era of international cooperation. Today, this makes vivid the hope that the coming century can truly be one of peace.

1

Wonder-Weapons and Prison Camps

Rocketry under Stalin and Hitler

It was noon on a bright-blue autumn day in early October of the war year 1942. At Peenemunde, a missile center on Germany's Baltic coast, a rocket stood ready for launch. It looked like a classic spaceship: slim and tall with a body that tapered to a point, fins of elegant shape fitted to its tail. A band of white frost surrounded its liquid-oxygen tank, partly hiding the bold black-and-white painted pattern that divided its surface into quadrants. Electrical cables ran from an adjacent pole to a plug near its nose. Otherwise, it stood alone and in the open.

Suddenly a cloud emerged from its nozzle, laced with sparks from an igniter. The sparks rapidly gave way to a flame, then to a reddish-yellow exhaust. Smoke gathered near the rocket's base. The cables detached and fell downward. Quickly the flame became brilliantly yellow-white, clean in appearance, rivaling the sun in color and intensity. With the sun itself shining brightly, the rocket rose and began to accelerate rapidly upward.

Ernst Stuhlinger, a specialist in rocket guidance, would view such launches from as close as five hundred feet. "We felt the heat of the rocket on our faces as soon as the flames developed," he recalls. "The sound was tremendous. We didn't hear with the ears only, but with the whole body. It was as if the entire skin was an eardrum vibrating. The rocket rose, very evenly. It moved as if nothing else on earth existed. It gained speed, while the flame grew longer. It went up straight; then it began to tilt over, again very evenly. It went out over the Baltic, still as if no other motion existed."[1]

Then came a critical moment as the rocket powered through the speed of sound. It did not falter; it continued to fly true. Suddenly, a shock! A long white streak began to form behind the missile, standing out sharply against the blue sky. An explosion? No, a vapor trail, forming naturally as its exhaust condensed within the cold upper air. Powerful winds, pervasive at high altitude, quickly seized this trail and blew it into a zigzag shape.

Coming up on one minute after launch . . . Through binoculars, the rocket, now some twenty miles away, showed a tiny reddish dot where its nose had heated through atmospheric friction. Ground-control equipment sent a cutoff signal; the flame vanished, along with the vapor trail. Still the rocket showed a dazzling bright dot at its tail, as the nozzle continued to shine with white heat.

Walter Dornberger, the project manager, was watching along with Colonel Leo Zanssen, head of the Peenemunde center. "Taking a deep breath, I put down my binoculars," Dornberger later wrote. "My heart was beating wildly. I wept with joy. I couldn't speak for a moment; my emotion was too great. I could see that Colonel Zanssen was in the same state. His eyes were moist. He stretched out his hands to me. I grasped them. Then our emotions ran away with us. We yelled and embraced each other like excited boys."

Later that day, Dornberger gave a short speech to some of his colleagues. "We are the first to have given a rocket a speed of thirty-three hundred miles per hour," he declared. "We have thus proved that it is quite possible to build piloted missiles or aircraft to fly at supersonic speeds. We did it with automatic control. Our rocket today reached a height of nearly sixty miles.

"We have invaded space with our rocket; we have proved rocket propulsion practical for space travel. So long as the war lasts, our most urgent task can only be the rapid perfecting of the rocket as a weapon. The development of possibilities we cannot yet envision will be a peacetime task. Then the first thing will be to find a safe means of landing after the journey through space."[2]

For that first launch, a number of the people who would lead these developments were present or otherwise deeply involved in the project. These included Dornberger's technical director, Wernher von Braun. But Sergei Korolev, the man who would set the pace in pursuing those possibilities, was nowhere in Germany. He was working some 1,500 miles to the east of the launch site. He was a prisoner of the NKVD, Stalin's secret police.

For Korolev, as well as for Dornberger and von Braun, the paths leading to their respective situations had a common origin in the work of a German-speaking mathematics teacher, Hermann Oberth. Oberth had studied in Munich and Heidelberg but had then returned to his hometown in Romania, where he lived amid a community of expatriate Germans and taught math in the local *Gymnasium*. After World War I, he developed a strong fascination with rockets.

People had been building them since the invention of gunpowder, some seven hundred years earlier. The best of them had seen use in nineteenth-century warfare, providing "the rockets' red glare" that had inspired Francis Scott Key. However, the subsequent development of highly capable artillery had driven rockets into the shadows. Oberth's point of departure started from a simple but powerful principle: By burning modern fuels, such as gasoline and liquid oxygen, a rocket could achieve performance far surpassing that of any blackpowder missile. To emphasize this, he made recommendations concerning specific choices of fuel, calculated their performance, and offered suggestions for the design of the rocket craft that could use them.

He then went further, proposing that this approach would make possible manned space flight. Leaping forty years into the future, he proposed that a liquid-fueled rocket weighing 400 metric tons could carry two astronauts into orbit. Drawing on the ideas of a science-fiction writer, Kurd Lasswitz, Oberth then suggested that these spacefarers could go on to build an orbiting station, which could serve for high-altitude observations and satellite communications. It also could operate as a base for other rockets, which would fly to the planets. As a final touch, Oberth proposed the building of enormous solar reflectors, "to keep the shipping lane to Spitsbergen and the North Siberian ports ice-free by concentrated sunlight [and to] make large areas in the north habitable."

Brilliantly original, Oberth's work fitted well with the spirit of the time. In recent decades, the internal-combustion engine had opened the way to aircraft and automobiles. The steam turbine, its contemporary, was spurring the rise of the electric-power industry and had brought about dramatic leaps in the size and speed of warships and passenger liners. Oberth now was pointing to another new engine, the liquid-fueled rocket, as a new goal for inventors. He also was offering his own sweeping vision of far-reaching applications.

In 1923 he wrote up his work in a book, *Die Rakete zu den Planetenraumen (The Rocket into Interplanetary Space)*, and had it published largely at his own expense. It was little more than a pamphlet—paper-covered and with less than a hundred pages—but it quickly began to draw the attention Oberth wanted. Among the places where it won this attention was Moscow, where some were aware that a countryman, Konstantin Tsiolkovsky, had nurtured similar ideas.

Like Oberth, Tsiolkovsky was a provincial math teacher. He had worked in the town of Kaluga, a hundred miles from the capital. Born in 1857, he developed a strong early interest in astronomy. He later tried to invent an airship, pursuing a line of research that led him to build Russia's first wind tunnel. Then, after 1895, he combined these two fields of interest and set out to design a spaceship.

He had a good background in physics, which enabled him to do a creditable job. He decided that such a craft would need rocket propulsion; he wrote of using fuels similar to kerosene, but found that he could get even better performance by using liquid oxygen and liquid hydrogen. Here was a bold step indeed; Britain's Sir James Dewar was only then preparing the world's first samples of liquid hydrogen, and no rocket would fly with that propellant until after 1960. Even so, Tsiolkovsky's work also pointed clearly to a far-reaching realm of possibility.

He then set out to publish his findings. During his airship days he had received valuable support from Dmitri Mendeleyev, who had discovered the periodic table in chemistry. But that didn't help him now. In 1898 he sent off his paper to a professional journal, *Science Survey,* whose editor sat on it for five years before getting around to publishing it. Nor did it take the world by storm; the journal had little readership, even within Russia. Undismayed, Tsiolkovsky continued his research, and shortly before World War I he published two other papers in a more widely read journal. These won the attention of one of the country's top science writers, Yakov Perelman. He proceeded to make Tsiolkovsky's name known, at least within Russia.

Then came Oberth's book of 1923. Officials in the new Bolshevik government, eager to claim that their country was at the forefront, dusted off Tsiolkovsky's 1898 paper and reprinted it along with a lament, "Do we always have to get from foreigners what originated in our boundless homeland and died in loneliness from neglect?" Certainly his work would receive no further neglect, for within Moscow and other cities, rocket societies would soon take form.

Leading the way was Fridrikh Tsander, a Latvian engineer and an early Tsiolkovsky devotee. In 1924 he launched a program of public lectures on spaceflight, which quickly led to the formation of an astronautics club, the Society for Study of Interplanetary Communication. (The word *astronautics* would come from France, entering the language in 1930.) Tsander's group proved to be little more than a gathering of enthusiasts, but a follow-on organization, the Interplanetary Section of the Association of Inventors, went further by sponsoring a public exhibition in Moscow during 1927. It amounted to an early version of a Star Trek convention, featuring works of Jules Verne and H. G. Wells rather than Gene Roddenberry, but it helped to spread the word.

Into this spirited group of amateurs came Sergei Korolev, born in the Ukraine. Aviation had captivated him at an early age, for he had found in it a goal and purpose that offered direction in a life that was often joyless and lonely. His mother had entered into an arranged marriage while still in her teens, and the marriage did not last, dissolving before she was twenty. She sent her baby to live with her parents, and she did not reclaim him until she remarried. Sergei's father died while he still was very young, and it was fortunate that his mother's new husband adopted him as a stepson.

Attending high school in Odessa, he joined a local aviation club. He participated eagerly, designing a glider that flew successfully. He learned math and aeronautics through self-study, but found no path to a university. Instead, he entered the building trades. Then, when he did succeed in entering college, it was in Kiev, though he knew that Moscow offered much more. But he found his opportunity in 1926, at age nineteen, as his mother and stepfather moved to Moscow. He left Kiev to rejoin them and won admission to one of the country's leading institutes, the Bauman Higher Technical School in Moscow. Here he would come into his own.

He worked in an aircraft factory by day, taking classes at night. He also learned to fly and won a pilot license. His native talent won the attention of Andrei Tupolev, one of the Soviet Union's top aeronautical designers, and Korolev went on to design and build another airplane as his student. He also worked with another leading designer, Sergei Ilyushin, creating a sailplane that stayed aloft for over four hours and won a nationwide competition.

In addition, Korolev gained a deep and enduring fascination with space flight. He had arrived in Moscow in time for the 1927 exhibition. In time he met with Tsiolkovsky himself, again in Moscow. Following his graduation from the Technical School, in 1930, he fell in with Tsander, who now was hoping to build liquid-rocket engines. Tsander was organizing a new society of young rocketeers, the Moscow Group for Study of Reaction Motion (MosGIRD). Korolev proposed to install a Tsander engine in a glider. He won modest financial support from Osoaviakhim, a government organization that worked to encourage activity in aviation.

The group set up shop in a former wine cellar, a basement that was dark and damp but spacious, and went to work. However, they didn't get very far. The level of support was modest, indeed; members of MosGIRD complained that their acronym really stood for a different Russian phrase, which translates as Moscow Group of Engineers Working for Nothing. Nor did it help that their principal equipment consisted of two obsolete machine tools. Nevertheless, Korolev would not merely take what he could get. If Osoaviakhim couldn't support MosGIRD in the manner to which he hoped to become accustomed, he would try for better things with the Red Army.

That army's chief of armaments, Mikhail Tukhachevsky, had a strong interest in rockets. He saw them as a means of overcoming the limits of heavy guns, by launching very large charges of explosive to any distance. In addition, at a time before the invention of the jet engine, he expected that rockets would lift "the ceiling of aircraft to the stratosphere," offering unprecedented performance. Tukhachevsky was widely viewed as having the finest mind in the Army, along with strong willpower. He also enjoyed high status; in 1935, following a military reform, he would receive the rank of marshal, one of only five generals to do so. And he was already funding a rocket-research group in Leningrad, the Gas Dynamics Laboratory.

The GDL had grown out of an earlier group that had reached beyond the traditional blackpowder rocket by working with smokeless powder, which offered more energy. This research center operated as an arm of the Army's Ordnance Department, with a rocket test stand at Leningrad's principal fort, and with an artillery officer as director. In 1929 a staff engineer, Valentin Glushko, had broadened GDL's program by initiating a liquid-rocket program. He went on to build the first such engine to fire successfully, producing forty-four pounds of thrust by burning gasoline with liquid oxygen.

Because Osoaviakhim had close links to the military, Tukhachevsky learned of MosGIRD quite early. At first he contented himself with endorsing the support of Osoaviakhim, but soon he decided that MosGIRD indeed needed more. Working directly with Korolev, Tukhachevsky arranged to fund MosGIRD directly, and to carry out a merger that would unite MosGIRD and GDL into a single strong organization. By August 1932, the Army's Department of Military Inventions was sending initial funds to MosGIRD. In addition, Korolev won the use of a rocket launch site on a nearby army base.

During the following year, as MosGIRD continued to work in its wine cellar, its members faced mishaps that involved more than the usual problems of rocket research. On one occasion, their truck overturned in wet snow. Another time, while carrying a rocket to the launch site, this truck struck a loose cobblestone and nearly overturned again. Finally, in August 1933, a rocket called "09" made its first flight. It reached 1,300 feet, but then tumbled and came down in a crash as a flange burned through. Still, Korolev was pleased. "She really flew," he declared. "We didn't do all that work for nothing."

Three months later came a similar partial success as a new rocket, GIRD-X, rose 250 feet and then turned sharply, falling to the ground 500 feet away. By then, MosGIRD and GDL had completed their merger into a new organization, the Scientific Research Institute of Reaction Propulsion (RNII). A former chief of the GDL, Ivan Kleimenov, became its head, with Korolev as his deputy, at least at first. The MosGIRD group moved out of their dank cellar and into a former diesel-engine plant on the outskirts of Moscow. It featured engine test cells that soon would see use in testing rocket motors. And Korolev, who had been a civilian, now joined the Red Army and obtained the rank of a commissioned officer.

Only one other country was building liquid-fueled rockets with military support: Germany. Indeed, that country's activities had developed along much the same lines as those of the Soviet Union, with the prophetic writings of Hermann Oberth sparking a wave of interest in rockets and space flight, a wave that led first to active experimentation and then to direct military sponsorship. However, in Germany a particular spur to rocket development came from a science-fiction movie, *Frau im Mond (The Girl in the Moon).*

Its producer, Fritz Lang, was more than a filmmaker; he was a leader of his country's art and culture. The science writer Willy Ley would note that at one of his premieres, "it was an unwritten but rigid rule that one had to wear full evening dress, not just a dinner jacket. The audience comprised literally everybody of importance in the realm of arts and letters, with a heavy sprinkling of high government officials." In 1926 Lang had released the classic *Metropolis,* in which the leading role was that of a robot. Now he would do the same for space flight.

His wife, the actress Thea von Harbou, wrote the script, drawing heavily on Oberth's writings. Lang hired Oberth himself as a technical consultant, and Oberth then convinced him to underwrite the building of a rocket, a real one. After all, there would be splendid publicity if such a rocket were to fly on the day of the premiere. Lang staked Oberth to a modest sum, which allowed him to hire the beginnings of a technical staff.

An early interviewee was a hard-faced man who dressed meticulously and showed a precise military bearing. "Name is Rudolf Nebel," he declared. "Engineer with diploma, member of oldest Bavarian student corps, World War combat pilot with rank of lieutenant and eleven enemy planes to my credit." He indeed had decent engineering credentials, though he had never worked as a designer. Instead he had worked as a salesman, selling ball bearings and then burglar alarms. Even so, Oberth hired him on the spot.

Frau im Mond would win lasting appreciation among rocket buffs; when Dornberger's rocket flew successfully in 1942, it displayed the movie's logo, showing a young woman sitting on a crescent moon. In addition, as a dramatic device, Lang introduced the countdown: "Five, four, three, two, one—FIRE!" But the film itself did rather poorly at the box office, and the reason was ironic: It hadn't kept up with technology. It was a silent film, and by the time of its release, late in 1929, most German towns were showing the new talkies.

The rocket project also flopped. It had attempted to build a liquid-fueled rocket that would climb to twenty-five miles, but while Nebel managed to build small engines for testing, he built nothing that even attempted to fly. Still, the effort resulted in the construction of bits and pieces of a liquid-rocket project, if only a small one. After Lang's support ran out, Nebel decided to continue on his own.

Oberth had headed a group of rocket buffs, the *Verein für Raumschiffahrt* (Society for Space Travel). It now provided a legally registered office through which Nebel could do business. But he needed money if he was to get anywhere, and his search for it led him to Colonel Karl Becker, chief of the Army Weapons Board. Like Tukhachevsky, Becker knew his artillery. And like his Soviet counterpart, Becker was nurturing his own thoughts on military rockets.

Within a hidebound Prussian officer corps that continued to cherish its horses, Becker stood out for his willingness to move in new directions. He held a Ph.D. in engineering; his professor, Carl Cranz, had authored Germany's standard textbook on ballistics. Becker himself had contributed to its 1926 revision, adding a lengthy section on rockets that included a discussion of the liquid-fueled type. And while artillery had been the standard weapon of World War I, Becker envisioned that rockets might play a major role in the next war. Like his fellow officers, he awaited it with eagerness.

The Treaty of Versailles, which had ended the recent conflict, had placed stringent limits on the size of Germany's army and on its weapons. This treaty had forbidden the country to develop heavy artillery—Becker's specialty—or to produce poison gas. But it carried no prohibition against rocket research, and Becker saw this as a loophole he could exploit. Battlefield rockets might offer a new type of artillery, and in addition could deliver poison gas.

Becker was well aware that the Army needed discretion more than it needed loopholes. Its leaders held the Versailles Treaty in utter contempt, secretly flouting it whenever they could, as they actively prepared to rearm. Already the firm of Krupp, Germany's great weaponsmith, had designs in hand for the next generation of tanks. Nevertheless, simple prudence demanded that these leaders investigate legal armaments as well as the illegal ones. In 1929 the minister of defense authorized Becker to initiate a small program for the study of rockets. Its initial focus would involve commercial blackpowder types. In addition, he directed his deputy, Captain Ritter von Horstig, to take a close look at the potential of the new liquid-fueled type.

After Nebel met Becker, Army Ordnance provided a start-up grant of 5,000 marks, some $2,000. Nebel needed a test area and found it in an abandoned ammunition dump near Berlin. Part of it was swampy, but it had walls and revetments strong enough to stand up to a rocket explosion. Its buildings belonged to the Army, but with help from Becker's office, Nebel arranged to lease what he needed—and for only ten marks a year. He called it the *Raketenflugplatz*, Rocket Airport, and pronounced it open for business in September 1930.

However, Becker soon decided that Nebel was too flaky for his taste. He disliked Nebel's flair for self-promotion, while Nebel's penchant for writing sensational news articles clashed strongly with Becker's insistence on absolute secrecy. Nebel thus found himself cut off from further Army funds, at a time when he needed both equipment and skilled workers. However, he knew how to live by his wits. By drawing on his talent as a salesman, he would get both. After all, he was offering nothing less than to make *Frau im Mond* a reality.

Nebel started by visiting the State Railroad Office, where he helped himself to abandoned lumber from railroad cars. This wood patched up drafty walls and covered the floors. In similar fashion he found two old stoves for heat in winter. He bought a typewriter at a pawnshop and proceeded to pound out hundreds of letters, asking companies and government offices for help. In this manner he secured free power and light, liquid oxygen, aluminum alloy, pots of paint, a small lathe, even a drill press.

Nebel's persuasiveness extended to the tax office, which waived the excise tax on gasoline, cutting its cost for him by nearly two-thirds. In addition, Nebel knew how to turn odds and ends to his advantage. Wernher von Braun would later recall how he "talked a director of Siemens Halske A.G. out of a goodly quantity of welding wire by vividly picturing the immediacy of space travel. Our own use for such wire was extremely small, but Nebel offered it to a welding shop in exchange for the labor of a skilled welder—which we badly needed."

His workforce similarly came in impromptu fashion. Some were members of the Verein für Raumschiffahrt; among these was von Braun himself. In addition, the Raketenflugplatz offered work to capable men who had lost their jobs amid the deepening Depression. Nebel couldn't pay salaries, but he could offer rent-free camping in his buildings. For food, a nearby Siemens plant ran a welfare kitchen. Some of these people were receiving unemployment checks. All were glad for the opportunity to stay busy and to keep up their skills.

"The rocket airdrome consisted of a few starkly simple barracks and many workshops," a visitor later wrote. "The impression you took away was the frenzied devotion of Nebel's men to their work. Most of them were like officers living under military discipline. He and his staff lived like hermits. Not one of these men was married."[3] They were not just amateurs, but unemployed amateurs. Still, they went ahead with surprising speed.

By mid-May of 1931, after less than eight months, Nebel's group had a test rocket ready for flight. In that time they had built and qualified two types of engine, the more advanced of which featured a water-filled cooling jacket. The author Willy Ley later would describe how it took off "with a wild roar":

> It hit the roof of the building and raced up slantwise at an angle of about 70 degrees. After two seconds or so it began to loop the loop, spilled all the water out of the cooling jacket, and came down in a power dive. While it was diving the wall of the combustion chamber—being no longer cooled—gave way in one place, and with two jets twirling it the thing went completely crazy. It did not crash because the fuel happened to run out just as it pulled out of a power dive near the ground; actually it almost made a landing.[4]

This was Repulsor No. 1; Ley took the name from the science fiction of Kurd Lasswitz. Repulsor No. 2 was ready a few days later. As Ley later wrote, "It climbed to about 200 feet and then turned sideways, like a car taking a curve. In that position it raced across the whole place, still under full power." It flew about 2,000 feet before crashing into a tree.

With No. 3, the experimenters began to use parachutes. On the first attempt the chute deployed prematurely and deflected the rocket before being torn off. Arcing upward like a howitzer shell, it crashed into the same clump of trees where No. 2 had landed. But in August, a more advanced type "reached an altitude of about a kilometer," according to Ley. "It was only a small dot in the sky. Suddenly we saw a tiny white parachute against the blue background and the repulsor drifted gracefully back to the ground."

During subsequent months this group carried out such launches increasingly routinely, flying the rockets every few days. Meanwhile, Colonel Becker had been expanding his own activities. He had a penchant for picking close associates who shared his background, which combined an advanced academic degree with wartime artillery experience. His deputy, von Horstig, had this background in his own right. So did Walter Dornberger, who joined Becker in 1930. In December 1930 Becker secured a commitment for funding that would permit him to initiate tests of rockets at an artillery proving ground.

The limitations of field artillery were very much on Becker's mind. The biggest guns could lob one-ton shells as far as 25 miles, but these guns were very heavy, both in weight and recoil. Only battleships could mount them; even a cruiser wouldn't do. On land, the recent war had seen the 16.5-inch Big Bertha, with a range of nine miles. However, each such weapon required twelve railroad cars when disassembled and demanded twenty hours to make ready for action. At the other extreme, the famous Paris Gun of 1918 had shelled the French capital at a distance of nearly 80 miles. But each shell weighed no more than 230 pounds, of which only a small portion was explosive.

Liquid-fueled rockets promised to overcome these limits. These weapons would need no heavy steel barrel, only a light and mobile launch platform. If built like an airplane, they could be quite large. Accelerating fairly slowly, they might nevertheless reach great speed and correspondingly great range. Accordingly, soon after Dornberger came aboard, Becker gave him a goal for the future: "Develop a liquid rocket that can carry more payload than any shell we presently have in our artillery, over a distance that is farther than the maximum range of a gun."

During 1932, as work at the Raketenflugplatz continued apace, Becker decided that its efforts merited a second look. Nebel himself remained persona non grata, not only because he wouldn't do things the Army way but because he had a most annoying propensity for promising more than he could deliver. A renewal of funding thus remained out of the question. But Becker now was ready to set up his own liquid-rocket program, and to carry it out within Army facilities. Nebel's associates were right there in Berlin, and Becker could expect to hire the best of them with ease. After all, he could pay salaries and Nebel couldn't. Becker started by recruiting one of the most outstanding, Wernher von Braun.

Wernher was the son of Baron Magnus von Braun, a Prussian Junker whose roots ran deep. A von Braun had fought the Mongols in 1245; seven centuries later, the baron served the Kaiser as a high-ranking civil official. After World War I he became a banker and developed close ties to Field Marshal Paul von Hindenburg, who had commanded the wartime armies and who went on to become president of the Weimar Republic. During 1932, amid this republic's last months, the baron served as minister of agriculture. Though deeply reactionary, he was no Nazi, and his attitudes carried over to his son. Wernher would work closely with Nazis, yet would avoid becoming one of them.

Wernher's mother was an enthusiastic amateur astronomer, and her son developed an early fascination with this subject. "For my confirmation," he later declared, "I didn't get a watch and my first pair of long pants, like other Lutheran boys. I got a telescope. My mother thought it would make the best gift." Astronomy led him to space flight, which quickly seized his imagination. He bought a copy of Oberth's book—and was shocked to find it contained a hefty array of mathematical equations. He disliked math; he had even flunked it in school. But as he later would recall, "I decided that if I had to know about math to learn about space travel and rocketry, then I'd have to learn math." He did just that, learning it so well that he wound up helping to tutor fellow students in both math and physics when he went to boarding school.

Most young aristocrats viewed university study as an opportunity to pursue social contacts while gaining a patina of classical knowledge; in their subsequent careers, hired staffs would do the real work. Von Braun chose instead to study engineering, at the Technical University of Berlin. He also joined Nebel's group, participating joyously in its rocket firings. Following successful launches, he often hosted parties at a downtown nightclub. When Dornberger met him in 1932, he was struck "by the energy and shrewdness with which this tall, fair young student went to work, and by his astonishing theoretical knowledge." Unlike Nebel, von Braun proved quite willing to accommodate himself to military practices. When Becker gave him the opportunity, he joined the Army's rocket program as the top civilian specialist. He was twenty years old.

Becker proceeded to place von Braun on a fast track that a few years later would raise him to the post of technical director at Peenemunde, with overall responsibility for its long-range missiles. Secrecy was of the essence, and von Braun accepted this even while continuing to pursue his university studies. Bypassing the second half of his undergraduate curriculum, he transferred to the University of Berlin and began to work on his Ph.D. His subject was liquid rocketry. His professor, Dr. Erich Schumann, headed Becker's principal research group, and von Braun continued to work on experimental rockets, though at the Army's proving ground rather than at the Raketenflugplatz.

Graduate students often took years to earn a Ph.D. at a major German university. Von Braun received his after only eighteen months. His dissertation work was so secret that even its title was classified, but his professor awarded him high honors. The entire affair amounted to having von Braun carry out a task of military research by working as a grad student, but the effort indeed created the Army's first successful test rockets.

At the outset, in 1933, there was the Aggregat-1, or A-1. As von Braun would describe it, "it took us exactly one-half year to build—and one-half second to blow up." A major redesign was in order, which led to the A-2. Like its predecessor, it would use an engine designed by von Braun, with 650 pounds of thrust. Two A-2s flew successfully in December 1934, reaching altitudes of a mile.

By then the Nazis were in power, ready to repudiate the Versailles Treaty outright. The legality of rocket research now would offer no advantage; Becker's projected super-rocket would have to stand on its military merits. As events would shortly disclose, those merits looked rather considerable, and one can appreciate them by considering the point of view of a German militarist.

Over a century earlier, Napoleon had declared that Prussia "was hatched from a cannonball." This was even more true of the Nazi state. Hitler had written, "We will have arms again!" and he would soon show himself as a man of his word. Funds for rearmament would flow in a powerful tide, encouraging generals to dust off their wish lists and to contribute to the buildup. As the Fatherland began to form its awesome fist in preparation for further battle, Becker's fellow artillerists were at the forefront.

World War I had been a war of artillery. Its acolytes, who dominated the Army, were well aware that advances in this area had helped the cause powerfully. Big Berthas, a product of prewar development, had smashed a strong ring of Belgian forts and opened the way for the invasion of France. At sea, heavily armed battleships had slugged it out with the Royal Navy at the Battle of Jutland, forcing the British to retire and demonstrating fatal weakness in their battle cruisers, a major class of warship. Becker's proposed rocket, overcoming the limits of conventional artillery, might point the way toward even more sweeping victories.

In addition, every artillerist understood the catastrophic effect of a barrage of heavy guns. These did more than blow soldiers' bodies to pieces while mangling flesh and bone in nearby survivors; they also produced shell shock. This was a terror that went beyond fear, which came of lying in a trench under an incessant fall of high explosives, unable either to hide or escape. What might happen, then, if Becker's rockets could carry out such a barrage against Paris or London?

German forces had shelled Paris with their eponymous gun and had used Zeppelin dirigibles to drop bombs on London. The attacks had produced little effect. But the mere sight of a 98-ton Big Bertha, pulled by three dozen horses, had shocked the citizens of Liège, Belgium, into mute consternation, a consternation that increased sharply when the firing of a single shell broke most of the windows in nearby buildings. Similarly, a massed barrage of Becker's long-range rockets might actually destroy Paris's or London's will to resist.

What was more, these weapons might be obtained both quickly and inexpensively. No one yet envisioned the high cost and long development time that would characterize future weapons programs, but everyone knew of the rapid and dramatic progress of recent aviation. Becker's rocket might come forth largely as a new type of fighter plane, designed in reasonably straightforward fashion and then produced inexpensively, in overwhelming numbers.

In March 1936, only fifteen months after the successful A-2 tests, Dornberger and von Braun were ready to respond to Becker's stated goal with a specific set of requirements. They already were working on the experimental A-3, with an engine of 3,300 pounds of thrust. The next in the series, the A-4, would be the real thing. They had already decided that its engine would achieve a thrust of 25 metric tons, or 55,000 pounds. Dornberger, describing himself as "an old long-range artillerist," now proposed that the A-4 should carry a ton of high explosive to a distance of 160 miles. That would match the war load of Big Bertha and of the heaviest naval cannon, while shooting to twice the range of the Paris Gun. Six and a half years later, on that bright autumn day in 1942, the launch of this rocket would drive Dornberger to his paroxysm of joy, as it made its first successful flight.

Meanwhile, in Moscow, Sergei Korolev was pursuing his own work, with support from his own army, though with goals that were nowhere so far-reaching. He was not envisioning a rocket barrage aimed at London; instead he followed the lead of Marshal Tukhachevsky and became interested in rocket-powered aircraft. In 1934 Korolev wrote a book, *Rocket Flight in the Stratosphere,* which the Red Army published. He also continued his experimental work at the newly formed RNII, which was seeking to merge the Moscow and Leningrad rocketeers.

The merger was not working well. The former GDL had come up within the Leningrad Military District, which Tukhachevsky had commanded. By contrast, MosGIRD had its roots in Tsiolkovsky's visions of space flight. The Leningraders, a hardheaded group, tended to view them as a bunch of dreamers. Nor did it help the Muscovites that GDL was the senior partner in the merger, with its former director, Ivan Kleimenov, now heading the combined organization. Korolev, who served initially as Kleimenov's deputy, held the post for only a few months before surrendering it to another GDL artillerist, Georgi Langemak. He would continue to work as a senior engineering manager, but would not direct the overall activities of RNII.

Soon Korolev would have far more to worry about than interoffice politics. A madness was gripping the Soviet Union, in the form of a state-sponsored terror that was engulfing the entire country. It would reduce Korolev to no more than a small pebble crushed beneath a very large boot. And the boot was Stalin's.

The roots of Stalin's terror lay in the unprecedented destruction of World War I, which had extirpated Russia's accustomed forms of social and political order. From its ashes had risen the Communist state, wielding absolute force in the service of naked power. After Stalin took over, he used this force to bring about the collectivization of agriculture. His policy amounted to nothing less than a ruthless war against his own people, a war that claimed some ten million lives. The creation of collective farms required far-reaching expropriation of privately owned land, and the imposition of serfdom upon a hundred million peasants who had thought they were free.

So sweeping an upheaval, so extensive an exercise of state power, produced violent opposition as peasants destroyed their cattle and foodstocks rather than surrender them. Stalin viciously crushed this opposition, first by attacking the peasantry and then by destroying major portions of his own Communist Party, portions whose leaders had dared to criticize him in prior years. After that, Stalin turned on the Red Army. He knew very well that his regime rested solely on force, for it lacked the rule of law and lacked such standard forms of legitimacy as popular consent and long-established custom. He understood that absolute power would flow to anyone commanding forces that could physically seize the Kremlin, with the Army's leaders being prime candidates for such an attempt. The generals might profess undying loyalty, but that didn't matter to Stalin. In his mind, they faced a potentially irresistible temptation: to seize power in the name of the peasants, then win their loyalty by abolishing the collective farms and restoring the lands expropriated by the state.

Tukhachevsky was one of the most powerful of these generals, and Stalin prepared his downfall with care. The charges would state that he had committed treason by working with the Nazis, and Stalin could make those charges appear convincing by relying on contacts that already existed between the Nazis and the NKVD, his secret police. Hitler and his senior officials willingly cooperated, knowing that Tukhachevsky's destruction would weaken the Red Army.

Accordingly, German officials arranged for a forger to create a file of letters that supposedly had passed between Tukhachevsky and members of Hitler's general staff. Tukhachevsky's signature came from a 1926 agreement between the two nations' armies; the signatures of the German generals came from bank checks. In May 1937 this dossier reached Stalin's desk. He acted swiftly; the NKVD arrested and executed several other top military leaders, along with Tukhachevsky. The NKVD also shot his two brothers and, later, his wife. His mother went to a concentration camp, where she died. Four sisters and two ex-wives also went into the Gulag. His daughter was allowed to live freely until she turned eighteen; then she too went to prison camp.

Tukhachevsky now stood condemned as a traitor. His protégés, including the leaders of RNII, were regarded as accomplices of a traitor. They too would be shot. Kleimenov met this fate, as did Langemak, his deputy. It didn't matter that Langemak had directed the development of an important weapon, the "Katyusha" artillery rocket, which would become a mainstay in the coming war. Beaten and tortured, both these officials were executed—after first being made to denounce their colleagues.

The NKVD then turned to other leaders of RNII, arresting the rocket designer Valentin Glushko. He had built the country's first successful liquid-fueled engine, and Korolev spoke out on behalf of his trusted friend. But under NKVD interrogation, Glushko was denouncing Korolev as well. In June 1938 the secret police came for Korolev and placed him under his own arrest.

They threw him into the prison Butyrki, a dank stone pile that dated back to the time of the Bastille. There they subjected him to a form of interrogation called the "conveyor," in which a group of officers, each working a shift of only a few hours, questioned him continually for days and nights without letup. Three months later he went before the prosecutor, Vasili Ulrikh, who had presided over Stalin's show trials of Communist leaders, and who had condemned Tukhachevsky and his fellow generals the previous year.

The historian Robert Conquest describes Ulrikh as "a fat man, with dewlaps like a bloodhound and little pig eyes. His shaven head rose to a point. His neck bulged over the collar of his uniform. His voice was soft and oily." In twenty minutes he reviewed Korolev's case and pronounced sentence: ten years of slave labor in the mines of Kolyma, deep in Siberia.

Amid this century's massive acts of murder, Kolyma stands out as one of the worst killing fields. Some three million *zeks* (political prisoners) died there under Stalin: some by gunshot, many more from disease and starvation. Deeply chilled during a winter that lasted much of the year, beset by dense swarms of mosquitoes in summer, the prisoners were denied both warm clothing and proper boots. They received inadequate rations but were under orders to fulfill their allotments of work, which demanded quotas that were out of reach. Then, as punishment for failing to meet these work norms, the guards would cut their rations severely. The zeks would weaken and die, but new prisoners would replace them—and would receive the same treatment.

"There was no investigation in the proper sense of the word," Korolev later declared. "I was bluntly accused of sabotaging research in a new technology. I could not imagine a more absurd and incredible charge." But the NKVD was highly skilled at viewing its prisoners in the worst possible light. "Our country doesn't need your fireworks," an interrogator told him. "Or maybe you're making rockets for an attempt on the life of our leader?"

Korolev did not go to Kolyma immediately; he spent the next eight months in prison in Novocherkassk, a city in the south. Meanwhile the NKVD received a new director, as Nikolai Yezhov, its head, died before one of his own firing squads. Yezhov's replacement, Lavrenti Beria, issued new orders, and one result was that Ulrikh now directed the NKVD to return Korolev to Moscow. But just then he was out of reach: Two weeks earlier, he had begun his journey to Kolyma.

Korolev crossed the country by rail, in a crowded and dirty cattle car and then endured an ocean voyage across the Sea of Okhotsk in the hold of a prison ship. The zeks stank with filth; there were no toilets or bathrooms. In August 1939 he reached his destination: the Maldyak mining camp, well inland. Soon he was slowly dying. Scurvy gripped him; with his gums bleeding, he lost over a dozen teeth. An earlier beating had broken his jaw; now a new blow from a guard scarred his head. And it was only November, with the worst of the winter still ahead.

At that point the new NKVD order reached Maldyak, and Korolev could retrace his steps down the long road that had led from Moscow. The initial leg of the route would recross the Okhotsk Sea, and Korolev was listed for passage aboard the *Indigirka,* an NKVD ship. Amid stormy weather, it went off course and ran onto rocks near Japan. The vessel's commander refused to open the hatches of the prison bay, deliberately drowning over a thousand zeks. Korolev was not among them. He had failed to arrive in the port of Magadan in time to board.

Korolev spent the winter in Magadan, repairing shoes and doing other odd jobs. Another vessel carried him to Vladivostok, the Pacific terminus of the Trans-Siberian Railroad. In Khabarovsk, still in the Far East, he received medical care and began to regain a semblance of health. The NKVD then brought him back to Moscow's badly overcrowded Butyrki prison, where in mid-1940 he received a new sentence: eight years. However, Korolev would not return to Kolyma. Instead he would benefit from another of Beria's decisions.

"The Gulag, the system of prison camps, was a major source of slave labor," notes the historian Aleksandr Gurshtein. "Beria decided not only to have slaves for physical work, but to also have slaves for intellectual work." He set up a series of highly capable aeronautical design groups that would operate behind barbed wire, with zeks in the key posts. They would work in standard engineering offices, carrying out the usual tasks of their specialties, but would not go home in the evening to their families. Instead they would sleep in barracks, with guards accompanying them at all times. These institutes were called *sharagas,* from a slang word meaning "scam," "con game," or "group of worthless people." Within them, Beria expected his inmates to design new aircraft that could win the war. If they didn't—well, there was always Kolyma.

The inmates now included Korolev's old mentor, Andrei Tupolev, who had supervised his design of an airplane at the Bauman Higher Technical School. Tupolev had run afoul of Stalin during the Spanish Civil War, when German fighters had achieved superiority over their Soviet counterparts. Tupolev had designed bombers, not fighters, but that didn't matter; Stalin had tossed his leading aircraft designers into jail willy-nilly. But now Beria directed Tupolev to set up his own NKVD design group, the Tupolevskaya Sharaga, and he remembered his old student Korolev, who soon joined him. A fellow engineer, Leonid Kerber, also joined the group and eventually wrote about it:

"We were taken to the dining room. Heads turned to our direction, sudden exclamations, people ran to us. There were so many well-known, friendly faces. At the tables we could see Tupolev, Petliakov, Myasishchev, Korolev, and many others—the elite, the cream of Russian aircraft technology. It was impossible to conceive that they had all been arrested, that they were all prisoners—this meant a catastrophe for Soviet aviation!"[5]

Actually, it didn't. Beria might be ruthless and wicked, but he was no fool, and he knew how to make it possible for these people to work. Even so, Kerber would remember Korolev as "a cynic and pessimist who took the gloomiest view of the future." Korolev did not disguise his contempt for the regime. He fully expected to be shot and often muttered, "We will all vanish without a trace. They'll just wipe us out and the newspapers won't even notice."

He found relief through his work. A fellow prisoner later would recall hearing a violin concerto on the radio, which made them both very homesick: "Tears ran down my cheeks, and I looked round to see Korolev standing beside me with tears in his eyes, too. I began to cry most bitterly. He went back into the office, and when I returned, he was sitting at his desk absorbed in his task."

Working in Tupolev's Moscow offices, his group of zeks redeemed their leader's good name with a successful twin-engine bomber, the Tu-2. Then the Nazi invasion, in June 1941, forced a general dispersion of aviation activities. Tupolev and his sharaga went to Omsk, well to the east, where their hard toil built a factory for production of the Tu-2. At times, Korolev succeeded in getting a stolen bottle of the plane's alcohol-based brake fluid, which was much in demand. It tasted horrible, but it was the only alcoholic drink available. By using it to bribe a guard, Korolev at times could arrange to spend a night with a woman.

Meanwhile, the RNII's Valentin Glushko had also survived imprisonment and now was running his own sharaga in Kazan, where he was building liquid-fueled rockets. These rockets were small units, intended to assist the takeoff of heavily loaded bombers. Korolev was well aware that Glushko had been one of his former colleagues who had betrayed him; as he had written in a letter from prison, "I have been foully slandered by the institute director, Kleimenov, his deputy, Langemak, and engineer Glushko." Still, Korolev was past the point of taking the treachery personally. Learning of Glushko's work, he asked the NKVD for a transfer into his group. He received this transfer in the fall of 1942. Though still a prisoner, he would spend the rest of the war involved in this new effort.

Korolev's contributions proved sufficiently valuable to win him his freedom, in mid-1944, when the NKVD released him from his prison sentence. He nevertheless continued to work with Glushko, for that was his duty in time of war. He had left a wife and daughter in Moscow, but he did not see them again until August 1945, four years after departing for Omsk and seven years after his arrest.

By then the world of rocketry was agog over Germany's wartime achievements of the previous ten years. At the outset, the Luftwaffe had matched the Army's interest in rockets by pursuing a plan for a rocket-powered interceptor. The invention of radar still lay well in the future, and Luftwaffe officials feared that an enemy might catch them unseen, striking with high-altitude bombers. Conventional fighter aircraft would need too much time to climb to altitude, but rocket-powered versions might get up there very nimbly, engaging the bombers on short notice.

With rocket concepts now burgeoning and with a national effort in the offing, the armed forces needed a major new facility for development and testing. The old Army proving ground wasn't big enough; in addition, as von Braun later noted, he could not concentrate on a calculation when troops nearby were testing new machine guns. Von Braun set out to search personally for a suitable site, which would offer thoroughgoing secrecy along with a clear firing range. The Baltic coast seemed to be the best bet.

He found a fine location on the island of Rugen, but the Nazi Strength through Joy movement had already claimed it for a vacation spot. Then came Christmas 1935, which he spent at home with his family. "Wernher," said his mother, "speaking of out-of-the-way places, your father used to go duck-hunting on Usedom, up near Peenemunde. . . ." This remote island of Usedom, adjoining the seacoast north of Stettin, proved ideal. Soon the Luftwaffe and Wehrmacht, the new and powerful army, were contributing to a pool of eleven million marks to fund the initial buildup of the Peenemunde rocket center.

The center would grow into a fantastic place, with no counterpart anywhere in the world. Test Stand 1, an enormous facility that towered over nearby trees, would accommodate rocket engines with over 200,000 pounds of thrust. Test Stand 7, a hundred feet tall, could hold a complete A-4 rocket in position for a static firing of its engine, then roll back for its launch. Measurement House, a center for advanced instrumentation, pursued the development of guidance systems with automatic control. There was a liquid-oxygen plant, a supersonic wind tunnel, a radar site, and a launcher for unmanned cruise missiles, as well as production facilities that would turn out A-4 rockets, each as tall as a four-story building, on an assembly line. And . . . a camp for prisoners of war, who served as workers.

Peenemunde would need these research laboratories, for in pursuing the A-4, its engineers and scientists were breaking new ground in a host of areas. Its rocket motor offered an early focus for effort; it would produce more than 800,000 horsepower, and like any engine, it had to be cooled. Water-filled cooling jackets were too heavy, but Walter Thiel, who was responsible for rocket development, had a different approach: regenerative cooling. It called for the fuel, alcohol, to flow through thin tubes that would line the outside of the thrust chamber and nozzle. The fuel then would carry away heat; being heated strongly, it also would burn with more energy.

Nevertheless, Thiel's tests showed that this approach by itself was inadequate. Regeneratively cooled engines would burn through in their hot areas, producing spectacular explosions. The answer lay in drilling small holes near the nozzle's throat, its narrowest and hottest region. Alcohol, seeping through these holes, would form a cool protective insulating layer over the nozzle's sheet metal. This approach was known as film cooling. It worked, and made it possible to build a reliable engine.

There also was the problem of delivering propellants to this engine at the required rate, 275 pounds per second. Pressurized tanks had worked for small rockets, but on the big A-4, tanks capable of resisting the pressure would be excessively heavy. The alternative was a turbopump, a pump driven by a lightweight steam turbine, and the requirements seemed daunting.

Von Braun visited a pump factory and presented his specifications. The turbopump had to achieve high pressure, 300 pounds per square inch, while pumping upward of fifty gallons per second. Yet it would need to have a simple construction, for high reliability. It should get up to full speed in only a few seconds, and then it had to maintain its delivery pressure and flow rate with almost no variation. Furthermore, it had to be very light.

He expected the factory's designers to protest that he was asking the impossible. Instead they told him that he was asking for a firefighter's pump. The requirements were nearly the same: simple construction, fast action, very high flow rate, and constant delivery pressure to allow a fireman to hold the hose steady. Existing pumps could offer the basis for what von Braun wanted.

The steam turbine that would drive the pump was another matter. The turbines themselves would not be hard to develop; the problem would be to produce the powerful flow of hot, high-pressure steam to drive it. Hydrogen peroxide gave the answer. With proper care, a nearly pure solution could be stored safely, and when made to decompose, it would instantly yield the desired flow of steam. In turn, the steam generator would be far lighter, more compact, more powerful, and faster acting than any boiler. Here was a technical solution that was not only feasible but elegant.

Guidance and control presented greater difficulties. To keep a rocket erect as it balances on its exhaust jet, immediately after liftoff, is like keeping a broomstick upright as you balance it on the back of your hand. You have to see or sense the beginning of any tilt, then apply a correcting movement. In rocketry, gyros would act as sensors and would generate inputs to a control system that would deflect the rocket's exhaust to provide a correcting force. However, rocket guidance had to be capable of coping with gusts of wind. It had to tilt the missile for long-range flight, and keep it steady while doing this. It also had to accommodate the changing weight of the rocket as it burned its propellants, the shifting location of the missile's center of gravity, as well as aerodynamic characteristics that would change as the missile passed the speed of sound.

Aircraft autopilots that came into use during the 1930s offered a way to start. But no one had built such guidance systems previously, and it took several years just to set up a mix of in-house, industry, and university labs that could address the problem. The first attempt, on the A-3 rockets of 1937, failed dismally; the guidance could not cope even with brisk wind. Later work showed that the problems were more difficult than anyone had thought; von Braun responded by setting up his own guidance lab. Although there were mathematical equations whose solution could help in designing the guidance, they were extraordinarily difficult to work with. An important advance came when a staff scientist built an early version of an electronic analog computer that could solve the equations. Guidance development proved so difficult that von Braun built an entirely new type of rocket, the A-5, largely to test experimental systems. Even when the A-4 entered flight testing, it relied at first on interim guidance systems that would require additional development.

The first test rocket reached the launch pad in June 1942. It flopped; so did the second. The third time was the charm: The rocket soared into the blue October sky and reached a range of 118 miles, encouragingly close to Dornberger's goal of 160. But that was the flight people would have to remember for a while. The next six months brought eleven more shots, but only one of them matched the performance of the earlier one. Not until the spring of 1943 did the A-4 begin to fly to full range with any kind of consistency.

By then the German military situation was becoming desperate. A major turning point, early in the year, was the defeat at Stalingrad, which ended any hope of victory on the Eastern Front. In the air, Allied bombers were striking with increasing force; the industrial Ruhr was a particular target. "The Ruhr will not be subjected to a single bomb," Hermann Goering, head of the Luftwaffe, had boasted back in 1939. The Royal Air Force nevertheless inflicted twenty-six heavy raids during the spring of 1943. In addition, in a particularly brilliant feat, a force of specially equipped bombers breached two of the country's largest dams, the Moehne and Eder, releasing a third of a billion tons of water to thunder down the Ruhr. And things were rapidly getting worse. July would bring the great tank battle at Kursk, in which the Soviets would take the offensive and keep it. And late that month, heavy British air raids on Hamburg touched off a firestorm and burned the city to the ground. Hamburg was the first city to burn in this fashion; it would not be the last.

For years, Hitler had been funding an assortment of costly development projects involving advanced weaponry. Four of them had now reached advanced stages of development and appeared ready for production: a jet fighter, the Messerschmitt Me 262; a rocket-powered interceptor, the Me 163; the A-4 ballistic missile; and a jet-powered cruise missile, the Fieseler Fi-103. These wonder-weapons now offered an enticing prospect indeed. If Germany could build them in sufficient quantity, the aircraft might turn back the Allies' growing air armadas. In turn, the missiles might provide a new German sword, striking at London with great power and renewing the offensive. The A-4 appeared especially attractive. It would fly at five times the speed of sound, and no defense against it existed.

Still, when the A-4 gained approval for mass production and operational use, this was not because it had proved itself on its merits, but rather because Hitler and his associates were increasingly willing to try almost anything. Dornberger has given a picture of Hitler's attitude as of that July, when he and von Braun met with the Führer to describe the current status of the A-4. During that meeting, Hitler went beyond hope to enter a realm of unreality.

"If we had had these rockets in 1939 we should never have had this war," Hitler said, ignoring just who had started the war in the first place. "Europe and the world will be too small from now on to contain a war. With such weapons humanity will be unable to endure it." He called for fortified bunkers as firing sites and added, "These shelters must lure the enemy airmen like flies to a honeypot. Every bomb that drops on them will mean one less for Germany."

Acknowledging that the A-4 would carry a one-ton war load, Hitler asked if his guests could not raise it to ten tons. Dornberger replied that a ten-ton warhead would call for a much larger rocket, which would take years to develop. Hitler's eyes flashed as he replied, "But what I want is annihilation—annihilating effect!" When Dornberger responded, "We were not thinking of an all-annihilating effect," Hitler turned to him in a rage. "You! No, you didn't think of it, I know. But I did!"[6]

What, after all, had Dornberger brought to that meeting? What was this weapon that lacked an all-annihilating effect but still would make war too terrible for humanity to endure? Hitler already had heard that the atomic bomb might someday become feasible. A year earlier he had told Field Marshal Erwin Rommel of a future explosive that could "throw a man off his horse at a distance of over two miles." Certainly the A-4 could make a strong impression on people far less prone to fantasy than Hitler. When Dieter Huzel, an associate of von Braun, first saw some of these weapons up close, he thought that "they must be out of some science-fiction film—*Frau im Mond* brought to earth."

Indeed, for years after the war, the A-4 would inspire thoughts of powerful spaceships of the future. Hitler's hopes ran in a different direction, toward all-annihilating weapons. Yet for him, as for so many others, the rocket evoked images of the shape of things to come. As it would inspire a generation of would-be astronauts, so it brought similar inspiration to this megalomaniac who dreamed of building armaments powerful enough to destroy the world.

Hitler gave his blessing to all four weapons. The A-4 and the Fi-103 cruise missile both received high priority; the latter promised a particular advantage because it was inexpensive and could be built in large numbers. Joseph Goebbels, the propaganda minister, then gave the projects his characteristic touch by christening them *Vergeltungswaffen,* Vengeance Weapons. The cruise missile would enter service as the V-1. The A-4, more amazing still, would become the V-2, suggesting a progression of increasingly deadly arms.

Production of the weapons now became a matter of first importance. A British air raid on Peenemunde, in August 1943, made it clear that the V-2 effort needed new facilities. The production manager, Gerhard Degenkolb, found what he needed in an underground oil-storage depot in the Harz Mountains near Nordhausen. To prepare for mass production, and to carry it out, the Nazis turned to slave labor. The SS controlled the slaves of the Third Reich, and an SS colonel, Hans Kammler, became a key man in the V-2 program. His qualifications were superb. He had taken an active role in the destruction of the Warsaw ghetto and later was responsible for designing and building the crematoria at Auschwitz.

The underground facilities needed expansion, and Kammler's prisoners set to work with pickaxes, choking on dust that contained anhydrous ammonia. They lacked showers and latrines; they relieved themselves where they stood, and the place stank like a sewer. Inmates lived in a concentration camp called Dora, a branch of Buchenwald, and they faced beatings, malnutrition, and exhaustion from overwork. Pneumonia, typhus, dysentery, and tuberculosis were rampant, and while Dora was not an extermination camp, the SS carried out frequent hangings. At first the dead went to Buchenwald for cremation; later Dora qualified for its own crematorium. Some twenty thousand people died there. When British officials arrived at war's end, they found stretchers heavily saturated with blood, along with a large pit full of human ashes.

In the face of this, what did von Braun know and when did he know it? He knew plenty; he visited frequently, and his brother Magnus worked in this Nordhausen plant as a manager. However, von Braun was the technical director at Peenemunde, which meant that he managed a highly skilled workforce that did not rely on slaves. Though he knew of the horrors of Nordhausen, he was not responsible for them and avoided personal involvement. This absolved him of liability for Kammler's crimes against humanity. Kammler was shot by his adjutant in 1945, a fate he preferred to being captured.

The first V-1s flew in June 1944, aimed at London. But they proved vulnerable to British air defenses, and by summer's end, England had largely neutralized the threat. This outcome contrasted with the Allied bombing campaign, which continued to gain in strength.

The V-2 was different; no air defense could shoot it down. But it did not enter service until September 1944, at a time when the American and British armies had taken Paris and were driving hard for the Rhine. The Germans successfully fired some 3,225 V-2s over the next several months, launching more against Antwerp than London. Though many of them missed their targets, the resulting explosive load of over 3,000 tons corresponds to the bomb load of some 1,500 Flying Fortress bombers, the standard warplanes of the European theater.

However, these figures represent the total effect of the A-4 over the full course of the war. Peak production never topped 700 a month, which would have corresponded, at most, to a single good-size air raid every few weeks. By contrast, without benefit of wonder-weapons, the Allies' bomber forces carried out 363 such raids on Berlin alone. Winston Churchill would note that for both V-1 and V-2, "the average error was over ten miles. Even if the Germans had been able to maintain a rate of fire of a hundred and twenty a day, the effect would have been the equivalent of only two or three one-ton bombs to a square mile per week." Lord Cherwell, Churchill's science adviser, believed for a time that the V-2 represented an enormous bluff, for he could not accept that Germany would squander resources on a weapon of such little effectiveness.

Hitler's other marvels, the advanced aircraft, were even less effective. The Me 262 was a true jet fighter, good enough to have fought alongside America's postwar jets in the Korean War. But if the 262 offered a vision of the next conflict, it, too, proved ineffective in the one that counted, which was World War II. It was unmatched in the air, but it lacked endurance and spent little time in the air, and on the ground it was a sitting duck. Indeed, it was Me 262 bases that lured the Allied airmen like flies to a honeypot. At war's end the U.S. Army Air Forces concluded that the 262 had shot down only fifty-two of its bombers, along with ten fighters. The combat record of the German rocket interceptor, the Me 163, was even worse: nine kills, including two probables.

Were these results inevitable? Certainly all these weapons would have been more effective if they had reached the field at earlier dates, or if Germany had succeeded in prolonging the war. Yet the Me 163 entered test flight as early as 1941, reaching a speed of 624 mph at a time when few fighters could top 400. The Me 262 flew in July 1942, the A-4 a few months later. From this perspective, what is pertinent is not that Hitler had launched a serious rearmament program as early as 1935 or that the war began in 1939. The pertinent time was the spring and summer of 1943, when the fortunes of battle turned decisively against the Germans. The weapons might indeed have been ready earlier, but for that to have happened, they would have had to obtain priority earlier. Such a course of action would have conflicted strongly with the natural tendency of all war leaders, which is to fight with arms that are well proven and readily available.

It is a reckless leader indeed who will hazard his country's fate on the unfamiliar and the unproven. Hitler did it, and granted priority to his Vergeltungswaffen, only out of desperation at a moment when outright defeat was less than two years away. Those two years, rather than the six years of the entire war, defined the available time in which Hitler's missiles and advanced aircraft could make their mark. It was not enough.

Yet if the V-2 appeared ineffective in the spring of 1945, this was no longer the case in the summer. The sudden advent of the atomic bomb raised the immediate prospect of a true wonder weapon, a long-range ballistic missile with a nuclear tip. This prospect greatly increased the importance of the Interdepartmental Technical Commission, a high-level group of Soviet rocket specialists who already were in Germany to seize what they could of that country's missile technology. In September 1945 Korolev went to Berlin as one of their members.

Members of the commission took part in three V-2 launches as observers, while British troops fired the weapons far out into the North Sea. By then Korolev was a lieutenant colonel, but he was still too junior to attend at the launch site; he had to watch from behind a fence. Nevertheless, this mission set him on the path to leadership of his own country's missile development.

"We Russians have a proverb," the historian Gurshtein recounts. "I'm a rooster; my job is to crow. If the sun doesn't rise, that's not my problem." Korolev would make damn sure the sun rose, whether it wanted to or not. He had long been blunt and plainspoken in upholding his points of view during arguments with colleagues. Now he took the initiative in outlining a missile program that could build on the German work. He also took increasing responsibility for his commission's activities.

Just then, Stalin was directing his minister of armaments to set up such a missile program. The minister, Dmitri Ustinov, had gained his advantage through the wide swaths that Stalin's purges had cut in the senior ranks of the government. He had risen quickly by filling the empty posts and had become a cabinet minister in 1941, at the age of thirty-two. He then had directed the massive production programs that equipped his armies for battle.

Now, in 1946, Ustinov traveled to Germany, met Korolev, and decided he liked him. Though only recently in prison, Korolev was no toady; he would set forth arguments and defend them, even if Ustinov disagreed. His fellow engineers knew that he would reproach a man severely for being careless or inattentive. Yet he did not stay angry for long or hold a grudge; he was quick to forgive. In addition, he had won his colleagues' loyalty because he readily gave credit to those who deserved it. And he already had a plan for a missile organization that would carry out Stalin's directive.

Ustinov appointed Korolev to head a department within a new research institute, NII-88, that would pursue a broad range of activities in missile development. Korolev's specific responsibility lay with long-range missiles. He was coming in as a middle manager, with a technical staff of only fifty-two engineers and with four levels of management between him and Ustinov. Even so, he was in a position to parlay his limited domain into a much larger one, for Stalin already had his eye on long-range rockets. And Korolev certainly was no longer a zek. While in Germany he had acquired a taste for that country's powerful cars; he now had one. People called him *Korol,* "King."

Korolev's immediate effort centered on a Soviet-built V-2. It offered a focus for missile production and for the training of operational rocket troops, but had no promise as a genuine weapon. At a Kremlin meeting in April 1947, Georgi Malenkov, one of Stalin's associates, reminded the aviation minister: "We are not going to fight a war with Poland. We have to remember that there are vast oceans between ourselves and our potential enemy."

The following day, as the meeting resumed, Stalin received a briefing on prospects for intercontinental missiles. "Do you realize the tremendous strategic importance of machines of this sort?" he responded. "They could be an effective straitjacket for that noisy shopkeeper Harry Truman. We must go ahead with it, comrades. The problem of the creation of transatlantic rockets is of extreme importance to us."[7] In creating them, Korolev would play a leading role.

2

Ingenious Yankees

The Rise of America's Rocket Industry

IN ADDITION TO Hermann Oberth and Konstantin Tsiolkovsky, there was a third prophet of space flight: Robert Goddard of Worcester, Massachusetts. He came up as an academic, earning a Ph.D. in physics at Clark University in his hometown. Clark's physics department was small but excellent; in 1907, while he was there, his department chairman, Albert Michelson, became the first American to win a Nobel Prize in physics. Goddard then stayed on at Clark, becoming a professor. But while Oberth and Tsiolkovsky contented themselves for the most part with writing about high-performing rockets, Goddard took long steps toward making them a reality.

He first began dreaming of flight to Mars in 1899, at age seventeen, after reading H. G. Wells's *War of the Worlds.* His physics studies led him into rocketry and into a program of experiments, which soon became too costly to pursue on his salary of $1,000 per year. But he knew that the Smithsonian Institution had a long record of supporting research for flight. He applied for and won a grant of $5,000. When the first thousand-dollar check came in the mail, his mother exclaimed, "Think of it! You send the government some typewritten sheets and some pictures, and they send you $1,000 and tell you they are going to send four more." Goddard's relationship with the Smithsonian would last the rest of his life.

An early consequence of that relationship came in 1920, as that institution published his early thoughts on rockets in a sixty-nine-page pamphlet titled *A Method of Reaching Extreme Altitudes.* Somewhat as an aside, he included a brief discussion of a rocket that would fly to the moon and would mark its arrival by exploding a charge of flash powder. The Smithsonian played up this space mission in a press release. Major newspapers played it up still further, including the *New York Times,* which ran a headline: AIM TO REACH MOON WITH NEW ROCKET. Goddard was appalled, for he loathed sensationalism. He was shy by nature, reticent and reserved; now he became much more so. During subsequent decades he would retreat not only from the public, but even from the company of engineering professionals whom he might have greeted as peers.

Goddard's early work had greatly improved the performance of powder rockets. Now he turned to liquid fuels, and in March 1926 he launched the world's first liquid-fueled rocket, five years ahead of the Germans and seven years prior to Korolev's "09." It flew a distance of 184 feet, with part of the nozzle burning off, but this short flight was quite enough to establish his priority. Characteristically, though, he issued no announcement.

His test area was a strawberry farm in nearby Auburn, owned by a distant relative, "Aunt" Effie Ward. She was tall, with gray hair that she wore in a bun; she lived as a spinster, and had a houseful of cats. She also had spirit, welcoming Goddard with pleasure and allowing him to store equipment in an empty henhouse. The local fire marshal was less accommodating; following a similar rocket flight in 1929, he told Goddard to shut down. Goddard then found another launch site, at the Army's Fort Devens. With help from the Smithsonian, he obtained permission to use it.

Meanwhile the aviator Charles Lindbergh was developing his own ideas about rocket-powered space flight. Goddard's 1929 test had brought a brief flurry of publicity, which Lindbergh noticed. Soon after, Goddard told his wife, "I had an interesting call from Charles Lindbergh." "Of course, Bob," she replied. "And I had tea with Marie, the Queen of Romania."

Lindbergh soon led Goddard to Harry Guggenheim, heir to a fortune in the copper industry and director of a foundation that was actively promoting new developments in aeronautics. The support that resulted allowed Goddard to leave Clark University and set up a serious program of rocket research in New Mexico, where there was plenty of land, along with all the privacy he could wish for. He would work alone except for help from his wife, Esther, and from a small group of mechanics who had worked with him in Massachusetts. Yet in this meager fashion, he would go surprisingly far.

His first important success came in December 1930 when a rocket reached 500 mph, faster than any airplane of the day, and soared to 2,000 feet. Its engine used no cooling, relying instead on a heat-resistant inner lining of ceramic. In addition, it had no guidance system, accelerating initially along tracks within a launch tower and then relying on fins to give it the stability of an arrow. In his next round of experiments, he introduced both cooling and gyroscopic guidance.

His guidance system anticipated features of the later German designs, for it used a gyro to sense deviations from the vertical, correcting them by causing heat-resistant vanes to dip into the rocket exhaust. These vanes, deflecting the exhaust, then would restore the rocket to its vertical path. To cool the engine, he anticipated the work of Germany's Walter Thiel by inventing film cooling. With these refinements, Goddard's rocket reached 7,500 feet in 1935. "It was like a fish swimming upward through the water," Esther declared.

So far, Goddard had pressurized the propellant tanks to feed his rocket motors. Now, seeking a path toward truly large rockets, he developed lightweight turbopumps that again anticipated those of the Germans. He also set out to build a film-cooled engine nearly twice as large as his earlier one, with a thrust of 700 pounds. He didn't quite succeed; his best flight, in August 1940, tilted immediately after liftoff and crashed to the ground only 400 feet away. Then, after the United States entered the war, Goddard left New Mexico to join a Navy rocket project in Annapolis.

"He could have had us sitting around his knee," says Robert Truax, his boss, who had initiated the Navy's work with rockets after receiving his commission at the Naval Academy. "But he would never talk rockets. I think he was afraid someone would steal his stuff. And his stuff wasn't that good. He was a physicist; his ability to do engineering was very limited. He had the vision, certainly. And there was almost nothing that was later developed successfully that he didn't try, at least once. But he was very much a loner. He had failure after failure. I'm amazed he even got a rocket out of the launch tower."[1]

Even so, working with nothing more than his out-of-the-way machine shop, his results could bear comparison with the best of Germany's entire Wehrmacht prior to 1942. His rocket of 1940 was twenty-two feet long, putting it in a class with von Braun's A-3 and A-5. His guidance system was far simpler, for he needed only to maintain a vertical course. By contrast, the German effort emphasized guidance along a precisely calculated trajectory aimed at a target such as London. But Goddard succeeded repeatedly with guided flights during 1935, whereas von Braun failed repeatedly with the guided A-3 late in 1937. Goddard's 1935 flights also successfully used a film-cooled engine, while the later A-3 and A-5 relied entirely on regenerative cooling, which was easier to develop but offered only limited use for the much larger V-2. And in using turbopumps rather than pressurized tanks, Goddard introduced an advanced feature that the Germans would only use on the V-2 itself.

A captured V-2 reached Annapolis in 1945, and Goddard and his staff examined it. "I was amazed," said one of his associates. "Of course it was more elaborate and much larger than the rockets we'd worked on, but it seemed extremely familiar." Another staffer said, "It looks like ours, Dr. Goddard," and he replied, "Yes, it seems so." At that moment he had only a few weeks to live, for he was ill with cancer. He died in August 1945 at age sixty-two.

In his lifetime Goddard was a prophet with only modest honor, for he rarely addressed meetings of technical societies and published very little. Following the publication of his 1920 monograph, he kept his silence until 1936, when he issued a second Smithsonian report. It outlined his work from 1920 through 1935—in all of ten pages. For further details, interested parties were referred to his patents.

These were already becoming both voluminous and important. By the time Goddard reached Annapolis he already held forty-eight rocket patents. Several of them anticipated key features of the V-2, including film cooling, gyrostabilization, and a turbopump suitable for use with liquid oxygen. During the war years he filed applications that led to thirty-five more patents, including one for a variable-thrust rocket engine and one that had a complete description of his 1940 rocket. Then, after his death, his widow organized his personal papers for later publication and secured over 130 additional patents. The last one was granted as late as 1957.

Many of these inventions represented his thoughts and ideas, rather than apparatus that he had built in practical form. Even so, the Germans made good use of them. In 1945 a Luftwaffe general, under questioning by an American, suggested, "Why don't you ask your own Dr. Goddard?" In 1950 von Braun declared that he was "virtually overwhelmed by the thoroughness of [Goddard's] work and found that many design solutions in the V-2 rocket were covered by Goddard patents."

His biographer, Milton Lehman, would salute him with lines from the poet Robert Browning:

> That low man seeks a little thing to do,
> Sees it and does it:
> This high man, with a great thing to pursue,
> Dies ere he knows it . . .

Still, Goddard's professional reticence prevented him from playing a leading role in building the organizations and institutions that by 1945 were already laying solid groundwork for the subsequent growth of American rocketry. One of the more important of these was a professional association from which he remained largely aloof: the American Rocket Society.

This society sprang to life in April 1930 as a gathering of writers who contributed to Hugo Gernsback's *Science Wonder Stories,* one of the first sci-fi pulp magazines. Its founder and first president, David Lasser, was a reporter at the New York *Herald Tribune* but wrote for Gernsback on the side. The group started by publishing a mimeographed newsletter, gathering every two weeks for meetings at the American Museum of Natural History.

The initial name of the group, American Interplanetary Society, reflected its origins in science fiction. Continuing this theme, early in 1931 Lasser arranged a public meeting at the Museum titled "By Rocket to the Moon." His prime attraction was an English-language edition of *Frau im Mond*. Editing out the love scenes, Lasser advertised it as "showing the actual flight to the moon of an imaginary but scientifically possible rocket" and announced the meeting by putting up dozens of placards in New York subways. Admission was free, and some two thousand people turned up. However, few of them joined the Society.

Lasser's friends included a fellow rocket enthusiast, G. Edward Pendray. Later in 1931, Pendray and his wife took a tour of Europe, with its rocket research very much on their minds. In Berlin they visited the Raketenflugplatz, where Rudolf Nebel gave them what Pendray would describe as "the most memorable experience of the entire trip": a static test of a small liquid-rocket engine. Pendray had never before seen such a demonstration, and on returning to New York, he resolved to build such rockets himself. He not only would follow Nebel's designs, but also would pursue Nebel's methods for achieving low-cost rocket development.

Pendray received considerable help from Hugh Pierce, a member who had been in the Navy—he had tattoos to show for it—and now had a job selling tickets to riders on the New York subway. The Navy had trained him as a machinist, and he was ready to build rocket components in a workshop located in the basement of his apartment building in the Bronx. Pierce worked without pay and had the use of the workshop rent-free. Part of his work would involve machining of aluminum castings, which he obtained as gifts from a manager at Alcoa.

Valves also cost nothing, after another member approached a supplier and asked for free samples. A cocktail shaker served as a water jacket, to cool the engine. A five-dollar piece of silk, bought at a department store, became a parachute for the rocket; its holder was a small saucepan. At the Air Reduction Company, which dealt with liquefied gases, an official sold them a secondhand liquid-oxygen container for fifteen dollars and arranged to fill it for free. These acts of generosity led Pendray to propose setting up an Air Reduction Extolling Corporation, "the members of which are to extol Air Reduction at all hours of the day and night, when not otherwise engaged."

The out-of-pocket cost of this first rocket came to less than fifty dollars, but Pendray would find that he was getting no more than what he'd paid for. Ground tests showed that the rocket was both fragile and excessively complex, thus requiring an extensive reconstruction. The remodeled rocket flew, or tried to, at Staten Island in May 1933. In Pendray's words, "It reached an altitude of about 250 feet, after firing about two seconds, and was still going well when the oxygen tank exploded." The engine shut off and the rocket fell into the adjacent bay, from which two boys in a rowboat fished it out.

Still, as Pendray would put it, "it was the first liquid propellant rocket any of us had seen get off the ground." It also marked an ongoing move by the Society away from its science-fiction beginnings and toward professionalism. Already it had abandoned its mimeographed newsletter in favor of an attractive magazine, *Astronautics.* Following the flight, at the annual meeting in April 1934, the group adopted the new name of American Rocket Society. Its experimenters then proceeded to prepare to launch an entirely new rocket.

Its engine featured a design by John Shesta, a tall, taciturn member with a degree in civil engineering from Columbia University. Shesta's rocket motor had four nozzles fed from a common thrust chamber. The test flight took place in September 1934, again in Staten Island, and a few seconds after launch, one of the nozzles burned out. The rocket tilted over, roared across the water at close to the speed of sound, and splashed into the bay a quarter-mile from the launch site, with Shesta's engine still firing vigorously.

Further research was obviously in order, but Pendray noted that "in the vicinity of New York, rocket shots or motor tests were not welcomed by neighbors, or approved by the police, and there was no way of obtaining permission to carry on such experiments in peace. As a consequence the Society performed many of these tests under some harassment, and found frequent and unannounced moving of the testing ground to be a wise and sometimes necessary precaution."[2]

Despite these difficulties, the ARS's experimental work found a focus in the efforts of a brilliant student at Princeton, James Wyld. In 1935, during his senior year, he learned of the Society, looked up its number in the Manhattan phone book, and quickly became deeply engrossed in its rocket activities. He graduated and got a job but worked on rockets in his spare time, seeking particularly to find the best way to cool an engine. He decided that regenerative cooling offered particular promise. He moved to Greenwich Village and set up a workshop in a pantry adjoining his rented room. Then in 1938 he lost his job and suddenly had more time for this project.

Wyld's finished engine delivered over ninety pounds of thrust; yet it weighed no more than two pounds. Tested in December 1938, it received far less damage from the heat of combustion than any previous rocket motor built by his colleagues. Then Wyld found a new job and put his rocket aside, but returned to it in 1941 and made improvements. Further tests, conducted that summer, showed that it was reliable and fully satisfactory.

Another member of the Society, Lovell Lawrence, was working in Washington and thought he might be able to win a government contract.

He tucked Wyld's engine into his briefcase—it was small enough to hold in his hand—and made the rounds at federal agencies. He aroused interest at the Navy's Bureau of Aeronautics, but learned that this service could not award contracts to private individuals. However, the situation would be quite different if Lawrence had a corporation. He promptly set one up, taking its presidency and naming Pierce, Wyld, and Shesta as the corporation's officers, directors, employees, and stockholders. Early in 1942, Lawrence received his contract.

Wyld christened the firm Reaction Motors, Inc., playing on the name General Motors. Start-up capital was $5,000. The group's initial shop was in the garage of Shesta's brother-in-law, but they soon moved to a former nightclub in Pompton Plains, New Jersey. The club's bar stools were still in place when they moved in. To test their engines, they found an isolated area in nearby Franklin Lakes and built a concrete blockhouse, complete with shatterproof glass.

The Navy was interested in rockets that could assist the takeoff of heavily loaded aircraft. Robert Goddard's wartime work at Annapolis addressed this issue; so did that of Reaction Motors. Its immediate task lay in developing larger versions of Wyld's engine, rated at 1,000 and later at 3,000 pounds of thrust, and the company carried out these tasks successfully. Then in 1945, employing all of thirty-five people, the firm received a new assignment: to develop an engine for a rocket plane that could break the sound barrier.

This plane, the X-1, was a project of the Army Air Forces and the firm of Bell Aircraft, in Buffalo, New York. A Bell engineer, Benson Hamlin, learned what Reaction Motors was doing for the Navy and realized that this company could build what he needed in a hurry. The engine that resulted, the XLR-11, drew on discussions over lunch and dinner at the Triangle Grill, a favorite restaurant of the Reaction Motors staff. A shop foreman later recalled that following such dinners, they would all go back to the office "and work like hell, sometimes to ten or eleven at night. Sometimes John Shesta would also come around with a clipboard and make modifications on the spot."

The engine emerged as a simple affair, featuring four Wyld-type units, each of 1,500-pound thrust. It could not be throttled, but the X-1's pilot could light the chambers one by one, for 25, 50, 75, or 100 percent of the full thrust of 6,000 pounds. It was supposed to have a turbopump, but this took time to develop. As a stopgap, the X-1 relied on thick-walled pressurized propellant tanks that added a ton to its landing weight. Even so, the engine had power to spare. The Air Force's Chuck Yeager used it to ride through the sound barrier in October 1947, thereby becoming the first pilot to fly at supersonic speed.

The American Rocket Society, Reaction Motors, and the XLR-11 all would go on to better things. With the coming of war, the ARS abandoned its rocket experiments and became a society of professional engineers. It flourished amid the burgeoning missile and space programs of the postwar years, and in 1963 it merged with the Institute of the Aeronautical Sciences. The engineering society that resulted, the American Institute of Aeronautics and Astronautics, became the nation's leading aerospace association.

Reaction Motors also prospered after the war. The financier Laurance Rockefeller purchased control in 1947, thus providing plenty of working capital. The Navy also provided a permanent test area, at an ammunition depot in Lake Denmark, New Jersey. The company then went on to power most of the nation's rocket aircraft of the postwar decades. Its XLR-11 would continue to see use whenever anyone needed a simple, reliable engine of its size.

Two new and major projects marked Reaction Motors's postwar years: a 20,000-pound-thrust engine for the Navy's Viking research rocket and an engine of 57,000 pounds of thrust for the Air Force's most advanced rocket plane, the X-15. In 1958 the firm merged with the much larger Thiokol Chemical Corp. However, it failed to win any more large contracts and went out of business in 1972. But through the success of the X-1 just after the war, Reaction Motors made its name as a cornerstone of a nascent rocket industry.

Meanwhile, a professor of aeronautics, Theodore von Karman, was laying other cornerstones at California Institute of Technology. He had studied with Germany's Ludwig Prandtl, who founded the science of aerodynamics and developed some of its most penetrating insights. In turn, other students of Prandtl laid groundwork for jet engines, gas turbines, supersonic wind tunnels, and wings for high-speed flight.

Von Karman had come to Caltech in 1930. Its president, Robert Millikan, was the second American to win a Nobel Prize in physics. The faculty Millikan recruited included several luminaries: Albert Einstein; the chemist Linus Pauling; Robert Oppenheimer, who later directed development of the first atomic bombs; and Charles Richter, who invented the earthquake scale. Here von Karman would make important contributions to supersonic flight and would earn a reputation as the No. 1 man in American aeronautics.

Von Karman had no consuming vision of a flight to Mars or of science fiction becoming a reality. Rather, he came to view rocketry as one of a number of interesting issues at the frontiers of flight. Still, the work of his colleagues would give rise to two major institutions: the Jet Propulsion Laboratory and the firm of Aerojet General, which in time would surpass Reaction Motors as a builder of rockets and their engines.

These developments started in 1936 when a graduate student, Frank Malina, approached von Karman and arranged to design and test rockets for his dissertation. With him were two other enthusiasts from Pasadena: Edward Forman, an experienced mechanic, and John Parsons, a chemist and expert on explosives. They couldn't fire their rockets at Caltech itself, which was as cloistered as a monastery. But Malina found the open space he needed a few miles from campus at the Arroyo Seco, which means "dry gulch."

Malina visited Robert Goddard in New Mexico, but while Goddard greeted him cordially, he was not about to share his expertise merely to help a graduate student obtain his Ph.D. Malina returned to Pasadena and proceeded in the usual catch-as-catch-can fashion, studying the meager technical literature, scrounging for funds and equipment with help from his fellow students. Then in May 1938, von Karman received a visit from General Henry "Hap" Arnold, who had recently taken over as chief of the Army Air Corps.* He believed that the future of air power lay in research, and he knew that Caltech was the place to find it.

Malina's rocket work fascinated Arnold; like his Navy counterparts, he saw that rockets could assist the takeoff of aircraft. In the fall of 1938 he called a meeting of an advisory committee that included von Karman, whose members chose from a list of research projects. Rocketry was one of them; another was the de-icing of aircraft windshields. "We'll take the problem of visibility," said Jerome Hunsaker, head of the aeronautics department at Massachusetts Institute of Technology. "Karman can take the Buck Rogers job."

Hunsaker wasn't the only one who viewed rockets in that light. Arnold arranged for Caltech to receive an initial grant of $10,000, but even his own staff members were dubious. His assistant visited von Karman late in 1939 and asked, "Do you honestly believe that the Air Corps should spend as much as ten thousand dollars for such a thing as rockets?" There was reason for this skepticism, for while Malina was seeking nothing more than to build solid-fuel units for takeoff assistance, even this modest goal seemed to lie out of reach.

The solid-fuel rockets that people knew about resembled skyrockets, which burn in a sudden burst. By contrast, Malina's units would have to burn slowly, giving sustained thrust for as long as twenty seconds. Experimental tests produced repeated explosions, but von Karman's strong suit lay in mathematical theory, and in 1940 he and Malina devised a set of

*This air arm had three different names in less than a decade. It was the Army Air Corps until 1941, then the Army Air Forces during the war. In 1947 it separated from the Army and became an independent service, the United States Air Force.

equations that showed them how to proceed. Encouraged, the Air Corps doubled the grant, to $22,000 for the fiscal year of 1941.

With this sudden wealth, Malina arranged to lease several acres of the Arroyo Seco from the city of Pasadena. Here he put up a few small wooden buildings with roofs of corrugated metal and with unheated, drafty, and cramped interiors. Another recent graduate, Martin Summerfield, joined the group and set out to build a liquid-propellant engine. But office space was so limited that for a time he worked out of the backseat of his car.

The group's initial task was to build reliable solid-fuel units, and for this purpose they could make full use of existing knowledge concerning propellants and explosives. They had an initial success in August 1941, when they mounted a group of test rockets to a 750-pound Ercoupe, one of the lightest aircraft available. "The plane shot off the ground as if released from a slingshot," von Karman later wrote. "None of us had ever seen a plane climb at such a steep angle." But these units proved unsuitable for military use, for they deteriorated in storage, forming cracks that led to explosions.

Further research found no combination of standard propellants that could do the job. Then John Parsons, the chief chemist, had a stroke of intuition. Most of these substances were powders, inherently brittle and prone to cracking. What if he were to use a fuel that wouldn't crack, such as asphalt or paving tar? Parsons melted a batch in a kettle, then stirred in a substantial quantity of potassium perchlorate, which would serve as an oxidizer. It worked! The propellant resembled stiff asphalt; it gave good thrust and would not crack. Indeed, in hot weather it tended to soften and flow, requiring that operational units be stored nozzle side up.

Here was a major advance in the field of solid propellants. In addition, Martin Summerfield was forging ahead in his work with liquid propellants, which also were designed to assist the takeoff of aircraft. The Navy's investigators, Goddard and James Wyld, had been following standard practice by using liquid oxygen. But this supercold liquid cannot be stored because it evaporates readily as it absorbs heat, and the Army wanted storable propellants. Summerfield thus set out to build a rocket engine that would use such substances.

Parsons told him to try red fuming nitric acid as an oxidizer—that is, commercial nitric acid with an admixture of nitrogen dioxide. Summerfield hoped to burn this acid with gasoline or kerosene, which were familiar fuels. He couldn't get the combinations to work, for his rocket engines experienced violent chugging. Amid rapidly pulsating pressures, the engines either would shut down or blow up.

No one at Caltech knew what to do, so Malina traveled to Annapolis to meet with his friend and fellow rocketeer, Robert Truax, who was running the Navy's rocket program. Truax, who was Goddard's boss, proved quite responsive, and one of his chemists suggested burning aniline with the nitric acid. The combination would ignite spontaneously when mixed, with no need for a separate igniter. Malina hoped that this rapid ignition would promote quick combustion, smoothing the engine's performance.

Summerfield tried it and in von Karman's words, "The results were spectacular. Nitric acid and aniline took to each other beautifully; the flame in the engine was absolutely steady." This breakthrough with liquid rockets matched Parsons's breakthrough in solid fuels. Malina and Summerfield not only had made the important invention of a liquid rocket with storable propellants, they had also shown how spontaneously igniting fuels could allow such an engine to run in a steady, reliable fashion.

By then, the United States had entered the war, and the Navy became interested in this work as well. Early in 1942 it offered a contract whereby Malina and his colleagues were to develop rocket units to help aircraft take off from carriers. Clearly, the armed services now would offer a ready market for production versions—and equally clearly, such a commercial venture could find no home amid the ivory towers of Caltech. But this problem would disappear if the rocket researchers set up a company that was separate from the university.

Accordingly, Malina suggested that von Karman take the lead in founding such a firm. General Arnold agreed that it would be a good idea, and Karman sought help from Andrew Haley, a friend who was also a lawyer. They met repeatedly in Karman's living room, along with Malina, Summerfield, Parsons, and Forman. The new firm needed a name, and Haley and Malina suggested Aerojet. The company would be selling rockets, not jets, but the name played to a growing fascination with jet propulsion while at the same time promising to steer clear of the Buck Rogers aura that was still attached to rockets. Haley filed the incorporation papers in March, with von Karman taking the post of president. For office space, the group settled on an auto salesroom in downtown Pasadena.

The venture was no more than a start-up, but it had clout in high places. This became clear when Haley was called to active duty in the Army as a legal officer. Von Karman lacked business know-how, and Aerojet suddenly found that with no one at the helm who knew about contracts and business law, they would receive no new War Department orders. The solution called for regaining the services of Haley, but only General Arnold could detach him from his military service. Karman phoned Arnold, and six hours later Haley was a civilian again, ready to take over as Aerojet's president. As Karman later put it, "The wonders of military efficiency at times are quite striking."

Still, though doors would open for Aerojet in Washington, they would not open so readily at the local banks. Like Reaction Motors, Aerojet needed working capital, and as with that New Jersey firm, bankers in Los

Angeles regarded rocketry as too dubious a business to justify a line of credit. Haley took care of this problem as well. His law firm's clients included General Tire and Rubber, of Akron, Ohio, and Haley knew that its executives were interested in acquisitions. He arranged for a buyout, and under its new name, Aerojet General, Haley's company continued its work.

By then, von Karman's rocketeers were running an engineering office in the Pasadena auto salesroom, a production plant in the nearby town of Azusa, and research facilities in the Arroyo Seco. You reached the Arroyo by driving over a dirt road that washed out when it rained, and if you got there, you would find rocket test pits lined with railroad ties. One employee declared that the corridors in the office building were so narrow that "you were liable to lose your teeth" if anyone opened a door while you were walking down a hallway. But though built with wartime haste, these facilities soon would grow in importance.

During 1943, Army intelligence analysts learned that Germany was building long-range missiles. Von Karman proposed a development effort that could yield a missile with a range of seventy-five miles. Captain Robert Staver of Army Ordnance, a liaison officer at Caltech, gave this proposal a strong endorsement and sent it upward through channels to Colonel Gervais Trichel, who headed Ordnance's rocket branch. In January 1944, Trichel asked Karman to take responsibility for a complete guided-missile program, which would include a guidance system. Karman agreed and Caltech soon had another contract.

However, as with Aerojet's production of rocket units for assisted takeoff, this new effort could not go forward as a project of the university. Caltech's aeronautics department could readily take on such research topics as the development of new propellants and the means to make an engine run smoothly when burning them. It would be another matter altogether for this department to carry out the focused development that could create anything so specific as a battlefield weapon. The work would go forward at the Arroyo Seco, and during 1944 the facility reorganized as a center for guided-missile research, taking the name Jet Propulsion Laboratory.

During 1944 and 1945 the Army undertook a buildup that added $3 million in new facilities, including laboratories, rocket test stands, a new office building, and a supersonic wind tunnel. This made the members of Pasadena's city council unhappy, for the city had agreed to lease the Arroyo Seco site only for the war's duration, and the new installations looked quite permanent. Their barracks-like appearance and the loud roars from rockets under test caused a major clash with upscale suburbanites who were building new homes in nearby Altadena and Flintridge.

When the Army sought to renew the lease, Pasadena would not go along; the city manager declared that the facilities violated "the first principle of proper zoning in residential territory." Army officials responded by threatening to condemn the land under eminent domain, and the city gave in. In time the rocket tests would go elsewhere, but JPL would remain.

Meanwhile, this center's focus on battlefield missiles offered new promise for liquid rockets. Aerojet and Reaction Motors had both gotten started with work on assisted takeoff, but this effort had slim prospects for the future. The Navy's carriers would soon have steam-driven catapults, able to launch aircraft without rocket assistance, while the Air Force would dispense with rocket boost by building longer runways and more powerful jet engines. Reaction Motors would find a niche by supplying engines for a small number of rocket-powered aircraft used in research on high-speed flight. But battlefield missiles constituted a separate path, which would guide the fortunes of JPL during the following decade.

Still another path involved sounding rockets, which would carry instruments to the unprecedented heights of a hundred miles and more. After the war, many scientists hoped to do research with these instruments. Astronomers wanted to view the sun in the far ultraviolet, whose wavelengths are absorbed in the lower atmosphere. Physicists expected to detect cosmic rays, which the atmosphere would absorb as well. Meteorologists looked forward to measuring temperatures and pressures in the upper atmosphere. And plenty of people were curious as to how the earth would look when photographed from these altitudes, which were much higher than any airplane or balloon could reach.

The war had hardly ended when the Army set the stage for a substantial effort in this area by bringing the best of Wernher von Braun's people to this country, along with a large quantity of key documents and equipment. As early as March 1945, Colonel Trichel in the Pentagon sent a request to Colonel Holger Toftoy, chief of Army Ordnance Technical Intelligence in Europe, asking for a hundred operational V-2 rockets that could be test-fired in the United States. To assist Toftoy, Trichel sent Robert Staver, now a major, who had midwifed the birth of JPL a year and a half earlier. He was to gather blueprints and documents, and find the project personnel.

Complete, ready-to-fire V-2s were virtually nonexistent, but Toftoy rounded up enough components to substantially carry out Trichel's request. Staver, in turn, located most of the top people of Peenemunde, along with a trove of documents. Yet von Braun and the others were not prisoners, for they had not been charged with crimes. The United States was not about to seize Germany's military scientists through outright capture; indeed, under occupation law, they were free to stay in Germany if they wished. Toftoy therefore met personally with the rocket experts, arranging to provide for their families and negotiating contracts whereby they would come to work in the United States. During subsequent months, von Braun led a contingent of 115 chosen specialists that crossed the Atlantic and traveled to a temporary home at Fort Bliss, Texas, in the desert near El Paso.

The scientists' immediate task was to assemble the V-2s from their components and fire them as sounding rockets. The test area, the new White Sands Proving Ground, was close to Fort Bliss and covered a tract of New Mexico land approximately the size and shape of Vermont. But while V-2s would indeed top the hundred-mile mark, at least occasionally, they were not well suited to this purpose.

The Army did not intend to put the V-2s back in production; when the limited supply from Germany was gone, there would be no more. In addition, the guidance system exercised control only during the powered boost. Following engine cutoff, there was no way to point the vehicle in a particular direction, such as the sun; indeed, it often would tumble or somersault rather than fly straight. What was more, it had been designed to carry a one-ton warhead, and it needed that much weight in the nose or the guidance system wouldn't work even during the boost. The scientists' instruments could not come up to that mark, so the V-2 would carry lead weights for ballast. As one project manager put it, "The more weight, the less altitude. We could see the miles of potential height drifting away as we poured pound after pound of lead—1,100 pounds in one case—into a V-2 nose."

But at JPL, plans for a small sounding rocket fitted in neatly with the lab's overall missile program. It had gotten its start late in 1944 with the Private, a missile eight feet long that burned John Parsons's solid fuel, with its mix of asphalt and potassium perchlorate. Next on the agenda was Corporal, thirty-nine feet long with a range of over sixty miles, flying with a guidance system and burning Summerfield's propellants, red fuming nitric acid and aniline. This proposed program led the chief of Army Ordnance research, Major General Gladeon Barnes, to ask von Karman how high up in rank he intended to go. "Certainly not over colonel," he replied. "That's the highest rank that works."

But to develop Corporal would represent a big jump, and Malina argued that JPL should carry out an interim project: the building of a smaller liquid-fuel rocket that could use an existing engine and would require no guidance. This project would lay the groundwork for the full-size Corporal; it also would fly as a high-altitude sounding rocket, fulfilling a hope that Malina had cherished since beginning his work in 1936. After Colonel Trichel gave his approval, the new rocket needed a name. It would be a smaller, slimmer sister of Corporal. The men at JPL called it the WAC Corporal, from Women's Army Corps, although the Army later declared primly that WAC meant "without attitude control."

The WAC Corporal stood sixteen feet tall and weighed 655 pounds when fueled. That made it only about the size of Goddard's rocket of 1940, but it was far simpler, lacking both guidance and turbopumps. However, while Goddard's rocket had drawn on his one-man design effort and never flew successfully, the WAC Corporal drew on the wartime work of Caltech and JPL. It made a brilliant flight on its first try, in October 1945. It got up to speed within a tower; then, relying on fins for arrow stability, it soared to forty-four miles.

But while the V-2 was too clumsy for effective use as a sounding rocket, the WAC Corporal was too small. It could carry only twenty-five pounds of payload. This limitation caused concern at two naval centers, the Naval Research Laboratory and the Applied Physics Lab at Johns Hopkins University. The latter had somewhat the same relationship to its parent university as did JPL to Caltech: that of a military lab operating under contract. At APL, two senior managers asked a staff scientist, James Van Allen, to look at the available sounding rockets and to see whether a new one might be worth developing.

Van Allen's inquiries led him to Aerojet, which was already building the WAC Corporal. He invited that firm to prepare a proposal, and in May 1946 the Navy's Bureau of Ordnance gave Aerojet a contract to build a new sounding rocket, the Aerobee. It closely resembled the WAC Corporal and would also fly from White Sands, but it would be somewhat larger, carrying 150 pounds of instruments to seventy miles. The first shot, in November 1947, was only partially successful but nevertheless approached the WAC Corporal's altitude. The second, the following March, reached seventy-three miles and scored a complete success.

For the rocketeers of White Sands, these flights provided bright moments in a dreary existence. The missile center was nothing more than an outpost amid the sagebrush, adjacent to a two-lane highway that extended across the flat desert to Alamogordo. A single large Quonset hut served as a hangar. Most other buildings were single-story barracks, and you could walk across the entire base in five minutes. Hot, windy weather prevailed, and at times rations dwindled to beans and cheese. Some people passed the time in all-night poker games; others crossed the border for a night in Juarez. But soon they would have company, for another sounding-rocket project was under way: the Naval Research Laboratory's Viking.

The project name was suggested by a staff scientist, Thor Bergstralh. The Viking itself was an earnest of the hope at NRL that the Navy might someday enter the field of long-range rockets by launching them from ships. Like carrier-based aircraft, these missiles then would add their range to that of their vessels. Milton Rosen, an NRL scientist with a strong interest in guided missiles, conceived the project and took the lead in winning the necessary approvals. His background was in electronics, and at the outset he hardly knew a rocket from a hole in the ground (of which the White Sands flights would soon be making plenty). A seven-month course in the workshops of JPL, where he learned his new trade through hands-on experience, gave Rosen a useful introduction. But like so many others in this field, he would learn what he needed as he went along.

His initial goal was to carry five hundred pounds to a hundred miles, which called for a rocket that could approach the performance of the V-2. However, the Germans had built their missile of heavy steel, to survive the heat of atmosphere entry and come down in one piece. Viking would save weight by being built of aluminum and came in at five tons when fully fueled, barely one-third the weight of a V-2. At Reaction Motors, John Shesta took the lead in designing its engine, which gave twenty thousand pounds of thrust.

Viking also broke new ground with its guidance and control. In designing any missile guidance system, a persistent problem was the determination of appropriate values for system parameters. A new method of analysis developed by Albert Hall, a professor at MIT, made it much easier to do the mathematics. In Rosen's words, "We could predict what would happen" during an actual flight.

This mathematical method made it possible to advance beyond the V-2's method of in-flight steering, which dipped heat-resistant vanes of carbon into the rocket exhaust to deflect it. It was wasteful, costing the V-2 up to 17 percent of its thrust. The Viking introduced the gimballed rocket engine, which would hang within pivoted supports, physically swinging to point its exhaust in an appropriate direction. Hall's analysis made it possible to calculate the dynamics of the rocket and its gimballed motor together, and ensured a successful design. After Viking, almost all large liquid-fueled rockets would use gimballed engines.

Viking also set a mark that far fewer subsequent projects would meet, for the entire program cost no more than $5 million. This sum paid for development of a completely new engine and guidance system, for two versions of the rocket, and for twelve flights. The second version, which carried more fuel, lifted 825 pounds of instruments to 158 miles on its best flight, in May 1954. And Viking itself proved fairly reliable. Following three initial launches that all featured premature engine cutoff, the program's remaining nine launches delivered seven strong successes and only one complete failure.

"There were only twelve men on the Viking launch crew," Rosen would write. "The Martin Co., which built the rocket, had no more than about two dozen engineers involved in the design. There were never more than about 50 men involved in building the vehicle. Finally, the government project group consisted of two men at first, and was never more than four. These people wrote the specifications, negotiated the change orders, analyzed flight and test data, and wrote all of the final reports on Viking. That is why you could buy a Viking vehicle for $250,000 and conduct an entire Viking operation for $500,000."[3]

These sounding rockets broke new ground in other respects. Carrying cameras, sometimes with color film, they protected the exposed film in strong containers that could survive the fall to earth and took astronauts'-eye views long before there were astronauts. High over El Paso, with the horizon over a thousand miles away, Viking cameras reached deep into Mexico and showed the Pacific, beyond Baja California. Aerobee did even better. In 1954, a hundred miles up, it photographed a well-developed tropical storm over southern Texas that resembled a hurricane. Ground-based observers knew that the weather was stormy, but they had failed to understand why. This incident gave dramatic proof that high-altitude photos could be of great value to meteorology, and it pointed toward the weather satellites of the future.

The V-2 and WAC Corporal also made their mark. They combined to form the first two-stage rocket, with the small one riding atop its booster. Rocket staging was technically demanding, for the upper stage would have to stand up to the stress and acceleration of the booster before igniting at high altitude, far from technicians. But the WAC Corporal used a pressure-fed engine, along with propellants that would ignite on contact. It needed no igniter, turbopump, or guidance system; all it had to do was open its valves at the proper time. In this fashion, in February 1949 it reached 244 miles, up where spacecraft would eventually fly, and set an altitude record that would stand until 1956.

Thus, an American community of rocket specialists, inspired by Goddard and growing out of the self-taught amateur experimenters of Caltech and the American Rocket Society, had drawn on federal funds and gained a modest but meaningful foothold in the areas of sounding rockets and battlefield missiles. Still, this work was a far cry from building all-annihilating missiles that could carry a nuclear bomb. Many people were ready to speculate about such weapons, in the wake of Peenemunde and Hiroshima. But Washington wasn't interested. Vannevar Bush, who had directed the wartime Office of Scientific Research and Development, made this clear in Senate testimony in December 1945:

> There has been a great deal said about a 3,000-mile high-angle rocket. In my opinion such a thing is impossible and will be impossible for many years. The people who have been writing these things that annoy me have been talking about a 3,000-mile high-angle rocket shot from one continent to another carrying an atomic bomb, and so directed as to be a precise weapon which would land exactly on a certain target such as this city.

> I say technically I don't think anybody in the world knows how to do such a thing, and I feel confident it will not be done for a very long period of time to come. I think we can leave that out of our thinking. I wish the American public would leave that out of their thinking.[4]

No one could doubt that missiles of this type would call for far more than the few millions of dollars that would bring forth Viking. The atomic bombs of 1945, both in their uranium and plutonium versions, weighed five tons. A rocket to carry one bomb to Moscow would have to weigh several hundred tons—compared to fourteen tons for the V-2. Its nose cone, entering the atmosphere at four miles per second, would burn up like a meteor from aerodynamic heating. And at intercontinental range, no guidance system existed that could ensure a hit. The missile would indeed deserve its name, for it would miss. Even if its warhead could somehow survive the plunge through the atmosphere, it would waste its energy on the cows and peasants of nearby collective farms, leaving Moscow unscathed.

Yet Bush's point of view drew on a larger argument: The United States had no need of such rocket weapons. The Twentieth Air Force had proved itself the master of air power, burning Japan with vast raids of B-29 bombers, then administering the coup de grâce at Hiroshima and Nagasaki. And the United States was already reaching well beyond the B-29. Its B-50 would amount to a B-29 with engines of far greater power. The B-36, mounting six of these engines compared to the B-50's four, would soon enter flight test. It had the range to fly across the Atlantic from the East Coast, drop a plutonium bomb on Moscow, and then return, all in a single unrefueled flight.

Over the next several years, new developments would continue to play into the hands of advocates of bombers. In-flight refueling would stretch the range of these aircraft at will. Jet bombers would offer higher speed and altitude, making them harder to shoot down. Overseas bases, close to the Soviet borders, raised the prospect that jet fighters would escort the bombers, providing further protection. And above all was the simple fact: We had the bomb and the other fellows didn't. Until they did, and until they gained the means of presenting a serious strategic threat, long-range missiles would continue to await their day.

Yet while the Air Force had no immediate intention of building such missiles, it appeared only prudent to understand what specifically was required to create them. In October 1945, five weeks prior to Vannevar Bush's Senate appearance, the Air Technical Service Command sent letters to officials at America's main aircraft companies, inviting them to prepare proposals for a ten-year program of development involving four categories of missiles. These would extend from a low range of twenty miles to a high of five thousand miles. At the firm of Convair in San Diego, the study that resulted drew particular attention from a specialist in aircraft structures, Karel Bossart, whom everyone called Charlie.

Bossart had grown up in Belgium, and after earning a degree in mining engineering at the University of Brussels in 1925, he had come to MIT on a fellowship. The school introduced him to aeronautics. He went on to make his name in the field of structures, which demand considerable strength for the lightest possible weight. He spent the war years as an aircraft man, and when he learned of the V-2, he decided it was a big waste of money. But when the contract for the Air Force study arrived, Bossart saw a challenge he couldn't resist. He went to his chief engineer and got the job of running it.

Bossart's group proposed three projects, all aimed at building missiles of 5,000-mile range. Design A would be a cruise missile, a pilotless jet aircraft that would fly its one-way mission by relying on an advanced autopilot. Design B would be a test rocket that somewhat resembled the V-2; it would permit Convair to gain experience in the new field of large liquid-fueled rockets. Design C then could take shape as the long-range rocket for which Vannevar Bush had expressed such disdain. These concepts needed names, and Bill Lester, a propulsion engineer, supplied them. He christened Missile A the Teetotaler, because it was the only one of the three that would not use alcohol as its fuel. Missile B, the test rocket, received the name Old Fashioned, because of its similarity to the V-2. Project C, the intercontinental rocket, would carry the atomic bomb; Lester called it Manhattan.

The Air Force provided $1.9 million, and Bossart believed this would suffice to pay for some hardware. There was no hope of building Manhattan, but the smaller Old Fashioned lay well within reach. It now received a third name, MX-774, which was the Air Force designation for the overall study. Bossart now set to work with a will on what he saw as the main problem: to use his background in structures so as to reduce the missile's weight to a minimum. Other people might talk of powerful engines, but Bossart knew that the adroit use of his specialty might lead to a lightweight rocket that would fly to Moscow with particular ease.

Von Braun had built the V-2 of steel, stiffening it with a supporting framework and building its propellant tanks as separate containers within this outer shell. Bossart saw that he could save weight in all three areas. He would use aluminum, naturally. He would allow the outer skin to do double duty by holding the propellants, thereby dispensing with internal tankage. Then, going further, he saw that he could also dispense with the stiffening framework. Instead he would support the outer skin, and keep it from collapsing, by pressurizing its interior with nitrogen. His missile would then resemble a balloon, which needs only modest internal pressure to maintain its shape quite effectively. It would be two-thirds the size of a V-2, but little more than one-eighth its empty weight.

There was money in the budget for a simple guidance system based on an autopilot, but when it came to a rocket engine for the MX-774, Bossart would have to take whatever was available. Again it was Reaction Motors that stepped up to the plate, offering a version of its four-chamber XLR-11 that the builders of the X-1 research aircraft, operating on their own financial shoestring, had already picked for their own project. This version used a turbopump to feed propellants at a faster rate, stepping on the gas, as it were, to boost the thrust to 8,000 pounds. Bossart deliberately copied the shape of the V-2; this allowed his aerodynamicists to use German wind-tunnel data in their studies.

Then the Air Force, beset by budget cuts, canceled the Convair contract. Nevertheless, the company still had some unexpended funds, which it could supplement with money of its own. Bossart won permission to build three MX-774 rockets, which flew at White Sands during 1948. They were designed to reach hundred-mile altitudes, but all three experienced premature engine cutoff and fell short, reaching no higher than thirty miles. In the wake of the third flight, Bossart's engineers found the problem and hoped for another try that might bring full success. But it wasn't in the budget.

Convair had already lost the long-range cruise missile, Teetotaler, to Northrop Aircraft, the firm that would go on to build a similar pilotless craft called Snark. Convair tried to pitch its MX-774 as a sounding rocket, but it lost out to the larger and more capable Viking. Manhattan, in time, would take shape as Atlas, the nation's first true intercontinental ballistic missile (ICBM), whose design would draw heavily on that of Bossart's MX-774. But all this lay far in the future, beyond the reach of current Air Force plans. Bossart had made a gallant effort and had gained experience that eventually would prove almost priceless. But for now he was stymied by the Air Force's lack of interest in big rockets and would have to wait for better days.

The Air Force would soon be cultivating a growing interest in missiles, but these would be jet-powered cruise types, amounting to pilotless aircraft. The 5,000-mile Snark would be one; another was the Martin Co.'s Matador, with a range of 600 miles. They could fly with standard jet engines, needing no rockets, and they didn't pose the problem of burning up like meteors. They would need highly capable autopilots; the Snark would have to hold a steady course for ten hours or more, whereas the autopilots of World War II had kept their planes flying properly for only fifteen minutes or so. But in most other respects, the technical demands of these cruise missiles appeared fairly conventional.

Yet the opportunity existed to use this interest in cruise missiles as a springboard for a far-reaching research effort that could indeed lay a foundation for the ICBM. The firm that seized this opportunity was North American Aviation, and the man who led the effort was William Bollay. His work established the main line of American rocketry.

Bollay had been born in Germany in 1911. The name is Huguenot in origin; his ancestors had fled religious persecution in the France of Louis XIV, and his father was an officer in the German army. Then in 1924, amid the ravages of postwar inflation, his family left their home in Stuttgart and emigrated to Evanston, Illinois, where the senior Bollay found work running a coalyard. The younger Bollay, staying in his hometown, went to Northwestern University. Then in 1933, as his widow Jeanne recalls, "he came bounding across the campus one day, his eyes sparkling. He'd gotten a $300 scholarship to Caltech."

The couple was still a year away from getting married, but she followed him to the West Coast, enrolling for her senior year at UCLA while he entered graduate school. "I'd go over to Caltech every weekend, in the big red streetcar," she remembers. "At Caltech, there wasn't much time for courting. Mostly, I sat by while he studied." His work won the attention of von Karman, who picked him as an assistant. They did calculations for the dome of the Mount Palomar observatory; they tried to understand the hydrodynamics of oil wells. "Von Karman would start thinking about ten P.M., and they'd work till three in the morning," Jeanne continues. "Then Bill would come in bleary-eyed, and he'd have an eight A.M. class the next day."[5]

Bollay also fell in with Frank Malina. It was customary for the graduate students to give seminars, and Bollay gave one on building a rocket-powered airplane that would reach 1,200 mph. He also joined Malina in the Arroyo Seco, crouching behind a pile of sandbags to conduct an early rocket test. But despite his interest in rockets, he was not about to hitch his career to anything so speculative. His Ph.D. dissertation was a solidly von Karmanesque piece of research on the theory of wings, which dealt with nothing more futuristic than standard aerodynamics. Bollay stayed on for a while at Caltech as a young instructor, then went to Harvard as a junior faculty member. There he built a wind tunnel, in the basement of a campus building, using surplus electric motors from the Boston subway.

Nevertheless, he could hardly escape the looming prospect of war. The Nazis had an organization of U.S. supporters, the German-American Bund, and one of its members approached Bollay, hoping his German descent would make him recruitable. Jeanne recalls his response: "Bill was so outraged by the idea of helping them, he went out and followed his brother Eugene, who had earlier joined the Naval Reserve." Eugene was a meteorologist for whom naval service offered a career in the Weather Bureau. For William, the Navy would offer a path to a career in rocketry.

In September 1941 that service ordered him to active duty. He arrived in Washington—and found he would be dealing with metal fasteners such as nuts and bolts. That didn't appeal to him, but he was ready to pull strings. Some of his Caltech colleagues now were high-ranking naval officers, and Bollay asked them for help. Within two days, he was transferred to the Power Plant Development Branch of the Bureau of Aeronautics, in Annapolis.

This was the research center where Robert Truax was working on rockets with Robert Goddard, while simultaneously managing contracts with Reaction Motors and Aerojet. Truax and Bollay became close friends. Another manager had similar responsibility for ramjets. Bollay himself won the key position, in the field of jet engines. The center was unique in that it provided a small organization where a handful of people would direct developments involving all three types of engine, and Bollay followed the developments closely.

The ramjet, or "flying stovepipe," was particularly promising, for it was the simplest engine imaginable. It was just a carefully shaped length of tube fitted with fuel injectors. At high speed, air would ram in the front, then burn the injected fuel and become hot. The hot airflow would blast out the back, giving thrust.

Initial work with ramjets was going forward in the same impromptu fashion as the work with rockets, but it was bringing results. At the Applied Physics Lab in nearby Baltimore, one group used nothing more complex than the exhaust pipe from a fighter plane as their tube. A cluster of small solid-fuel rockets provided the initial boost, accelerating the ramjet to a speed where the air-ramming effect would come into play. In June 1945 this makeshift arrangement reached 1,400 mph, nearly twice the speed of sound and more than twice as fast as the hottest jet fighters of the day.

Bollay's own work focused on the design and production of turbojets. A British inventor, Frank Whittle, had built the first of these and Bollay was eager to get them into production. By the war's end, the Navy was seriously pursuing the development of jet aircraft. North American Aviation, in Los Angeles, was developing the FJ-1 Fury, which would become one of the first carrier-based jet fighters.

This fighter project led Bollay to North American, where he would build up the nation's premier rocket-development group. J. Leland Atwood, the president of the firm after 1948, recalls that the end of the war had brought a dramatic falloff in North American's prospects. During the war, the company had been a mainstay for the nation's wartime aircraft production. "We had 90,000 employees at the peak," Atwood declares.

But by the fall of 1945, amid sweeping production cancellations, the firm was down to only 5,000. At the nadir, it had a backlog of only two dozen aircraft on order.

Still there was some work, and it was largely in the new area of jet-powered fighters and bombers. To Atwood and to his boss, company president James "Dutch" Kindelberger, this work represented the way to the future. In Atwood's words, "It was quite apparent that the country would be needing new military aircraft, and we would participate."

A host of yeasty technologies had come out of the war: jet planes, rockets, radar and other electronics, automatic control, atomic energy. Kindelberger decided to bring in the best scientist he could find and have him build up a new company research lab staffed with experts in these fields. An executive recruiter, working in the Washington area, recommended Bollay.

He and his wife Jeanne, along with their infant daughter, Melodie, arrived in Los Angeles by Thanksgiving of 1945. They bought a big, rambling home on the bluffs of Pacific Palisades, across a street from the beach. Bollay set up shop in a red-brick company building close to the airport. It had been called the tooling building but now became the Aerophysics Laboratory.

As Bollay was settling in, his old mentor Theodore von Karman was pointing the way to the Air Force's future. At the request of its commanding general, Hap Arnold, Karman wrote a report titled "Toward New Horizons." He predicted that the future would bring supersonic jet fighters, long-range rockets tipped with atomic bombs, and satellites in orbit. Thoroughly pleased with the report, General Arnold told Karman that it would be used "for some time to come as a guide to the Commanding General in discharging his responsibility for research and development." There was little money for such efforts, but with Arnold's support, Bollay and his colleagues could go forward.

Their rocket research began in a company parking lot, with parked cars only a few yards away. The steel blade of a bulldozer's scraper shielded engineers in case an engine blew up. Some engines were so small that they seemed to whistle rather than roar, and Atwood recalls, "We had rockets whistling day and night for a couple of years." Bollay, who had built a wind tunnel at Harvard, now built a supersonic wind tunnel in this new lab. He also began to bring in specialists in gyros, electronics, and rocketry. These included John Parsons, the propellant specialist from Aerojet. "We were feeling our way," says Atwood. "We had no coordinated plan; but what we were picking up was all aimed toward propulsion, aerodynamics, or control. Those were the pillars or legs on which things would be developed."[6]

The small home-brewed rockets would give his people a basic introduction to the field, but Bollay knew very well that the world's best rocket experts were Wernher von Braun's specialists, who now were cooling their heels at Fort Bliss. Bollay recruited several, including Dieter Huzel, who had been von Braun's assistant. Soon the North American people were working with original V-2 blueprints, building and testing copies of its rocket engine.

It was the world's most powerful rocket motor, but Bollay could see plenty of room for improvement. A particular bone of contention lay in its arrangements for injecting and mixing propellants within its thrust chamber, for the immense V-2 had performed this critical task with nothing more than technology from the engine of the much smaller A-3. The A-3 had used a "burner cup," an inverted cup with a brass stud protruding from the center. Alcohol sprayed inward from small holes in the cup's wall; liquid oxygen streamed outward into the cup through similar holes in the central stud. This was suitable for the 3,300-pound thrust of the A-3 engine. But the much larger A-4 needed more.

It needed an injector, a flat circular plate mounted at the top of the thrust chamber and pierced with channels, like a big showerhead with its holes. Fuel and liquid oxygen, spraying from the channels like water from the showerhead, then would mix and burn. But injectors take time to develop, and Walter Thiel, head of engine development, had been in a hurry. To inject the A-4's propellants, he settled instead on arranging eighteen standard A-3 burner cups at the top of his thrust chamber. Each of them needed its own liquid-oxygen line, leading a colleague to describe this "eighteen-pot motor" as "a monstrosity and a plumber's nightmare." Thiel tried to build a true injector but ran into difficulties, and as one commander wrote, "The war is not going to wait for Dr. Thiel." Thiel died in a British bombing raid in 1943, and the war was nearly over before his engineers succeeded in building a workable version.

For Bollay, a suitable injector offered much more than a simpler engine layout. It could pave the way to rockets of much greater thrust, because injectors for such engines might take form as straightforward scale-ups of a basic design. But Bollay could not carry out the necessary tests in his parking lot; he needed a major set of rocket-test facilities.

"We scoured the country," recalls Atwood. "It wasn't so densely settled then—and we located this land at Santa Susana Pass, on the hilltop." It was owned by a family named Dundas, who allowed Hollywood film crews to use it for Western movies. The land was stark and sere, full of rounded reddish boulders, offering spectacular views of the adjacent San Fernando Valley. However, it was so rugged that Bollay's associates had to use a jeep to get around.

Early in 1947, North American leased the land and proceeded to erect a rocket test center, at a cost of $713,000 in company money. That was quite a lot amid the postwar stringency. But it amounted to a declaration that rockets represented the company's future, and that Atwood and Kindelberger intended to be at the forefront and to lead.

By then, Bollay was developing a highly original approach to the problem of designing a long-range guided missile. He was well aware of Air Force interest in jet-propelled cruise missiles, but he knew that these would fly below the speed of sound and might easily be shot down by interceptors or antiaircraft fire. A rocket-powered ICBM would fly so high and fast as to be invulnerable, but its technical problems put it out of reach. Bollay's approach promised to combine some of the most attractive features of the two types of missile.

He called for a winged rocket that would cruise at high speed using ramjets. It then would fly much higher and faster than if it used ordinary turbojets, and would not fall so easily to enemy fire. Yet its speed would not be so great as to make it burn up like a meteor. Its guidance system could continue to draw on the technology of aircraft autopilots, and might not be so demanding as the guidance of an ICBM. Its rocket engines would also be far less demanding—so much less, in fact, that work could proceed on the engines immediately. The concept was christened Navaho, reflecting a penchant at North American for names beginning with "NA." With strong Air Force support, Navaho would quickly emerge as the focus for postwar development of rockets.

Navaho's early design showed that Bollay's thinking continued to reflect the powerful influence of the V-2. It featured a large rocket resembling the V-2, with a single engine in the tail along with a pair of stubby wings and upper and lower vertical fins. Each fin mounted a ramjet at the tip, Buck Rogers–style. The Germans had built a similar winged V-2 in 1945, though without the ramjets. Gliding at over 2,600 mph, it was to stretch the V-2's range to 750 kilometers, enough to reach Glasgow. With the ramjets, Bollay expected that his Navaho would achieve a range of a thousand miles.

Bollay would not try to build the ramjets; the Air Force left that part of the problem to Wright Aeronautical Corp., a major builder of aircraft engines. But the rockets would be all his own. Lieutenant Colonel Edward Hall, who was funding the work from the development center at Wright-Patterson Air Force Base, states that "Navaho was the main line of things because, although it was a cruise missile, we could put into it a large enough rocket motor to allow us to turn out engines that would be near the size we would need for a ballistic missile." From the start, then, it was clear that Navaho would help lay groundwork for an ICBM.

With his Air Force counterparts, Bollay settled on a thrust of 75,000 pounds. That would improve on the V-2's 56,000-pound thrust. Still, during 1949 it became clear that the work was not going well. "Engineers have a tendency to redesign and redesign, over and over," recalls Jim Broadston, the test director. Bollay thus needed a manager who would make decisions and push the project. He found his man in Sam Hoffman, a salty aircraft-engine leader who had been a chief engineer and had then become a professor at Penn State. He had dealt with Bollay during the war, and his combination of industrial and academic experience closely matched Bollay's own.

"I came out to get a rocket engine for them," Hoffman would declare. "Bill had a group of brilliant young fellows with no practical experience—which probably helped them with the new things that were coming along. Bill wanted me because I knew how to build engines, had built them, and brought practical experience to this young group. They all were a generation after me. I'd had a career before, but these guys hadn't, except for being in the army and the navy."[7]

By March 1950 the first version of Bollay's engine was ready for testing at full thrust at Santa Susana. It appeared tiny, lost amid the stark propellant tanks and looming steel girders of its massive test stand. The stand was not beautiful, but it showed integrity and purpose. It contrasted sharply with the reddish rock formations of "the Bowl," a craterlike ravine that served as the test area. Several officials were present, including von Braun. If all went well, a bright sword of yellow-white flame would stab downward from amid the girders, shaking the rocks with its roar, a roar that would grow louder as it reverberated off the canyon walls.

The test conductor called, "Oxygen valve open!"—and the rocket blew up in a sudden explosion. A designer had specified that a key part be made of mild steel, not knowing that this commonly used metal would become brittle when chilled with liquid oxygen, and would shatter. Kindelberger, who now was chairman of North American, sought out the luckless engineer and chided him: "You should have made the blasted thing out of solid gold!" The specification changed to stainless steel, which stood up well when very cold. Later that spring, the engine was working well in tests.

By then the design of Navaho was changing dramatically. Bollay and the Air Force had established that by leaping beyond approaches based on the old V-2, they could anticipate vast increases in range, to 3,000 and even 5,500 miles. However, the new Navaho would not be a winged rocket with auxiliary engines. It would be an unmanned supersonic airplane of stunningly rakish appearance.

This missile would feature delta wings, small forward-mounted winglike canards, a double vertical tail, and a long, slender, pointed fuselage.

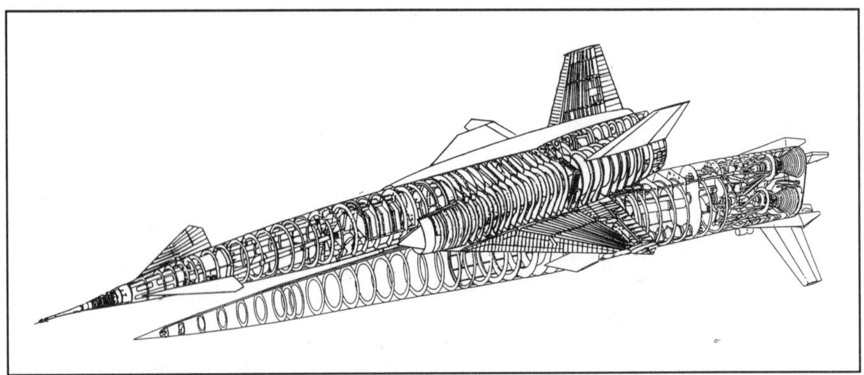

Navaho missile with its booster, in cutaway view. (Courtesy James Gibson)

Two of the largest ramjets then conceivable, each four feet in diameter, would provide the thrust. The guidance system would pick out selected stars even in daytime and use them for navigation; these stars would provide reference points that could keep the gyros aligned and pointing in their proper directions. The missile would cruise at 60,000 feet and 1,800 mph, and it might feint or zigzag along the way. Then, with its fuel burned off, it would approach the target at 80,000 feet. Using infrared sensors, it would feel for the warmth of a city under blackout.

This missile would need more rocket power for its boost than the 75,000-pound engine could provide. Bollay therefore decided to pursue an engine of 120,000 pounds of thrust. Two such engines, mounted at the base of a large booster rocket, would propel Navaho during launch. Rising into the sky, the ramjet-powered missile then would ride on the back of this rocket booster, much as the space shuttle rides its propellant tank today. Then, high in the stratosphere and at supersonic speed, the Navaho's ramjets would kick in and the missile would separate from its booster to fly on.

In developing this new engine, engineers drew on experience with the injector from the previous project. But a stress analyst, Matthew Ek, pointed to problems with the thrust chamber and nozzle. These were to follow earlier practice by being built as shells, one inside the other, with tubing for regenerative cooling in between. Ek showed that this standard approach would lead to stresses that would limit the thrust and chamber pressure, and hence the engine's power.

The alternative demanded nothing less than to build the engine entirely from this tubing, with the tubes brazed together. At Reaction Motors, the designer Ed Neu had pushed this approach strongly. No sheet-metal wall then would separate the cooling passages from the fierce heat of combustion. But with this form of construction, and with metal hoops surrounding the brazed tubes for structural strength, future engines could grow as large as anyone might wish, and could provide arbitrarily high levels of thrust.

With this newest advance, the 120,000-pound engine demonstrated the design of America's future rocket motors in definitive form. This engine burned alcohol and would give way to an advanced version that would burn more energetic fuels obtained from kerosene. Then this version, increasing in thrust, would power most of the liquid-fueled missiles that came to the fore during the 1950s, including the Atlas ICBM. It would provide thrust for launching satellites and spacecraft, beginning in the late 1950s and continuing to the present day. And by scaling up its basic layout, Bollay's engineers would build the F-1, with 1.5 million pounds of thrust. It would drive the Saturn V moon rocket of the 1960s, carrying astronauts to the moon.

By 1950 the Aerophysics Laboratory had grown into a major corporate enterprise, with over 1,600 employees. A few years later it would spin off two formal divisions of North American, Rocketdyne and Autonetics. Rocketdyne would become America's premier builder of liquid-rocket engines, overshadowing Aerojet General, its only serious rival. Autonetics grew into a leading builder of guidance systems. With these in-house divisions, North American would build on its Navaho experience to become the nation's principal builder of manned rockets and spacecraft.

In 1950, all this lay well in the future. Bollay meanwhile was maintaining and even strengthening his ties to the academic world, taking a part-time professorship at UCLA. "He was a quiet, direct person," recalls Paul Castenholz, a student whom Bollay hired in 1949. "He was not flamboyant; he was to the point. Bill was tall, slender, looked active. You saw him and you felt he was a dynamic person, ready to go somewhere." Jim Broadston remembers him as "a charming person. He looked well-groomed, as an engineer should." Jeanne Bollay uses the same adjective as Castenholz: "He was quiet; he thought a long time. Very amiable; the only problem I ever had with him was that he'd be off in some other world. He had a wicked sense of humor, in an understated, English way."

In his heart, Bollay remained an academic. As with von Karman, he delighted in attracting highly intelligent people with new ideas, nurturing them, and watching them grow. "Bill brought into North American an unusual bunch of highly gifted people," remembers his brother, Eugene. "And they were all like him. They all were pushing like mad to be the top in their particular field."

Chauncey Starr, who became a national leader in the field of atomic energy, was one of them. Others were Sam Hoffman, who built the rocket engines for the Apollo moon-landing program; Dale Myers, who became NASA's director for manned space flight; Paul Castenholz, who designed the main engines for the space shuttle; and John R. Moore, who became president of Autonetics. There were others as well, in guidance, control, aerodynamics, and other fields.

However, as the Navaho effort grew, Bollay increasingly found himself at odds with his boss Larry Waite, a corporate vice president. "Bollay was oriented to university-type approaches," says Tom Dixon, one of his rocket managers. "He wasn't interested in 'management systems.' Larry Waite was an adamant fellow; he disagreed with the way Bill wanted to run things, and Bill just couldn't get along with him." Lee Atwood, who was Waite's boss, has similar recollections: "Larry was from MIT and was technically able, but he shouldn't have held that job. Eventually we moved him over to Contracts and Pricing."

But in 1951, with their disagreements at a height, Bollay left North American and his Navaho project. He set up a new company, became deeply involved in battlefield missiles for the Army, and ceased to remain involved in rocket-engine development.

Just then the Air Force was only three years away from a commitment to build the Atlas ICBM, with the highest military priority.

3

Racing to Armageddon

The Superpowers Begin Their Missile Programs

AT WAR'S END, American air and sea power bestrode the world like a colossus. The bombers of the Army Air Forces, assisted by the U.S. Navy, had broken Japan's will and had forced that nation to surrender without a single American soldier setting foot on its shore. Stalin had the world's most powerful army, which had crushed the Nazis. But he had no navy or bomber force to speak of, and he feared that America would use the atomic bomb to launch a war against his own state. He needed both nuclear weapons and long-range bombers, and he felt their lack keenly. He wished passionately to break America's nuclear monopoly.

In pursuing this goal, Stalin would show the same single-mindedness with which he had built the industrial base that supported his army. "To slacken the pace would mean to lag behind, and those who lag behind are beaten," he stated after launching his first Five-Year Plan in 1928. "We do not want to be beaten. No, we don't want to be. We are fifty or a hundred years behind the advanced countries. We must make good the lag in ten years. Either we do it or they crush us."[1]

To make good the lag, prior to the war, he relied heavily on industrial espionage. Amtorg, a trading organization in New York, acted as a front for the NKVD in this effort. Because the Soviet Union had virtually no modern industry at the outset, the processes Stalin sought to uncover were often basic indeed, involving no more than production of common industrial chemicals. The appropriate processes were available for purchase under license, but Moscow lacked reserves of hard currency. Espionage saved the cost of paying license fees.

When the war came, Stalin's industrial spies gave him a strong base from which to ferret out the secrets of America's atomic-bomb effort. The NKVD acquired some ten thousand pages of pertinent documents.

Nevertheless, the head of the Soviet atomic program, Igor Kurchatov, failed to win the funding that would allow him to take the costly initial steps toward building the weapons. The prospects for an atomic bomb seemed too uncertain to justify a major commitment.

At the end of the war, then, Kurchatov knew what to do, for he had carefully studied the NKVD's trove of stolen reports. But he had nothing with which to do it. Then came Hiroshima and Nagasaki, which suddenly raised nuclear arms to the highest level of concern. Stalin summoned the appropriate officials and said, "A single demand of you, comrades: Provide us with atomic weapons in the shortest possible time. Provide the bomb—it will remove a great danger from us."

Stalin would need long-range bombers as well, and the planebuilder Andrei Tupolev was already addressing this issue. Like Kurchatov, he would proceed by copying American designs, but he could go Kurchatov one better because he already had in hand several examples of the B-29, the world's best bomber. A B-29 had made a forced landing near Vladivostok in July 1944. Later that year, two others fell into Soviet hands in the same manner. Similar aircraft had carried the first atomic bombs, and Tupolev's copy, the Tu-4, would carry such weapons as well.

"The B-29 looked like something from the future," the historian Steven Zaloga notes. "The gun turrets were unmanned and lacked the normal Plexiglas dome. They were remotely controlled from central gunners' stations. The aircraft had many electronic systems that had no Soviet counterparts, including navigation radars. Although Soviet engine designers had attempted to develop turbochargers for aircraft engines, none approached the complexity or performance of the General Electric systems on the B-29's engines."[2] To overcome these difficulties, Stalin gave the Tu-4 effort the highest state authority.

In pursuing the atomic bomb, Kurchatov's first nuclear reactor was virtually a duplicate of an American design of 1944, except that it required more uranium to make up for impurities in his own supply. His plutonium bomb posed the difficult problem of implosion, of surrounding the fissionable core with high explosives that could detonate in a precise manner, producing a spherical shock wave that would compress and implode the core rather than blow it apart. Here, too, Kurchatov relied on the experience of Los Alamos. His eventual design would duplicate the Fat Man weapon of Nagasaki.

In entering the field of long-range ballistic missiles, the Soviets quite naturally used the V-2 as their point of departure. Here, however, they would do more than merely make copies. At the outset, they would rebuild enough of the German effort to put the V-2 back into production within their zone of occupation in Germany. While they did this, the Ministry of Armaments set up an array of engineering and development centers that would allow them to build and launch the V-2 on Soviet soil as well.

Not much had been left of the German effort at the end of the war. Peenemunde lay abandoned, stripped of documents and equipment; retreating German forces had deliberately blown up its principal facilities. Nordhausen, site of Hans Kammler's production center, was in central Germany, within the Soviet zone. Moreover, it had continued to turn out V-2 rockets until foreign troops arrived. But the troops had been American, and the Yankees had stripped this center bare as well. A few installations remained ready for Soviet use, notably an engine test center at Lehesten, which was also in central Germany. But the U.S. Army had walked off with the key people, along with most of the documents and the available V-2 components.

The United States had not gotten everything, however, and the meager remnants left would suffice to serve the Soviets' needs. The top people might be on their way to the States, but plenty of lower-level staffers remained, with invaluable information, even if it was in bits and pieces. Some contractors were still in business, including a plant in Berlin that built missile control systems.

In addition, one high-level manager, Helmut Grottrup, cast his lot with the Russians. He had been deputy to Ernst Steinhoff, who had been responsible for guidance, control, and telemetry, and could easily have accompanied Steinhoff as part of the group recruited by Colonel Holger Toftoy. But Grottrup wanted to stay in Germany and disliked the terms of employment offered by Toftoy. This left him open to an offer from Boris Chertok, a member of Moscow's Interdepartmental Technical Commission who was trying to restart V-2 production at the Nordhausen center. Grottrup certainly had been no Communist, not in Nazi Germany, but he proved receptive when the Soviets offered him a chance not only to remain in his homeland but to take a leading role in the development of rockets.

The Commission set two immediate goals: to reconstruct a full set of V-2 production drawings, and to turn out these rockets on a pilot production line. Grottrup soon was granted responsibility for all missile development work being done in the Soviet zone by Germans. In addition, as early as September 1945 a test stand at Lehesten flamed to life again, as another German specialist, Joachim Umpfenbach, began training the rocket crews of Valentin Glushko in procedures for firing the V-2 engine. A year later the needed engineering drawings were in hand, while the pilot plant built fifteen complete V-2s, along with major components for fifteen more.

Also during 1946, the Soviet government began to establish the initial centers for an indigenous missile effort. The main installation, NII-88, sprang to life in the Moscow suburb of Kaliningrad, on the site of a former artillery plant. Its layout called for the German pilot production plant, transplanted whole, along with an engineering center, and a center for applied sciences such as aerodynamics and guidance. The engineering group featured several departments that would work on different missiles, with Sergei Korolev heading up Department 3, which dealt with long-range designs.

Other centers supported NII-88. Valentin Glushko, who had headed a wartime rocket-development sharaga in Kazan, now became director of a major engine-development center in Khimki, a few miles from NII-88. Separate engineering groups would develop guidance systems, gyroscopic instruments, and ground support equipment for missile launches. The directors of these groups all had participated in the work of the Commission in Germany, as had Glushko and Korolev.

Although Korolev was only a middle manager, he soon made it clear that he was the man to watch. He took the lead in setting up a working panel, the Chiefs' Council, whose members included Glushko as well as four other engineering-group directors. One participant, Vladimir Barmin, would recall that "the council's meetings were usually held in the design bureau whose field was the topic of discussion. If, for example, the discussion centered on engines, the meeting would be held in Glushko's bureau. But Korolev always presided." This council's members would play central roles in the rocket and space projects of the subsequent twenty years.

The immediate effort called for building V-2s. These missiles had short range and were too small to carry the atomic bomb, but as with Tupolev's Tu-4, the project would extend the limits of what Soviet industry could do during the early postwar years. Many parts and materials proved difficult to obtain, but the first made-in-Moscow V-2s were completed during 1947. Subsequent production of the V-2 involved eighteen factories along with thirty-five engineering and science centers. By contrast, von Braun in the United States was merely assembling his own V-2s from available components, with little industrial support.

Moscow's work also involved Grottrup's associates. In October 1946 Soviet troops carried out a large-scale roundup of technically capable Germans who had been working in the Soviet zone, and shipped them off to Russia. Grottrup's colleagues split into two groups, one of them accompanying him to NII-88 near Moscow, the second installed on the island of Gorodomliya in a lake some 200 miles away. Grottrup and his family were put up in a six-room villa, and he had a chauffeur. But the others did not fare so well.

Grottrup's Moscow staffers got one room for a family of three and two rooms for a family of four; university graduates had an extra room. These associates of Grottrup nevertheless were better off than Korolev's engineers, who were living in barracks and tents. At least half of Korolev's men were on a waiting list for a room for their families. At Gorodomliya it was worse yet. The island acquired fine research facilities, including a Mach 5 wind tunnel, a rocket test stand, and an electronics lab. But living conditions were appalling.

People lived in filthy huts. Grottrup's wife Irmgard would note their "raw floorboards with wide cracks. The dirt can be brushed straight into the cracks; on the other hand, you can't get rid of the bugs." Tap water was undrinkable; it contained dirt, seaweed, and tiny animals, and people hauled water from the lake. Sewage facilities were nonexistent. Meat, when available, was mostly bones, and meals ran heavily to cabbage soup. Stoves were constructed of clay and stones. And amid bitter winters and long northern nights, windows rarely shut properly and children lacked decent shoes.

Back in Germany, these engineers and technicians had used the V-2 to help the Russians learn the rocket business. Now, at NII-88 and Gorodomliya, they would go beyond the V-2 with design studies of more advanced missiles. However, they would not participate in these rockets' detailed design, let alone in their construction and development. The Germans were there so that Russians could pick their brains by having them work on ideas for Soviet use. They would not join in learning about Soviet advances, and as their expertise grew increasingly stale, Russians would shift them more and more to the margins. The NII-88 group, including Grottrup himself, would leave Moscow for the wilds of Gorodomliya. A few years later, their Soviet masters would return them to East Germany.

Still, with Grottrup's colleagues at hand to help, Korolev began planning seriously for major improvements on the V-2 even before the first of the German-built rockets were ready for launch. As early as the summer of 1946, while Grottrup's associates were still in Germany, these engineers had begun to propose improvements in the V-2 by drawing on concepts that had originated in Peenemunde. Then, after settling in Russia, they broke decisively with the V-2 by proposing an entirely new missile that would triple its range.

To Grottrup, the path to long-range missiles demanded the same innovations that his American counterparts were developing independently. High on Grottrup's wish list was a set of turbopumps that would dispense with hydrogen peroxide, driving these pumps with hot combustion-chamber gases. At North American Aviation, this "gas-generator cycle" became part of Bollay's 120,000-pound rocket engine. This arrangement saved weight. To save even more, Grottrup's new missile would dispense with the V-2's internal propellant tanks, allowing its outer skin to hold propellant directly. Further weight loss would come by separating the warhead in flight, for the rocket's body then could avoid requiring the robustness to survive re-entry. Within the engine, extra thrust would come by boosting the propellant flow rate.

As Grottrup pursued these concepts during 1947 and 1948, Korolev, working separately, developed a design that would feature similar advances in lightweight structure but would not push as far in propulsion. After all, Korolev's engines would come from Glushko, who was not ready to proceed to a gas-generator cycle but preferred instead to stick with the V-2's hydrogen peroxide. With the weight savings thus compromised, this rocket would achieve a range of six hundred kilometers, twice but not three times that of the V-2. It still could not carry the atomic bomb, and although its lightweight structure would offer an important advantage, its reliance on V-2 engine technology meant that it represented only a half step into the future. As early as April 1947, Korolev presented an initial design concept to Armaments Minister Ustinov. It would enter development as the R-2.

Meanwhile, amid these design studies, there remained the matter of actually launching some of the V-2s that were rolling off the production lines. The Soviet Union needed a test range, and Ustinov's officials found what they wanted near the town of Kapustin Yar, seventy-five miles east of Stalingrad. A railroad ran to the site, but the region was so desolate that no other town lay near its route. One visitor wrote of "bare, lifeless steppes; dry winds chased billowing dust and tumbleweeds about. There was essentially no water." Camels were common. The workforce lived in army tents, while the more fortunate project officers found housing in railroad cars. Industrial safety was minimal. Few people cared when one worker fell to his death from a scaffold, or when a steel beam broke loose and dropped onto a welder, killing him as well.

Ustinov now had two batches of V-2s, from Germany and from NII-88. Flight testing began in October 1947. The initial flight was sufficiently important to draw Ustinov to Kapustin Yar; when the V-2 lifted off successfully and flew downrange, he seized Korolev in a bear hug and danced him about. Mrs. Grottrup recalls that senior officials "jumped up and down like little children. Then they grabbed their vodka bottles and got drunk." Ustinov had reason to celebrate. Only two years after the war, he now was well on his way to matching the German achievement in rocketry.

Even so, Korolev's proposed plans far outstripped what he could actually build and launch. The first flight of an experimental R-2 lay two years in the future; it would not take place until September 1949. But during late 1947 and 1948, Korolev began to nurture thoughts of a 3,000-kilometer rocket designated R-3, both within his own design office and amid the complementary work of Grottrup's staff. Such a weapon still would lack intercontinental range and would not strike the United States. But it pushed the limits of what he might pursue.

As the Cold War deepened, the planebuilder Andrei Tupolev continued to offer only improved versions of his piston-powered Tu-4, which copied America's B-29. The new models promised longer range, but their slow speeds would leave them vulnerable to jet interceptors. Dissatisfied, Stalin called a meeting at the Kremlin in July 1949. Igor Kurchatov described recent work on the atomic bomb, which was only a month away from its first successful test. His report overshadowed that of Korolev, whose R-3 existed only in the form of design studies.

"We want long, durable peace," Stalin declared. "But Churchill, well, he's warmonger number one. And Truman, he fears the Soviet land as the devil's own stench. They threaten us with atomic war. But we are not Japan. That is why you, Comrade Kurchatov, and you, Comrade Ustinov, and you as well"—turning to Korolev—"must speed things up! Are there any more questions?"[3]

In December 1949 Korolev responded to Stalin with a formal proposal for his 3,000-kilometer R-3. He designed it as a single-stage missile, thus avoiding the problem of igniting a second stage. He also drew on studies by Glushko of a new engine, which would seek to achieve a thrust of 260,000 pounds. This was a bold leap indeed, far beyond the realm of design where experience with the V-2 could serve as a guide. Nor was there any guarantee that the means were at hand to turn this concept into an operational missile.

Still, there was new joy in Korolev's personal world. His marriage had faltered, particularly after a young and pretty divorcee, Nina Kotenkova, entered his life. She worked for him as a translator, and fell in love with him. His wife, Xenia, resided in downtown Moscow, while Korolev had a flat near his work, outside the city—in Nina's apartment building. Korolev divorced an embittered Xenia in 1948, then married his Nina the following year.

April 1950 brought a bright moment for Korolev professionally. He earned a promotion. His design group, Department 3, had operated within the engineering division of NII-88; now he took over as director of the entire division, with responsibility for missiles of all ranges. As the year progressed, he initiated two new projects that would build his country's first nuclear-capable missiles, the R-11 and R-5.

The R-11 grew out of dissatisfaction among Army officials with the V-2 and the forthcoming R-2. Both needed liquid oxygen, which might be hard to obtain during field operations, and which boiled away readily. It would not be possible to fuel a V-2 or R-2 in advance and leave it on its pad for any length of time, ready to launch, like artillery loaded and prepared to fire. But Korolev had an in-house rocket specialist, Alexei Isayev, who had been following up wartime German work on the use of nitric acid as an oxidizer. As at Jet Propulsion Laboratory in California, this approach promised truly storable propellants, and Isayev's first such engine ran successfully in ground tests during August 1950. The missile that would use nitric acid, the R-11, then took shape as a longer-range counterpart of the Corporal. NATO called it the Scud, as it went on to see wide use in wars of the Third World—including the Gulf War of 1991, in which Iraq's Saddam Hussein repeatedly fired it against the city of Tel Aviv.

The R-5 took shape during 1952, as it became clear that the 3,000-kilometer R-3 lay well off in the future. The range of the R-5, 1,200 kilometers, would double that of the R-2 and quadruple that of the V-2. It did this with an engine by Glushko that represented a final elaboration of V-2 technology. Yet it was the first long-range missile with strategic capability, for it could fly from Soviet territory to deliver a nuclear strike against American air bases in Italy and West Germany.

There remained the bothersome matter of the R-3, whose engines represented a particular bone of contention. Glushko tried to push V-2 technology to build his 260,000-pound engine, but during 1950 and 1951, test versions repeatedly blew up. This was not new in itself, but what made it worrisome was that the engine was reaching beyond the state of the art in metallurgy. It could not stand up to the heat and vibration associated with its high thrust. Nor was there any near-term prospect for advances in metallurgy that could overcome the difficulties.

For both Glushko and Korolev, it was back to the drawing board. Korolev found hope in new work of Alexei Isayev, which seemed to promise an engine of 140,000 pounds of thrust. Korolev responded with the R-3A, a modified R-2 that could test elements of the R-3 design. He also took advantage of the broad scope of his professional charter by raising a far-reaching question: Should long-range missiles perhaps be of the cruise variety, flying in the atmosphere rather than relying on rockets for ballistic flight? He knew that the U.S. Air Force was putting its money on the ramjet-powered Navaho, and he had his designers take a close look at whether a similar ramjet missile might better serve his country's needs. Navaho was still a long way from the flight-test stage, and Korolev in no way was ready to copy its design, as Tupolev had done with the B-29. Yet Korolev was quite prepared to believe that North American Aviation might be on to something.

But during 1952, Glushko's work took a turn for the better. He had not favored using numerous engines of modest thrust for the R-3, because multiple engines meant multiple chances of failure. Particular points of failure were in the turbopumps, which would continue to rely on hydrogen peroxide, V-2-style. However, no law required that each thrust chamber had to have its own set of pumps. Glushko saw that he could build engines of truly enormous thrust by having a single set of turbopumps feed propellant to several thrust chambers.

This approach was brilliant because it precisely addressed the problems of the moment. The turbopumps could be large and very capable; such designs indeed lay within reach. The thrust chambers could each be small enough to stand up to heat and vibration without straining the limits of welds and metallurgy. As Glushko's new concept took form, Korolev saw that he could use it to rescue the R-3. Better yet, he could leap beyond the R-3 to propose a ballistic missile of intercontinental range, able to strike the United States.

Just then, at the end of 1952, the Soviet strategic position was unpleasantly weak. The air war in Korea had made it clear that the B-29, which the Americans had used routinely, could survive against jet fighters only if escorted by its own jets. Accordingly, the Strategic Air Command was phasing out its B-29s and was rapidly replacing them with highly capable jet bombers. But Moscow's bomber force consisted almost entirely of B-29s, in the form of the Tu-4. Stalin had asked Tupolev to build a long-range jet bomber, but the engines he might use were far more fuel-hungry than their American counterparts. The most Tupolev could offer was a bomber with turboprop engines, the Tu-95. Stalin accepted it, but he still wanted a jet and turned to Tupolev's son-in-law, Vladimir Myasishchev. This senior designer proceeded to build a four-engine jet bomber, which NATO called the Bison. But he could not overcome the engine problem and his bomber lacked enough range to attack the United States.

Then in November 1952, American scientists detonated the first hydrogen bomb, with a yield nearly a thousand times greater than that of Hiroshima. The following February, three weeks before he died, Stalin signed an order directing NII-88 to develop a missile of intercontinental range. The immediate question was whether it would be of the cruise or the ballistic variety. In April the cabinet-level Council of Ministers modified Stalin's order by taking the cruise-missile effort away from NII-88 and turning it over to the aircraft industry.

Two aircraft design centers, headed respectively by Myasishchev and by Semyon Lavochkin, each would vie to build a ramjet-powered missile, and both organizations were strong. Myasishchev had won Stalin's personal favor and was leading the way with his jet bomber, while Lavochkin had taken over an antiaircraft-missile project in response to a direct order from Stalin. In turn, Korolev now would focus all his efforts on long-range rockets. But he had an ace up his sleeve, for he now could propose to use Glushko's new engine to build a true ICBM. Indeed, he would do more. He would recommend that the government stop work on the R-3, which could not reach the United States, and proceed directly toward intercontinental capability.

Korolev went to a meeting at the Kremlin to discuss the R-3. Officials expected him to move forward with it, then use its technology as the basis for an ICBM. This would continue the cautious, step-by-step approach that had produced the sequence from the V-2 to the R-2 and then the R-5, while avoiding such blunders as Glushko's early engine for the R-3. Korolev had other ideas. Facing his listeners, he boldly declared: "A new concept has emerged. The R-3 design with a range of three thousand kilometers will not answer for our long-term needs. Our collective has come to the well-founded conclusion that there is the opportunity to go over the R-3 and begin work on an intercontinental missile."[4]

The historian Asif Siddiqi notes that "all those present at the meeting were stunned." They were eagerly awaiting the R-3, even though it existed only on paper, and feared Korolev's bold leap. He met strong opposition, with the minister Vyacheslav Malyshev challenging him emphatically. With his cabinet rank, Malyshev was responsible for the nuclear-weapons program. He knew that Korolev was interested in space flight, a wild fantasy if ever one existed, and accused him of hoping to build a space booster that would have no military use. Malyshev insisted that Korolev proceed with the R-3.

"I refuse," Korolev declared. "This is an antistate approach to this matter."

"No! Really? He refuses," Malyshev responded. "People are not irreplaceable. Others can be found."

But Ustinov, not Malyshev, was in charge of the strategic-missile program, and Ustinov was at least cautiously in favor of Korolev's plan. He lacked the authority to accept it outright as a basis for policy, but he could set up a review board that would look into the matter.

At high levels of government, the convening of these boards represents an art. When done properly, such a panel will feature members of impeccable reputation and far-reaching authority, whose points of view will just happen to coincide with those of the chairman. As head of the commission, Ustinov picked Konstantin Rudnev, a deputy defense minister who had headed NII-88 and who knew Korolev well. His fellow panelists included Marshal Mitrofan Nedelin, the top commander of artillery, who also supported Korolev's proposal.

This ICBM concept called for a central core powered by a Glushko engine with four thrust chambers that would develop some 170,000 pounds of thrust. Four strap-on boosters would surround the core, each carrying propellant and fitted with a similar engine. All rockets would ignite on the launch pad, thus avoiding the problems of ignition while in flight. The total thrust, 900,000 pounds, would compare with the 100,000 pounds of the largest Soviet rocket flown to date, the R-5. Following liftoff, the strap-ons would fall away, leaving the core rocket to continue to push its nuclear warhead.

The R-3 studies had benefited from the close involvement of one of the country's leading mathematicians, Mstislav Keldysh. Keldysh now appeared before Ustinov's commission and asserted that his analyses showed that Korolev's new concept was sound. Further support came from Glushko, who declared that he could build his clustered engines using existing engineering practice. This panel made its report, which was favorable, and Ustinov won his point. On May 20, 1954, he received final approval to initiate the development of Korolev's ICBM.

It was designated R-7, though people would call it *Semyorka* (No. 7). However, it would not be the only intercontinental missile, for Myasishchev and Lavochkin would continue to pursue their ramjet-powered cruise projects. The three designers would compete; if one faltered, then the other two could carry on. But one way or another, Moscow was bent on attaining success.

Meanwhile, what was the United States doing? The country had rapidly demobilized after the war, with the defense budget falling to a low of $14.4 billion for 1949. The focus of the Cold War was Berlin, and Stalin expected to hold East Germany, keep eastern Europe as a buffer zone, impose Communist rule upon these lands in the face of widespread opposition, and counter any threat of invasion from the West. The Army he maintained featured a strong offensive threat because it was far larger than it had to be in order to do these things. Secretary of State George Marshall would note that the Soviets had over 260 active divisions, whereas his force in reserve consisted of "one and a third divisions over the entire United States."

Truman had sought to maintain a balance of forces by having the atomic bomb in reserve as an equalizer. But during 1949, Kurchatov detonated his own bomb and exploded that policy as well. At the same time, China fell to its own Communists. This event raised the specter of a global power wielding nuclear weapons, armed by the Soviet industry that had defeated Hitler, and deploying the limitless reserves of Chinese manpower.

Truman's initial response was to strengthen his nuclear advantage. He approved plans to expand supplies of weapons-grade uranium and plutonium. He stepped up production of atomic bombs, and launched a major program that would build the hydrogen bomb. Then in April 1950 he approved a National Security Council policy paper known as NSC-68. It stated that the United States would devote up to 20 percent of its gross national product to the military and would resist the Communist threat anywhere in the world.

Two months later the Korean War erupted, adding a sense of urgency to the buildup. The nation's nuclear stockpile, numbering 56 bombs in 1948, approached 300 in mid-1950, and reached 832 two years later. The Strategic Air Command, which would deliver them, grew apace. Because

Soviet air defenses were weak, the B-36 was considered a formidable bomber, even though it was piston-driven and relatively slow. In addition, the Air Force purchased over two thousand jet-powered B-47s. With midair refueling, their range could increase at will. And in 1952 the B-52 made its first flight. It would dominate the field of strategic bombing for decades, until the Cold War finally ended.

Amid this overwhelming emphasis on air power, America's first step toward long-range rockets amounted to small change. It took place at Huntsville, Alabama, a farm town known for watercress, cotton, and mosquitoes. Nearby, the U.S. Army had built a wartime arsenal for the manufacture of poison gases and chemical weapons. After the war, Army Ordnance stepped in. It needed a missile center and the Huntsville facilities, known as Redstone Arsenal, looked attractive. After all, if the good townspeople had put up with production of phosgene and mustard gas, then they probably wouldn't object to mere rockets, particularly when the alternative might involve picking cotton at a dollar per hundred pounds. Many citizens were living in slums—Boogertown for the whites, Honeyhole for blacks—and the local job market just wasn't that great.

The new Ordnance Rocket Center opened for business in June 1949. Its initial tasks involved battlefield missiles: Jet Propulsion Laboratory's Corporal, a solid-fuel rocket called Honest John, and the Nike series of antiaircraft missiles that were under contract at Bell Labs. The Secretary of the Army soon approved the transfer to Redstone of Wernher von Braun and his Peenemunde veterans, who had been languishing at Fort Bliss since the end of the war. They arrived in April 1950, and the question then arose of what to do with them. The answer lay in an ongoing study of possible V-2 follow-ons that had been under way at General Electric since 1946, without ever getting very far. In von Braun's hands, it would acquire a sharp focus and turn into the rocket called Redstone.

Redstone would carry a nuclear warhead, but its range of two hundred miles made it little more than a V-2 upgrade. Its engine was an Air Force hand-me-down, the 75,000-pound motor that William Bollay had developed at North American Aviation for his Navaho and then abandoned when that missile's thrust requirements increased. And when Army Ordnance sought a contractor to build the Redstone, its officials discovered that aircraft firms weren't interested. Their executives believed, correctly, that the Army would not stay the course in the field of large liquid-fueled rockets, and would offer no continuing flow of business. The contract went to Chrysler, a Big Three auto company that took on Redstone as a simple production job.

Redstone would be ready when the nation needed it, not for war but to launch America's first satellite in 1958 and to put its first astronaut, Alan Shepard, into space in 1961. In addition, it helped keep von Braun's

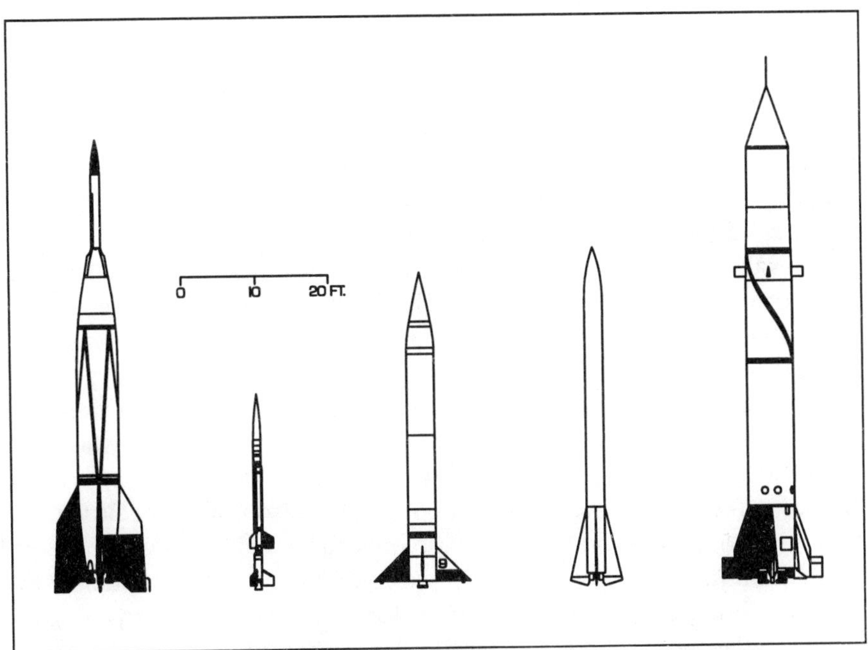

Postwar American rockets: V-2 with WAC Corporal; Aerobee; Viking; Corporal; Redstone. (Dan Gauthier)

associates busy for several years, while they waited for the opportunity to lead the country into space. But its significance paled when compared to that of the Air Force's bombers.

The bombers, indeed, lay at the root of the difference between the Air Force's reluctance to enter the field of long-range missiles and the Soviets' eagerness to build them as soon as possible. For Stalin and his immediate successors, the overriding strategic problem was to find a way to strike the United States, and they would allocate large sums to anything that might work: jet bombers, ramjet-powered cruise missiles, Korolev's ICBM. The Air Force faced no such urgency, for it had what the nation needed in the Strategic Air Command. These contrasting positions resembled those of 1943, when the Allies' low-tech bombers had forced Hitler to take his chances with Germany's jet interceptors and missiles.

The Air Force had flirted briefly with Convair's MX-774, but though it had abandoned this effort, the Rand Corporation, a Los Angeles think tank, had kept an eye on the field. In December 1950 it reported that long-range missiles now lay within reach. A month later the Air Force gave Convair a new study contract, inviting Charlie Bossart and his colleagues to take a fresh look at his earlier ICBM concept. In August 1951, Bossart christened it Atlas, after Convair's parent company, the Atlas Corp.

It would be a behemoth. It was to weigh 670,000 pounds, standing 160 feet tall and using seven of Bollay's new 120,000-pound engines. Atlas was thoroughly unwieldy and represented a basis for further studies rather than a design for a practical weapon. Still, it was a milestone. For the first time the Air Force had a concept for a rocket with a range exceeding five thousand miles, which it could pursue using rocket engines that were already being developed.

Navaho was not an approved project like the Redstone, but was at the stage of research and development. However, Atlas was at an even lower level, that of engineering studies. This meant that its engines had to be funded through a back-door arrangement known as bootlegging. Bootlegging has been aptly described as taking money that has been aimed in one direction and pointing it in a different direction. The man who controlled the funds was Lieutenant Colonel Edward Hall, who was running the Air Force's rocket activities from Wright-Patterson Air Force Base. "We had a hell of a time funding Atlas," Colonel Hall recalls. "Navaho became a line item in the budget very early; we could fund it. The developmental work on the rockets for Atlas, as well as for those of Navaho that would be used in Atlas, were all done under the Navaho program."[5]

Hall was quick to appreciate that Atlas engines would have to differ from those of Navaho. The Navaho engines could burn alcohol, which had served for the V-2. Alcohol had the advantage of mixing easily with water, to lower the temperature in the combustion chamber and make it easier to cool. But the 120,000-pound engine already had improved its cooling arrangements enough to allow it to burn alcohol at nearly full strength. During 1952, studies showed that this engine could be modified to burn another undiluted fuel: kerosene. Kerosene promised to deliver more energy than alcohol, and Hall knew that Atlas would need this energy to achieve its range.

He found his opportunity in plans for an advanced Navaho, which would need more performance in its rockets. Sam Hoffman was running the rocket project at North American, and Hall recalls: "I called him up. There were several of us on the phone. We said, 'Sam, we want you to lay out an engine for hydrocarbon use; we want to throw the alcohol.' Sam was shocked. And his reasons were good—a hydrocarbon is a mixed bag of cats, you don't really know what's gonna happen. Sam was unhappy, and he let us know he was unhappy. We said, 'No, you have to do it, and we want 120,000 pounds out of it.'

"There were several weeks when it was in the balance, and we told Sam that we much appreciated what North American had done on the 75,000-pound engine. But if his company would not take on the 120,000-pound hydrocarbon engine—we'd have to give it elsewhere. And Sam then collapsed and said OK." In January 1953 he initiated an effort called

REAP, Rocket Engine Advancement Program, to solve the problems that would arise in switching the Navaho engine from alcohol to kerosene.

The standard Air Force kerosene fuel was jet fuel: JP-4. But its specifications permitted so much variation in density that it would lead to unacceptable variations in the weight of a loaded rocket, which would be fueled by volume. In addition, test-stand firings with JP-4 showed that it could clog an engine's thin cooling passages with gummy residue or with carbon, leading to the engine's destruction. This stemmed from the presence in the fuel of compounds that had caused no problems in jet engines but that now would call for careful removal. The fuel that resulted, RP-1, free of contaminants and constant in density, went on to set its own standard.

The Atlas engine that came from REAP, planned initially for use as a Navaho engine, gave Hall more thrust than he had called for: 135,000 pounds. During the next few years this engine would see use not only in the Atlas but in most of America's other large liquid-fueled rockets. But for the moment, Atlas still was something the Air Force wouldn't buy.

Its original 1951 design had featured an 8,000-pound warhead. During 1952 the Air Force cut its estimated weight to 3,000 pounds, thereby achieving a welcome reduction in the projected size and cost of this missile. But there remained a sticking point: the rocket's accuracy.

Experienced Air Force bombardiers could place atomic bombs within 1,500 feet of an aim point, when using good radar bombsights to strike at night from combat altitudes at targets directly below. The view at the Pentagon was that Atlas would have to do as well when flying all the way to Moscow. This accuracy corresponded to hitting a golf ball a mile and having it fall right into the cup, making a hole in one without even rolling or bouncing. And Atlas would have to do this entirely through automatic control. But in no way could Atlas meet this requirement, not by a long shot. It would miss by miles, making it useless as a serious weapon.

Then on November 1, 1952, the first hydrogen bomb detonated at Eniwetok Atoll in the Pacific. "The thing was enormous," one observer said. "It looked as though it blotted out the whole horizon, and I was standing thirty miles away." The weapons designer Theodore Taylor described it as "so huge, so brutal—as if things had gone too far. When the heat reached the observers, it stayed and stayed and stayed, not for seconds but for minutes." It yielded ten megatons, compared to thirteen kilotons at Hiroshima. That detonation in 1945 had produced a fireball with a diameter of a third of a mile. At Eniwetok the fireball spread so far and fast as to terrify the distant observers, expanding until it was over three miles across. The explosion vaporized an entire island, leaving a crater two hundred feet deep and more than a mile wide.

This bomb was far too heavy to serve as a weapon. But within months two leading weapon designers, John von Neumann and Edward Teller, determined that lightweight H-bombs were feasible. Among knowledgeable Air Force officials, this immediately suggested a solution to the guidance problem. Atlas, armed with such a bomb, might miss by several miles and still destroy its target, by the simple method of wiping out everything that lay between. But for Atlas to proceed as a major program, the officials would need information on the size, weight, and explosive power of the hydrogen bombs that were likely to come along during the next few years.

Theodore von Karman was chairing the Air Force's Scientific Advisory Board, and he set up a panel to explore these questions. He recruited von Neumann as its chairman, who was one of the world's top mathematicians. Von Neumann was the man who invented computer programming. He had pioneered in carrying out the difficult calculations needed for the first plutonium bomb, had led the way in using the first computers to carry out the calculations, and then went on to address the even more intractable computations that the hydrogen bomb required. His fellow panelists included a cast of noteworthy bomb-builders: Hans Bethe; George Kistiakowsky; Edward Teller, who had invented the hydrogen bomb; and Norris Bradbury, director of the Los Alamos nuclear laboratory.

Their work took on new urgency in August 1953, as the Soviets detonated their own new weapon, with a yield of four hundred kilotons. It drew on the work of the physicist Andrei Sakharov, who had shown how a modest amount of thermonuclear material, incorporated in an atomic bomb, could release a flood of neutrons that would markedly boost its yield by causing more uranium or plutonium to fission. The name Sakharov derives from *sakhar,* sugar, and Sakharov's colleagues called his process "sugarization" because it would make their bombs much sweeter.

American scientists, studying fallout from this test, succeeded in determining its yield and in reconstructing the main features of its design. They concluded that while it was not a true hydrogen bomb, it nevertheless signaled a sizable advance over earlier Soviet weapons. In addition, Air Force intelligence had become aware of Soviet studies of long-range rockets. Taken together, these developments pointed inescapably to the likelihood that Moscow would soon build its own big missiles.

Von Neumann's panel had recently learned that projected ICBM warheads, weighing 3,000 pounds, would yield 500 kilotons. In September the Air Force Special Weapons Center added its own report, confirming that bombs with this yield would weigh as little as 1,500 pounds. Atlas then could come in at no more than 240,000 pounds, barely a third of the 1951 design weight.

This was a breakthrough. It changed all the prospects for rocketry, and for an ICBM. Furthermore, it took place at a time when the recently elected Eisenhower administration was adopting a defense policy called the New Look, which would strengthen the nation's reliance on nuclear weapons. Even so, the prospect of lightweight, powerful bombs would not trigger immediate approval of a major new program in strategic missiles, for the New Look sought specifically to cut the budget and to cancel existing programs if possible, rather than initiate new ones. But this prospect certainly would justify taking a fresh look at the nation's ongoing projects.

The Air Force secretary, Harold Talbott, had a special assistant, Trevor Gardner, with sweeping responsibility for research and development. In a service whose generals had mostly come up as wartime operational commanders, Gardner had made his name as a director of technical projects. Cultural clash was inevitable, for these generals did not relish the thought of taking orders from an upstart civilian. Nor did Gardner's personality help; others described him as "sharp, abrupt, irascible, cold, unpleasant, a bastard." But he had Talbott's strong support. And although the generals favored bombers, Gardner wanted to push ahead with missiles.

Once again his advocacy for such a policy would find its focus in a high-level review panel, the Strategic Missiles Evaluation Group, which Gardner christened the Tea Pot Committee. He recruited von Neumann to serve anew as its chairman, had Kistiakowsky join him, and then added a dazzling array of talent from Caltech, Hughes Aircraft, Bell Labs, and MIT. As Gardner put it, "The aim was to create a document so hot and of such eminence that no one could pooh-pooh it." The group met during the fall and winter of 1953 to 1954, and gave particular attention to the Navaho and Atlas. Like Dmitri Ustinov's review panel, which was working in Moscow at virtually the same time, the Tea Pot Committee would weigh the merits of a bold leap toward an ICBM, bearing in mind that a ramjet-powered cruise missile might offer a highly plausible alternative.

Gardner was lukewarm toward Navaho but enthusiastic for Atlas, and von Neumann's report, delivered in February 1954, strongly supported these viewpoints. To add to its authority, the report drew on work of the Rand Corporation, whose recommendations held weight.

Just then, early in 1954, Navaho existed in bits and pieces. North American was developing the rocket engines for its initial boost, while the firm of Wright Aeronautical Corp. was building test versions of ramjets. A Navaho guidance system was also in development. In addition, North American had built another unmanned aircraft, the X-10, powered by turbojets. It had the detailed shape of the planned Navaho missile and would test its aerodynamics at supersonic speeds.

The Tea Pot Committee nevertheless stated that "at this time the Navaho program would not benefit from a broad attempt at acceleration of the entire program." It recommended that the Air Force choose a second contractor to ensure that "ramjets of adequate performance and reliability will be available," and called for "a prompt decision" regarding the choice of a contractor to build "the difficult Navaho guidance components." This cruise missile would go forward, but would enter a stage of serious development only if Atlas were to falter.

In the matter of Atlas, von Neumann's group faced a dilemma. It could recommend an immediate go-ahead, inviting Convair to build it. Yet while Convair was a builder of bombers par excellence, it lacked the competence to proceed with an ICBM. It had engineers aplenty, but it needed the scientific talent that William Bollay had brought to North American. The comparison was instructive. Even with this talent, the company's Navaho still existed only as test components, with a flight of a complete prototype years away. Yet Navaho from the start had appeared less demanding than Atlas; that is why the Air Force had pursued it. How then could the less capable Convair succeed with the more difficult Atlas?

The report pulled no punches: "While much credit is due Convair for pioneering work, it is the conviction of the Committee that a radical reorganization of the project considerably transcending the Convair framework is required if a militarily useful vehicle is to be had within a reasonable span of time." It called for a new round of design studies, "adequately based on fundamental science." It also called for new management, "which we propose should be given directive responsibility for the entire project." The report then made specific recommendations: "The Committee judges that at this moment the project could not be effectively accelerated by heavy financial commitments, early freezing of the design, and production planning. However, the Committee expects that the new group referred to above will within a year be in a position to recommend in full detail a redirected, expanded, and accelerated program, which is likely to call soon for increased financial support and high project priority."[6]

Responding to the breakthrough in nuclear-weapons design, the report specifically referred to the allowable miss distance, or in Air Force parlance, circular error probable: "The military equipment on C.E.P. should be relaxed from the present 1,500 feet to at least two, and probably three, nautical miles." The Atlas warhead thus would deliver a yield of one-half megaton.

The Tea Pot report gave Gardner a basis for action. For Atlas to go ahead, and to win the necessary priority, it would need support from the Joint Chiefs of Staff, the National Security Council, and the president. However, recent events made it likely that these endorsements would come forth quickly and that Atlas would go forward at full throttle.

Back in April 1950, only a few months after the first Soviet A-bomb test, the CIA had estimated that two hundred such bombs could defeat the United States in a war, and that Moscow might have that many "some time between mid-1954 and the end of 1955." This made 1954 a year of danger. Curtis LeMay, head of the Strategic Air Command, described it in May 1950 as "the critical year at which time we must be prepared to meet, and effectively counter, the full military force of the USSR." The year of danger was now upon the nation, and the powerful Soviet bomb of August 1953 added weight to the warning.

Further danger lay in missile development. The R-5, with an initial range of a thousand kilometers, first flew in March 1953. A draft of the Tea Pot report, noting that "intelligence data have been made available to the Committee," stated that "most of the members believe that the Russians are probably significantly ahead of us in long-range ballistic missiles." The Russians also were building new bombers. The jet-powered Bison made its first flight in January 1953. In addition, Andrei Tupolev had won Stalin's assent to build a fleet of Tu-95 turboprop bombers, with enough range to strike the United States.

A CIA estimate in mid-1953 warned that "the U.S. is losing, if it has not already lost, its longstanding invulnerability to crippling attack, and with it the immense strategic advantage of being able to conduct the traditionally deliberate and extensive post–D-Day mobilization." A follow-up estimate, in February 1954, added that within three years, the Soviets could deploy five hundred Tu-95 bombers. Hence, if war came, America would have to fight with weapons already in hand.

In reviewing the policy decisions of the day, one finds apocalyptic attitudes that often jar the modern sensibility. For instance, in preparing the National Security Council paper of 1950 that became NSC-68, presidential adviser Paul Nitze had written that Stalin's regime is "animated by a new fanatic faith, antithetical to our own, and seeks to impose its absolute authority over the rest of the world." Calling for a "rapid and sustained build-up of the political, economic, and military strength of the free world," Nitze asserted that "budgetary considerations will need to be subordinated to the stark fact that our very independence as a nation may be at stake."[7]

This was not a speech by Senator Joe McCarthy; this was a sober policy paper requested by the president. Similarly, in an Eisenhower paper titled "Basic National Security Policy" (NSC-162/6), architects of the New Look stated bluntly that "in the event of hostilities, the United States will consider nuclear weapons to be as available for use as other munitions." Secretary of State John Foster Dulles, addressing the Council on Foreign Relations early in 1954, spoke of this policy as "massive retaliatory power," with which America would "respond vigorously at places and with means of its own choosing."

It is tempting to view such statements as reflecting the panic of frightened men who too readily confused the ruthlessly opportunistic Stalin with the recklessly aggressive Hitler. But we do this from the vantage of the thirty years of coexistence that followed the 1962 Cuban missile crisis, decades in which superpower leaders avoided bluster and threat while declining to challenge each other's key interests. Policymakers in the 1950s had to consider Moscow's capabilities, regardless of its professed intentions. They knew that the Cold War had not grown out of a diplomatic misunderstanding. They lived with the awesome conviction that if they blundered, the nation might be destroyed.

But America's breakthrough in hydrogen-bomb design offered the promise of an effective counter. This became dramatically evident during March and April of 1954, as several different types of bomb exploded at Bikini Atoll in the Marshall Islands. The 1952 test, named Mike, had used liquefied deuterium, which was much colder than liquid oxygen and more difficult to handle. The test designers had used it because it simplified the weapons physics. The new bombs would use lithium deuteride, a dry powder similar to salt, which could be stored indefinitely.

The first test weapon, Bravo, detonated on March 1, 1954. It had been designed to yield five megatons, but researchers at Los Alamos had missed discovering a key nuclear reaction in lithium that would markedly increase its explosive power. Bravo thus ran away and went out of control, yielding fifteen megatons, more even than Mike. A similar weapon, Romeo, ran away for the same reason and reached eleven megatons. This unexpected reaction meant that future warheads could be even smaller than expected, further reducing Atlas's required accuracy and making it more attractive yet. And another shot at Bikini, Nectar, showed that the lightweight bomb was already at hand. It weighed only 6,520 pounds but yielded 1.69 megatons.

In addition to Gardner and Talbott, supporters of Atlas included General Thomas White, the vice chief of staff, and Donald Quarles, assistant secretary of defense for research and development. Significantly, Quarles had come to the Pentagon from Bell Labs, and his background resembled that of Gardner. He and Talbott thus were positioned to win the support of the secretary of defense, Charles Wilson, who would push for approval from the National Security Council. Meanwhile, Gardner and White would gain endorsements from senior Air Force generals. These would include Nathan Twining, the service's chief of staff and a member of the Joint Chiefs.

White carried the ball to the Air Staff, which had responsibility for recommending approval of new programs. In the words of an observer, White told them that "ballistic missiles were here to stay, and the Air Staff had better realize this fact and get on with it." Then on May 14, having secured concurrence from Defense Secretary Wilson, White assigned Atlas 1-A priority, the Pentagon's highest. He directed that the program should accelerate "to the maximum extent that technology would allow." Gardner, whom Talbott had already designated as his direct representative in dealing with Atlas, declared that this meant "the maximum effort possible with no limitation as to funding."

Yet if Convair could not run this program, who could? The Tea Pot report had called for a new "development-management agency" that would place "over-all technical direction in the hands of an unusually competent group of scientists and engineers capable of making systems analyses, supervising the research phases, and completely controlling the experimental and hardware phases of the program." The Manhattan Project had built such a center for technical leadership at Los Alamos, and Gardner wanted something similar for Atlas.

The Air Force's own development center, Wright-Patterson Air Force Base, lacked the necessary smarts. The biggest problems would involve electronics, and Bell Labs, RCA, or IBM might have been suitable. But their directors would want to bid on major Atlas contracts and would face a conflict of interest if they also were managing Atlas as a whole. This conflict would arise from a temptation to make program-level decisions in ways that would produce more business for the home company.

Gardner proceeded to build his own Los Alamos by turning to two Tea Pot members, Simon Ramo and Dean Wooldridge. These scientists had made their names at Hughes Aircraft by leading the Air Force into a new realm of electronics, building radar fire-control systems that became standard installations within the nation's first-line interceptors. They had gone on to build an electronic guidance system, parlaying it into a contract for the Falcon air-to-air missile. Then in 1950 they scored the coup of defeating some twenty competing companies to win a major contract for the navigation and fire-control system to be used in the F-102 supersonic fighter. *Fortune* magazine would note that this gave Hughes "a virtual monopoly on the Air Force's advanced electronics requirements."

But Hughes Aircraft was the lengthened shadow of Howard Hughes, who owned it but who was becoming increasingly quirky. He would make bad executive decisions or no decisions at all, disappearing for weeks at a time and letting matters dangle. Then, when he reappeared, he could continue to ignore pressing matters while involving himself deeply in trivia, such as candy-bar sales in the company's vending machines. His trusted associate Noah Dietrich, who stepped into this managerial void, was a meddlesome hack. A buyout could have refreshed the company with serious leadership, and Howard Hughes received an attractive offer from Lockheed, but vowed he'd never sell. Ramo and Wooldridge both were corporate vice presidents, but now they'd had enough and handed in their resignations.

They set up a new company in a one-room office that would later become a barbershop, with a card table, a phone, a rented typewriter, and a secretary. But they quickly lined up financial backing from Thompson Products, a Cleveland auto-parts firm that was seeking to diversify. In addition, Trevor Gardner had known Ramo since 1938, when they had shared an apartment while working at General Electric, and greatly appreciated the work he and Wooldridge had done at Hughes. Within days after starting operation, the Ramo-Wooldridge Corp. received a contract to provide technical support to the Tea Pot Committee, and the company was suddenly profitable.

It was still several years before the company would formally merge with Thompson and create the firm of TRW. However, with Gardner's support, Ramo and Wooldridge soon could pursue a path of rapid growth. In addition, they attracted a host of gifted managers who would outshine those of William Bollay in providing future leadership. James Fletcher, who had worked with them at Hughes, came over to join them; he would become head of NASA. Louis Dunn, director of Jet Propulsion Laboratory, left it to join them, too. Other recruits included Richard DeLauer, who would become an undersecretary of defense; George Mueller, who would direct the Apollo manned lunar-landing program; Ruben Mettler, who would eventually head all defense and space work at TRW; and Albert Wheelon, who later became chairman of a revived Hughes Aircraft.

R-W would serve as a technical staff for the Air Force project office that would direct the development of Atlas. It would also give technical direction to the program contractors—including Convair, which would build these missiles under close supervision. By 1957 R-W would be coordinating the work of 220 such firms. This was a clear demonstration that the ICBM was indeed so difficult a project as to exceed the capacity of any single aircraft company, even with copious funding from the Pentagon. The ICBM demanded nothing less than the marshaling of much of America's aviation and electronics industry.

As R-W evolved during 1954, so did Air Force management. Gardner had proposed that a major general should lead the effort, but James McCormack, the man he had in mind, suffered a heart attack and soon retired from active duty. The job went instead to his intended deputy, Brigadier General Bernard Schriever. Schriever, wearing only a single star, took over this highest-priority project and began reporting to General White, who wore four stars and who stood at the elbow of the Chief of Staff.

Schriever had come up in the Army Air Corps as a bomber pilot during the 1930s. With this background plus an engineering degree from Texas A & M, he won a posting to Wright Field, which was already becoming famous as a development center. The Air Corps sent him to Stanford, where he received a master of science in mechanical engineering in 1942. He spent the war in the Pacific, flying combat missions in

B-17s, but his engineering studies now qualified him for a post at the Pentagon, where he would forge links between the Air Force and the scientific community.

Rockets captured his interest soon after the war. A key moment came in April 1953, as Edward Teller and John von Neumann formally advised him that lightweight hydrogen bombs now were possible. He then joined with General White in advising von Karman to convene the Scientific Advisory Board panel that laid out the thermonuclear breakthrough. *Action, dynamic,* and *capability* were three of his favorite words. "I hate to admit defeat in anything," he once said, as if that were the most natural of feelings. He came through the back door to take charge of Atlas, but another general said: "We didn't appoint him; Benny was born for this job. There wasn't another soul we knew who could handle it, so we just sort of nodded and said, 'OK, now,' and Benny walked in and took over."

In mid-1954 he slipped into Los Angeles to set up his project office, the Western Development Division, a name deliberately intended to be obscure. It set up offices in a former parochial school in Inglewood, close to R-W, and its Air Force staff wore civilian clothing to avoid standing out. One important task was to select contractors, and in doing this Schriever would show that like Sergei Korolev, he would stand his ground against even the highest of officials.

This official was Harold Talbott, the Air Force Secretary, a presidential appointee, and a civilian who outranked every one of the generals. Schriever had awarded a contract to a certain firm, and one of the losing bidders succeeded in persuading Talbott to reverse this decision. Unfortunately, Ramo and Schriever viewed that company as only marginally competent. Talbott nevertheless met with these men, bringing Gardner to the meeting, and he told Schriever what he wanted.

Schriever refused, and Ramo and Gardner backed him up. As Ramo would recall, "Talbott glowered, then lost his temper. It was scary to see him come apart, get red in the face, and with an ugly expression yell at Schriever, 'Before this meeting is over, General, there's going to be one more colonel in the Air Force!'"

Schriever kept cool. "I can't accept that directive," he replied, "because I have a prior and overriding order. On being handed this assignment, I was directed to run this program so as to attain an operational ICBM capability in the shortest possible time." He then asked if Talbott might care to issue a new order, in writing, lowering the priority of the earlier directive.

"The redness left Talbott's face and he turned pale," Ramo continues. "He made no comment and began to stare at the table, vigorously tapping a pencil on it and trying to pull himself together. In a few moments, his face returned to normal. With the fewest of words, he said to leave the contractor decision as it was. Schriever had passed a vital test. If he had given in, he would have retained his title but lost control of the project. He was now firmly in charge."[8] He kept his title and his rank as well.

Schriever would need this stubbornness to face the technical challenges, which were unprecedented in severity. Guidance was particularly difficult. In October 1954 the Atomic Energy Commission predicted that weapons weighing between 1,500 and 1,700 pounds could deliver a full megaton. Atlas thus could relax its miss-distance requirement still further, to five nautical miles. But if it were to use V-2 technology for its guidance, it would miss Moscow by 150 miles, which was five hundred times worse than a bomber's 1,500 feet.

This vast gulf in required accuracy would have shrunk only slightly if the Tea Pot Committee report had eased the requirement from 1,500 to 3,000 or 4,000 feet. A 4,000-foot goal would have been almost as far out of reach as one of 1,500 feet and would have kept Atlas on the back burner for years to come. The new five-mile goal appeared much more achievable, giving designers something well worth pursuing. Even so, in 1954 this goal constituted a topic for research rather than a goal that anyone could readily attain.

Why was guidance so inaccurate? It relied on the gyroscope, which featured a rapidly rotating wheel that weighed several pounds. The wheel's mountings had to be correspondingly robust, and these mountings would transmit forces to the wheel, causing it to drift from its proper orientation. To eliminate these forces, the point of departure involved finding a way to mount a gyroscope wheel using fine jeweled bearings, like those of a wristwatch. The man who did this was Charles Stark Draper, a professor at MIT.

Draper enclosed the heavy gyro within a canister and floated this sealed container within a heavy fluid. Within the canister, the gyro spun on a robust mount that left it unable to drift. That could have made the gyro useless as an instrument. But the fluid now made the container buoyant, thereby supporting its weight. It now could indeed turn on fine bearings, which would not have to carry its load, and it had plenty of sensitivity.

Draper had used such floated gyros during the war, in a stabilized fire-control system. But to achieve the precise requirements of Atlas, he would have to meticulously eliminate all sources of error. The fluid had to be heated, to expand slightly and become marginally less buoyant, in order to reduce to zero the canister's weight on the bearings. In addition, suitable gyros demanded the most scrupulous attention to eliminating dust.

Time magazine, describing a Falcon missile plant, noted that "no speck of dust can be tolerated. The air is changed by fans and filters every nine minutes, and positive air pressure is maintained inside the building so that any air leakage will be outward, not inward. Engineers in the drafting rooms are forbidden to tear paper or use pencil erasers (both make dust), and all employees must wear nylon smocks."[9] Women couldn't wear makeup to work, and people with peeling sunburns had to stay out of the clean room.

Moreover, a guidance system needed a computer, to receive signals from the gyro units and convert these signals into precise measurements of position and velocity. An analog computer would do the job, by processing the signals directly; one could avoid using a digital computer, which in 1954 required a roomful of vacuum-tube circuitry. Still, the question remained: Should the computer stay on the ground or should it ride along with the missile?

A ground-based computer could remain heavy and somewhat delicate, protected against a missile's vibration and intense noise. But it then would have to exchange data with the missile in flight using a radio link, demanding what Schriever described as "a very substantial ground installation which was highly vulnerable." An onboard computer would have to be robust and light in weight. But it would do away with the radio link and the ground installation, offering truly self-contained guidance, immune to interruption.

Radio guidance offered an interim system, because it was easier to build. But Draper strongly advocated the second approach: inertial guidance. The Air Force lent him a B-29; in 1949 he began making test flights from a base near Boston, with the flight crew standing by while instruments did the steering. Flights of 1,800 miles, during 1955, showed errors as low as two miles. The equipment weighed 2,700 pounds, but it showed dramatically what could be done. In addition, as a professor, Draper nurtured a "guidance mafia" made up of his former students, who agreed that his approach was best. Many of these people rose to high and influential positions. They included Schriever's guidance officer, Lieutenant Colonel Paul Blasingame.

Guidance represented one main problem; another involved re-entry. The warhead would have to enter the atmosphere at 16,000 mph, following its ballistic flight, and would have to survive to approach its target without burning up like a meteor. As early as 1953 two aerodynamicists, Julian Allen and Alfred Eggers, showed by mathematics that a nose cone could most readily avoid this fate if it had a blunt shape. This was surprising; a common view at the time was that nose cones would be sharply streamlined, resembling the needle-nose shapes of the latest jet fighters. But a warhead didn't need streamlining; it needed a heat shield. With this shield, it would slam back into the upper atmosphere with no more subtlety than a runaway freight train. Compressed air, piling up in front of the blunt nose, would act as an insulating blanket and would keep much of the heat from reaching it.

This was only a start, though, and researchers needed experiments that would directly simulate re-entry in their laboratories. At Cornell University, the physicist Arthur Kantrowitz led the way with a simple device called a shock tube. It consisted of a chamber filled with an explosive mix of hydrogen and oxygen and separated by a thin metal diaphragm from an adjoining chamber, holding a small nose cone and pumped down to vacuum. Detonation of the mix would burst the diaphragm, sending a pulse of gas past the model at the actual speeds and temperatures of re-entry. The flow would last only a thousandth of a second, but high-speed instruments would make the appropriate measurements. These measurements determined the conditions that the warhead would face.

Other tests sought to show that model nose cones could indeed stand up to their moments of crisis. In some of the simplest, researchers placed their models within the exhaust of a rocket. But these rockets weren't hot enough, and a better method used a new instrument, the arc tunnel. This was a supersonic wind tunnel in which the high-speed flow of air would pass through a flaming electric arc, with enough kilowatts to light a city. The air's temperature would soar as high as 14,000 degrees Fahrenheit, and it would slam into samples of materials.

To learn still more, Colonel Edward Hall arranged for test flights aboard a new solid-fuel rocket, the X-17. Hall had joined Schriever's staff as the propulsion officer and arranged for Lockheed to build the rockets. These were three-stage affairs, with the first stage thrusting them high above the atmosphere. Then, while they were falling downward, the second and third stages would fire, accelerating the nose cone to the highest possible speed.

This work led to initial concepts that resulted in designing the nose cone as a solid slab of copper, which would soak up the heat without melting. But these heat shields were heavy, and a much more elegant approach came from George Sutton, a physicist at General Electric. Ironically, he proposed that a nose cone should indeed burn up like a meteor, but in a controlled fashion. Instead of copper, a lightweight shell of reinforced plastic would cover the warhead. Its outer surface would burn and vaporize during re-entry, carrying away heat rather than absorbing it, with the vapor forming a cool protective layer. Such "ablative" nose cones became the standard type.

Schriever kept a close eye on these developments and made key decisions personally on Black Friday. This took place once a month, when staff members presented the status of specific projects and learned clearly, and often painfully, just who was falling behind. If an issue needed a decision then Schriever, having read the appropriate reports, would let people have their say and then issue his ruling. He had the authority to match his responsibility, and he used it.

To move forward at top speed, Schriever and Gardner broke with the usual practice of awarding major Atlas contracts to individual firms. Such arrangements would have selected North American Aviation as the sole-source contractor for the rocket engines, but this looked like putting their eggs in one basket. As early as August 1954, with the Scientific Advisory Committee expressing doubts, Schriever took steps to bring in Aerojet General as a second source.

The new practice, known as parallel development, drew on previous experience of nuclear-weapons development. It insured against major delays on the part of any single contractor, while encouraging competition between the contractors that could speed them both up. Schriever also chose firms for other major project elements: General Electric and Bell Labs, for radio guidance; A.C. Spark Plug and a team that included MIT, for inertial guidance; General Electric and Avco Everett, for nose cones; Burroughs and Remington Rand, for the computer.

Critics would describe this approach as wasteful duplication. But Schriever knew that the real waste would arise if Atlas were to depend upon a single contractor for some critical element and that firm encountered delays. If that happened, the whole program might have to stop in its tracks until the difficulty was resolved. Parallel development, by contrast, would ensure there was a second contractor that might still be surging forward. And their systems would be interchangeable; either could substitute for the other. Moreover, parallel development would broaden the base of firms that could qualify for future missile projects, thereby spurring the growth of their industry.

During 1955, parallel development would culminate in the approval of a second complete ICBM project, the Titan. Martin Co., the builder of Viking, won the contract, putting it on the same footing as Convair. Titan would share the same prime contractors that were building the principal systems for Atlas, and Schriever now divided them into two groups, one for each ICBM:

	Atlas	*Titan*
Airframe	Convair	Martin
Radio guidance	General Electric	Bell Labs
Inertial guidance	A.C. Spark Plug	American Bosch Arma and MIT
Rocket engines	North American	Aerojet General
Nose cone	General Electric	Avco Everett
Computer	Burroughs	Remington Rand

And so, during 1954 and 1955, the Soviets and the United States began a race toward doomsday, as they vied to build the intercontinental missiles that could blow up the world. Moscow had two ramjet-powered cruise missile efforts, from Myasishchev and Lavochkin, along with Korolev's ICBM, the R-7. The United States had two ICBM projects as well as the ramjet-powered Navaho. Both nations would take their cruise-missile projects to initial flight test and then drop them only when it became clear that the ICBM would succeed.

In addition, there was enough similarity between the basic layouts of Atlas and the R-7 to make them seem like parallel developments of a different sort. Both featured "stage-and-a-half" design, with all engines burning at liftoff. Both used a core vehicle with a single engine, flanked by booster engines that would drop off in flight. Both would rely on new engines that burned kerosene, and for both nations these represented the first large rocket motors to break with the use of alcohol as fuel, which dated back to the V-2. Each project had won approval in May 1954. Three years later, the Atlas and the R-7 would attempt their first flights within a few weeks of each other, and following initial failures both rockets would fly successfully later in 1957.

But there were differences as well, and the most important was that Korolev had not waited for a thermonuclear breakthrough in his own country. He had gone ahead as soon as he could, and it was no accident that the R-7, with five engines, would resemble the Atlas concept of 1951 with seven engines. The new Atlas would be modest indeed: a three-engine affair weighing 240,000 pounds, according to a design of late 1954. The R-7 would be two and a half times heavier. In addition, Korolev would pursue a development plan that would be notably more aggressive in flying the R-7 first to full range and then in using it to launch satellites. From these differences, important consequences would follow.

4

The Mid-1950s

Spacecraft, Planned and Imagined

THE SOVIETS had used their spies to good advantage in building their atomic bombs. A decade later, the United States prepared to return the favor, for the prospect of an ICBM spurred far-reaching thoughts of using space as a vantage point for military reconnaissance. Satellites in orbit could pierce Moscow's curtain of secrecy, disclosing its preparations for war. And while these plans were highly classified, there was strong public interest in the new rockets, as a number of writers vied to prophesy the conquest of space. One of the most prominent was Willy Ley, author of *Rockets, Missiles and Space Travel.*

One day in the spring of 1951, he had lunch with Robert Coles, chairman of the Hayden Planetarium in Manhattan. Ley remarked that in Europe, interest in astronautics was reviving strongly; an international conference, held in Paris the previous October, had attracted over a thousand people. However, none of these people had come from the United States, and this suggested that Americans should organize a similar conference. Coles replied, "Go ahead, the planetarium is yours."

Ley proceeded to set up a symposium that took place on Columbus Day. Admission was by invitation only, but some invitations went to members of the press, and the attendees included a few staffers from *Collier's,* a magazine with a readership of ten million. Two weeks later its managing editor, Gordon Manning, read a brief news item about an upcoming Air Force conference in San Antonio, on medical aspects of spaceflight. He sent an associate editor, Cornelius Ryan, to cover this meeting and to see if it could be turned into a story.

Ryan was no space enthusiast, though he was a meticulous reporter, as he would show with his book *The Longest Day.* But at the meeting he fell in with von Braun, whose salesmanship had swayed even Adolf Hitler. Over cocktails, dinner, and still more cocktails, von Braun delivered his pitch. It focused on a manned space station, and von Braun declared that it could be up and operating by 1967.

The concept of such stations went back at least to 1897, when Kurd Lasswitz, a founder of science fiction, described them as the key to space travel. Hermann Oberth, in his book of 1923, *The Rocket into Interplanetary Space,* outlined their uses:

> With their powerful instruments they would be able to see fine detail on earth. Since they, provided the sky is clear, could see a candle flame at night and the reflection from a hand mirror by day, they could maintain communications between expeditions and their homeland, far distant colonies and their mother land, ships at sea. The station would notice every iceberg and warn ships. The disaster of the Titanic in 1912 would have been avoided by such means.
>
> The observing station could also be a refueling station. If hydrogen and oxygen are shielded against solar radiation they'll keep for any length of time in the solid state. If we now connect a large sphere of sodium metal which was assembled and filled with fuel in the station's orbit with a small solidly constructed rocket which draws fuel from it, we have a highly efficient apparatus which should easily be capable of flying to another planet.[1]

Six years later, in 1929, an Austrian engineer who wrote under the name of Hermann Noordung described how a space station might look. He introduced the classic shape of a rotating wheel, slowly spinning to provide artificial gravity, with an airlock at the hub. To provide electric power, solar mirrors would focus sunlight onto boiler pipes, to operate a type of steam engine.

Von Braun started with these ideas and went much further. Noordung's station was to have a diameter of a hundred feet; von Braun's would be two and a half times larger and carry a crew of eighty people. These would include astronauts operating a major telescope. Meteorologists, looking earthward, would study cloud patterns and predict the weather.

To serve the needs of the Cold War, von Braun emphasized the use of the station for military reconnaissance. He also declared that it could operate as a high-flying bomber, dropping nuclear weapons with great accuracy. To build it, he called for a fleet of immense manned cargo rockets, each weighing seven thousand tons, the equivalent of five hundred V-2s. Yet the whole program—rockets, station, and all—would cost $4 billion, only twice the budget of the Manhattan Project.

Once completed, the station could serve as an assembly point for a far-reaching program of exploration. An initial mission would send a crew on a looping flight around the moon, to photograph its unseen farside. Later, perhaps by 1977, a fleet of three rockets would carry as many as fifty people to the Bay of Dew, to stay for six weeks while ranging widely in mobile vehicles. Eventually, perhaps a century in the future, an even bolder expedition would carry astronauts to Mars.

By the end of the evening, von Braun had won a new convert in Ryan, who now believed that manned spaceflight was not only possible but imminent. Returning to New York, Ryan persuaded Manning that the topic merited not merely an article, but an extensive series that in time would fill parts of eight issues of the magazine. Manning invited von Braun to come to Manhattan for interviews and discussions, together with several other specialists. These included Willy Ley; the astronomer Fred Whipple of Harvard, who knew about the moon and Mars; and Heinz Haber, an Air Force expert in the nascent field of space medicine.

In preparing the articles, *Collier's* placed heavy emphasis on getting the best possible color illustrations. Its artists included Chesley Bonestell, who had founded the genre of space art by presenting imagined views of planets as seen close up from nearby satellites. Von Braun prepared engineering drawings and sketches of his rockets and spaceships, which Bonestell and the other artists used to create working drawings for his review. They would execute the finished paintings only after receiving von Braun's corrections and comments.

The first set of articles appeared in March 1952, with the cover illustration showing a cargo rocket at the moment of staging, high above the Pacific. "MAN WILL CONQUER SPACE SOON," blared the cover. "Top Scientists Tell How in 15 Startling Pages." Inside, an editorial noted "the inevitability of man's conquest of space" and presented "an urgent warning that the U.S. must immediately embark on a long-range development program to secure for the West 'space superiority.'"

The series appeared while Willy Ley was bringing out new and updated editions of his own book, *Rockets, Missiles and Space Travel.* In addition, during 1951 Arthur C. Clarke published *The Exploration of Space,* which became a selection of the Book-of-the-Month Club. But the *Collier's* articles set the pace. Late in 1952, *Time* magazine ran its own cover story on von Braun's ideas. Then Walt Disney got into the act, phoning Ley from his office in Burbank, California. He was building Disneyland, his theme park in nearby Anaheim, and would advertise it by showing a weekly TV program of that name over the ABC network. With von Braun's help, Disney went on to produce *Man in Space,* an hour-long feature. It ran in October 1954, with subsequent reruns, and audience-rating organizations estimated that forty-two million people saw it.

The influence of von Braun's scenario would echo powerfully during subsequent decades, for the *Collier's* series would define NASA's eventual agenda for manned spaceflight. Cargo rockets such as the Saturn V and space shuttle, manned moon landings, a large space station, eventual manned flight to Mars—all these ideas would dominate NASA's projects and plans. Yet in an important respect the space station, which was the centerpiece of the whole, would soon be obsolete.

The concept had taken root in the writings of Oberth and Noordung, at a time when the only useful electronic technology was radio. It emphasized the view that while space offered such useful activities as reconnaissance, weather observation, communications, and astronomy, these would rely on astronauts for their achievement. It then made sense to have the astronauts live together in the station, whose design would necessarily focus on their comfort.

In fact, the uses of space would rely almost entirely on unmanned spacecraft, with astronauts filling only the most marginal roles. Von Braun didn't expect this. He was quite prepared to envision an unmanned instrumented satellite, a "baby space station," which would carry a TV camera along with rhesus monkeys, to test the medical effects of weightlessness and cosmic rays. But this would merely be a prelude. The real effort would revolve around the space station, with all eighty of its crew members hard at work.

But while von Braun's far-reaching manned program was appearing on magazine covers and on TV, several much quieter efforts were laying the groundwork for the unmanned spacecraft. Within the Air Force, the Rand Corporation emerged immediately after the war as a major center of activity. This think tank came into being late in 1945 and quickly began issuing classified reports that would exert great influence. In May 1946, only a few months after Vannevar Bush had declared that an ICBM was an impossibility, a group of Rand researchers considered that rockets more powerful than an ICBM might succeed in launching satellites.

The report, "Preliminary Design of an Experimental World-Circling Spaceship," focused on the prospect for orbiting a 500-pound package of instruments. Such a spacecraft would be far too small either for manned operation or as a carrier of an atomic bomb, which raised the question of its uses. These could include such nonmilitary applications as communications, observation of cloud patterns, and biomedical research in zero gravity. In addition Louis Ridenour, who later became the Air Force's chief scientist, noted that even blurry TV images could have military value: "Perhaps the two most important classes of observation which can be made from a satellite are the spotting of points of impact of bombs launched by us, and the observation of weather conditions over enemy territory."

Another scientist, David Griggs, added a prophetic prediction that merits being quoted at some length:

> Although the crystal ball is cloudy, two things seem clear:
>
> 1. A satellite vehicle with appropriate instrumentation can be expected to be one of the most potent scientific tools of the Twentieth Century.

2. The achievement of a satellite craft by the United States would inflame the imagination of mankind, and would probably produce repercussions in the world comparable to the explosion of the atomic bomb.

The nation which first makes significant achievements in space travel will be acknowledged as the world leader in both military and scientific techniques. To visualize the impact on the world, one can imagine the consternation and admiration that would be felt here if the U.S. were to discover suddenly that some other nation had already put up a successful satellite.[2]

As with Convair's MX-774, this early interest in spacecraft quickly died. Late in 1948, Defense Secretary James Forrestal mentioned an Earth Satellite Vehicle Program, but added that he was following a recommendation that "current efforts in this field be limited to studies and component designs." At Rand, studies of satellites continued apace. In October 1950 another researcher, Paul Kecskemeti, issued a report that raised a critical question: Even if we could launch such spacecraft, would the Soviets let us?

Kecskemeti realized that reconnaissance would provide a strong rationale for a satellite program and emphasized that Moscow would view the orbiting cameras as a major threat: "Fear of loss of secrecy is constant and intense. A picture of the outside world as engaged in penetrating Soviet secrets is likely to be highly anxiety-provoking." This attitude was already clear in the area of commercial aviation. Most nations freely granted overflight rights to airliners. The Soviets did not.

If the United States launched a satellite to fly over the Soviet Union, the Kremlin might view it as a form of aggression. Stalin then might respond with military activity, directing force or threats against nearby nations that were providing bases to the U.S. Air Force. Hence, before proceeding with such launches, the United States would first have to establish the legal right to carry out the overflights. This right would amount to "freedom of space," akin to freedom of the high seas. It would allow a spacecraft to overfly other countries without the countries' permission.

"Perhaps the best way to minimize the risk of countermeasures would be to launch an 'experimental' satellite on an equatorial orbit," Kecskemeti concluded. The spacecraft would be small, for it would carry no camera. It also would steer clear of the Soviet landmass, which would remain well to the north of its orbit. By pursuing research in science rather than military objectives, it might overfly a number of countries without drawing complaint. In doing so, it would establish freedom of space as a legal principle, thus providing cover for the reconnaissance satellites that would follow.

As events would demonstrate, the pursuit of freedom of space would clash strongly with the insistence on being first into orbit. Kecskemeti's small scientific spacecraft might well succeed in establishing his legal principle, but no one could deny that if Russia launched the first such satellite, it would immediately establish the principle par excellence. To the degree that Soviet security indeed demanded thoroughgoing secrecy, the achievement might actually undercut that security, even while winning worldwide renown. This in fact would happen, though the events would take more than a decade to play themselves out.

Meanwhile, as the Rand Corporation was stirring the pot, several civilian scientists were preparing for a satellite effort of their own. One evening in April 1950, James Van Allen, the man behind the Aerobee sounding rocket, hosted a gathering of friends at his home in Silver Spring, Maryland. The featured guest, Sydney Chapman, was from Oxford University; he was a leading expert on the behavior of the atmosphere. Other guests included Lloyd Berkner, head of Brookhaven National Laboratory, and Fred Singer, a young physicist at the University of Maryland.

The talk turned to whether scientists in a number of countries might coordinate their efforts so as to advance in their studies of the upper atmosphere. Everyone knew that the International Meteorological Organization had set up two such coordinated programs called the International Polar Years. These had taken place in 1882 and 1932 and had considerably enhanced scientific understanding of the Arctic. Berkner was a polar explorer of long standing, having taken part in the 1932 activities. He now suggested that the time might be ripe for a Third Polar Year, to be held in 1957 or 1958, a quarter-century after the last one. The other scientists responded with enthusiasm.

A transnational organization, the International Council of Scientific Unions, offered a forum that could make the necessary arrangements. Berkner was rising rapidly within it and would soon become its president. He and Chapman presented a proposal to a commission and parlayed its members' support into approval by the full ICSU. Indeed, as the topics of interest broadened, Chapman saw that this would be no mere polar year and proposed the name of International Geophysical Year. It would run for eighteen months, from mid-1957 through the whole of 1958, a period that would coincide with a peak of sunspot activity as well as with several eclipses. In time it would involve some sixty thousand participants from sixty-six nations.

The ICSU set up a special IGY committee, with Chapman as president and Berkner as vice president. In Washington, the National Academy of Sciences soon took note. It was America's member organization within the ICSU, and it set up its own committee. The chairman, Joseph Kaplan, had organized the 1951 San Antonio conference on space medicine, and had contributed to the *Collier's* series.

Also in 1951, a number of national rocket societies combined to form the International Astronautical Federation. At its 1953 conference in Zurich,

Fred Singer presented the first unclassified proposal for an initial scientific spacecraft. Talking with two British specialists, Val Cleaver and Arthur C. Clarke, he christened it MOUSE (Minimum Orbital Unmanned Satellite of Earth). No mouse would fly in it, let alone a rhesus monkey, but it would weigh a hundred pounds and carry a valuable array of instruments. Addressing a subsequent symposium at the Hayden Planetarium in May 1954, Singer emphasized that MOUSE was "within range right now, should be built now, and can be built, or at least started, now with available technical facilities."

The president of the IAF, Frederick Durant, was a former naval aviator and rocket specialist. He learned that a fellow aviator who now was in the Office of Naval Research, Commander George Hoover, wanted to launch a satellite like MOUSE. Late in June, Durant phoned von Braun: "I just had a very interesting talk with a man in the ONR; he wants to get rolling on a space vehicle. Do you want to meet him?" Von Braun came a-running, and two days later he met with Hoover and Durant. Singer and Whipple were also at this meeting.

"Everybody talks about a space satellite, but nobody does anything about it," Hoover declared. "I'd like to get the ball rolling with a small satellite carried into orbit by a combination of existing rockets." Von Braun, ever helpful, already had a plan in mind to do just that. For some time he had been thinking about how a cluster of small solid-fuel antiaircraft rockets, mounted atop a Redstone missile, could yield the boost to get to orbit. Now he asserted that he could put up an inflatable balloon weighing five to seven pounds. It would carry no instruments, not even a radio transmitter, but it would appear in the sky like a new star, rising in the west as it flew along its orbit. Whipple, the astronomer, assured the group that this star would indeed be bright. And it could fly merely by mounting existing rockets atop an existing booster.

The attendees agreed that even this modest effort would be valuable, and the Chief of Naval Research authorized further discussions. Then in September 1954, Singer decided to get Berkner to agree that satellite launches should take place within the IGY. Berkner was dubious; he knew that Singer was young and impetuous. But he agreed to consult with a group of advisers and to have a look at the matter.

On the eve of a meeting of his ICSU committee in Rome, Berkner invited these advisers to gather in his room at the Hotel Majestic. Singer was there, as was Joseph Kaplan, head of the National Academy IGY committee. As one participant would note, they had to consider whether talk of satellites was real or just a "pious resolution" such as Sir Isaac Newton might have presented in the seventeenth century. But despite words of caution, the group became caught up in the excitement of the idea. When one man noted that electric batteries might break down amid weightlessness, another one banged his fist and shouted, "Then we'll get batteries that won't!" The session lasted six hours and concluded with unanimous support for IGY satellites. A few days later, the ICSU committee endorsed the proposal.

Now it was Kaplan's turn to take the lead. A few months later, in January 1955, he set up a panel on rocketry, with a subcommittee on satellites. The members of the subcommittee included Milton Rosen, the head of the Viking effort. He had studied von Braun's formal proposal, called Project Orbiter, and he knew he could offer something better. Some months earlier, the Air Force's General Schriever had asked him to propose an advanced version of Viking that could serve for flight tests of experimental nose cones, and Rosen knew that this improved Viking could offer a better booster than the Army's souped-up Redstone.

The standard Viking used a Reaction Motors engine that had topped 21,000 pounds of thrust in recent flights. This wasn't enough, but General Electric, a leading builder of jet engines for aircraft, had a rocket motor under development that promised 27,000 pounds. In addition, Aerojet General, builder of the Aerobee, was proposing an uprated version called the Aerobee-Hi that would boost the achievable altitude from 75 to 150 miles. Rosen had proposed to put the General Electric engine within his Viking, place an Aerobee-Hi on top as a second stage, and use this two-stage rocket to launch the nose cones. Now he declared that by adding a solid-fuel third stage, he could put up a satellite.

In turn, the satellite would improve substantially on von Braun's simple balloon. It would weigh thirty pounds and carry instruments along with a radio transmitter, making it a true mini-MOUSE. In addition, the Viking effort had already developed and used a high-quality system for tracking and receiving telemetry from its sounding rockets. A new version called Minitrack could also serve the satellite effort.

In March the National Academy's IGY executive committee endorsed the satellite decision, noting that either Project Orbiter or Viking could serve the purpose. Kaplan shot letters off to Detlev Bronk, head of the Academy, and to Alan Waterman, director of the National Science Foundation, who was working closely with Bronk on preparations for the IGY.

The National Academy, in operation since 1863, was an elite society for the nation's top scientists, some of whom would regard membership as a way station en route to the Nobel Prize. The NSF, in business only since 1950, had a modest budget with which it made grants to university researchers. But neither organization possessed the means to put up a satellite. Only the Defense Department could do that.

Waterman and Bronk approached the defense secretary, Charles Wilson. They proposed that he put a spacecraft into orbit in the name of the Academy, an outfit as divorced from the military as the Nobel Foundation itself. Wilson bucked the issue over to Donald Quarles, his assistant secretary for research and development. And just then, in the spring of 1955, Quarles was deeply concerned with strategic reconnaissance.

General Curtis LeMay, head of the Strategic Air Command, was ardently in favor of doing much more in this area. His bomber crews had to be ready to fly to the attack, and they needed photos of radar images of targets as well as information on Soviet air defenses. He knew that reconnaissance overflights were legally acts of war; Truman had banned them for that reason. LeMay had gotten around the ban by sending B-45 bombers to Britain; the Royal Air Force had flown them over the Soviet Union and had shared their findings with SAC. By 1954 LeMay was flying his own aircraft for the purpose of reconnaissance, with Eisenhower's permission, and for a while he found air defenses that were weak indeed. He would later recall a time when "we flew all of the reconnaissance aircraft that SAC possessed over Vladivostok at high noon." Vladivostok was Moscow's principal Pacific naval base.

In addition to targeting information, Pentagon analysts wanted badly to learn the state of Soviet preparations for war. The 1953 nuclear test, with its 400-kiloton yield, had caught everyone by surprise, while additional worrisome news came from Helmut Grottrup and his colleagues, most of whom had returned to Germany between 1951 and 1953. The Soviets had kept them in the dark concerning the work of Korolev and other Russians, but had called on them to carry out detailed studies of long-range missiles. American specialists, questioning these returnees, could easily conclude that Moscow was preparing to build the missiles on its own.

Bronk and Waterman both were members of a presidential science-advisory committee, with Waterman serving as deputy to its chairman, Lee DuBridge, the president of Caltech. In March 1954 Eisenhower had met with this committee and had warned them that he feared a surprise attack, a new Pearl Harbor that would destroy cities rather than battleships. The CIA had already issued its 1954 estimate, predicting that Moscow would have five hundred turboprop Tu-95 bombers in 1957. DuBridge's initial steps aimed at setting up a high-level commission, the Technological Capabilities Panel (TCP). It would recommend new policies that could meet this danger.

Within weeks, new events heightened the concern. On May Day, at a public air show, Moscow's air force showed its new jet bomber, the Bison. Here was another surprise—a Soviet jet bomber! It was all the more worrisome because no one had known about it until the Kremlin displayed it openly. A week later, one of LeMay's B-47s flew deep into the USSR on a reconnaissance mission and ran into a MiG-17 jet interceptor. The B-47 got away with a fuel leak and made it back to England, but the point was clear: America's own jet bombers were now at risk.

One response would revolve around Atlas, which garnered its 1-A priority during the same month of May. Clearly, though, a further response would require a much stronger commitment to reconnaissance. LeMay had been freelancing in this field, but to Eisenhower this situation was unacceptable. It meant that the Air Force would control the assets used to measure the Soviet threat as well as the forces with which to respond, and it raised the prospect that its generals would slant their intelligence estimates so as to support their own demands for new weapons. Ike wanted to carry out reconnaissance within an outfit that LeMay could not control: the Central Intelligence Agency.

The new commission, the TCP, took form in mid-1954; Eisenhower recruited James Killian, the president of MIT, for its chairman. A subpanel, Project 3, dealt with technical means for surveillance, and among the people who learned of it was Clarence "Kelly" Johnson of Lockheed, one of the country's top aircraft designers. His facility, the Skunk Works, had designed and built America's first operational jet fighter, the P-80, during 1943 and 1944. At that moment, in the wake of the Korean War, Johnson was developing the F-104, which would reach Mach 2. Earlier in the year he had prepared a design for a spy plane and had pitched it to the Air Force, without success.

But Johnson had received strong encouragement from Trevor Gardner, and he and Gardner now approached Project 3 for a new try. The subpanel's chairman, Edwin Land, had invented the Polaroid camera and was president of the Polaroid Corp. Johnson could move quickly when he had to; he had built the first P-80 from scratch in only 143 days, and he now promised to put his spy plane into test flight within eight months of signing a contract. Land and Killian took the proposal to Ike and convinced him to accept it. In keeping with the CIA's penchant for names that disclosed little, it would receive the name Utility-2.

The man at the CIA who would manage the U-2, right from the outset, was Richard Bissell. Tall and lean, he looked the part of a patrician and had the background to match: Groton, Yale, London School of Economics. Soft-spoken, courteous without excessive formality, and with a taste for three-piece suits, he had broad social connections that might by themselves have marked him for advancement. However, what set him apart was his mind. He had taught economics at Yale; his students had included McGeorge Bundy, who would be JFK's national security adviser. After the war, he took the lead in pushing for the Marshall Plan that helped rebuild Europe. The historian Thomas Powers would describe him as having "lucidity, a capacious memory, and a confidence, unsettling in its coolness, that the only relevant question to be asked of a system was whether it worked. He had a vast appetite for detail, and once he set himself to understand how something worked, he made it his own."[3]

Bissell had been with the CIA for less than a year when its director, Allen Dulles, told him one morning that he was to head the U-2 program. That afternoon, Bissell met with an Air Force group in Gardner's Pentagon office. He later would recall that "it became readily apparent that although a top-level decision had been made, nothing had been worked out. When the question of money arose, everyone looked in my direction." Bissell agreed to put up $22 million from a discretionary fund that Dulles controlled. This would allow the project to go forward without Congress approving it or even being aware that it existed.

The Air Force would remain closely involved, supplying engines for the aircraft while training their pilots. These pilots would be "sheep-dipped": transferred out of the service and into the CIA while receiving new identities as civilians working for Lockheed. In addition, LeMay in time would receive his own U-2s. But project management would be the exclusive province of Bissell. His staff set up shop in offices set off from the rest of the CIA and wrote its own contracts, kept its own financial records, and handled its own administration and security. Indeed, nearly one staff member in seven was a security officer. Bissell kept its cable traffic private; even Dulles couldn't read it.

Bissell also gave Johnson a free hand. They transacted much of their business by telephone, though once a month, Johnson would write a progress report of five or six pages. Still, within the CIA this could amount to voluminous documentation. Other activities of similar sensitivity, such as plans to assassinate Fidel Castro, would rely entirely on verbal discussion, with nothing whatever in writing.

In mid-February of 1955, the full Killian Committee made its report, entitled "Meeting the Threat of Surprise Attack." Killian would note that a presidential aide "carefully orchestrated all the plans for the presentation of the report and put those of us who made the presentation through a rehearsal." They proceeded to address a session of the National Security Council, held within the White House itself.

The report declared: *"We have an offensive advantage but are vulnerable to surprise attack"* (emphasis in original). "Because of our vulnerability, the Soviets might be tempted to try an attack." The nation needed radars to give early warning of approaching Soviet bombers. In addition, the report stated: "Our defense system is inadequate; therefore SAC is vulnerable." Until the nation could make good these deficiencies, it would indeed lie open to a nuclear Pearl Harbor.

But Killian's group also noted that new initiatives, particularly the Atlas ICBM, seemed likely to strengthen the United States. It called for more such initiatives, with the gathering of intelligence at the top of the list. In Edwin Land's section of the report, he wrote: "We *must* find ways to increase the number of hard facts upon which our intelligence estimates are based, to provide better strategic warning, to minimize surprise in the kind of attack, and to reduce the danger of gross overestimation or gross underestimation of the threat." To do this, the CIA would require "the extensive use of the most advanced knowledge in science and technology."[4]

The U-2 would help, but it could amount to only an interim system. It would cross the entire Soviet Union at an altitude of 70,000 feet, far above the reach of interceptors or antiaircraft missiles, but not beyond the reach of radar. The view within the CIA was that within a year or two after beginning these overflights, Moscow would have radars capable of tracking the U-2 reliably. Its leaders then could file diplomatic protests with supporting evidence, generating enough political pressure to halt the missions.

But satellite reconnaissance would be another matter, and with its customary prescience, the Rand Corporation had already laid important groundwork with classified studies. In April 1951 its researchers had issued two new reports, "Utility of a Satellite Vehicle for Reconnaissance" and "Inquiry into the Feasibility of Weather Reconnaissance from a Satellite Vehicle." Then, with CIA sponsorship, Rand undertook a major study called Project Feedback. Its highly secret reports came out on March 1, 1954, the same day as the Bravo hydrogen-bomb test. The studies were detailed indeed; the summary alone, "An Analysis of the Potential of an Unconventional Reconnaissance Method," filled two volumes.

Principal topics of discussion involved methods for transmitting images to a ground station. One technique, investigated at RCA, called for use of a TV camera, which would store its images on videotape. Then, when the spacecraft flew over a ground station, it would send down the tape's contents using a radio link.

Another method, studied at Eastman Kodak, CBS Laboratories, and Philco, promised finer detail by using a photographic camera. Its film would pass through an onboard processor for development into negatives. The developed film then would roll past a scanner, which would capture each image in the manner of a wirephoto, as a succession of closely spaced lines. The resulting data then would again be stored on tape for later transmission.

This was no proposal for anything as simple as Fred Singer's MOUSE, let alone von Braun's orbiting balloon. These were detailed concepts for automated imaging labs that would accomplish by electronics what von Braun believed would require astronauts in his space station. The Air Force was now ready to move quickly. In mid-March of 1955, only a month after the Killian Committee issued its report, the Wright Air Development Center issued General Operational Requirement No. 80, requesting proposals from industry for a "strategic reconnaissance satellite weapon system," designated WS-117L. In time it too would enter the expanding empire of Richard Bissell, but while in development, it would remain an Air Force project.

Not long after GOR-80 went out, Detlev Bronk and Alan Waterman approached Defense Secretary Wilson with their proposal for an International Geophysical Year satellite. One must not think of them as naive intellectuals entering a realm of worldly cynicism; the scientists were presidential advisers in their own right. But when Wilson bucked the issue over to Assistant Secretary Quarles, the matter at hand was no straightforward choice between Project Orbiter and Viking. Lurking in the background was WS-117L, and its needs would take precedence. In particular, the IGY satellite would have to serve those needs—by establishing the principle of freedom of space, in the name of science pursued for all humanity, so that other satellites might fly for the CIA.

Could von Braun do this? No way; his satellite would have Army written all over it. His project center would be Redstone Arsenal, the chemical-warfare plant that had become a facility for military rockets. His booster, the Redstone, was a weapon in its own right, able to carry the atomic bomb. Against this nakedly bellicose background, the IGY would represent too thin a veil. The world would see von Braun's satellite as a mere prelude to an invasion of space by military force.

But Milton Rosen's proposal was something else entirely. His booster would derive from Viking and Aerobee, which had become known as research rockets launched for scientific purposes. The Naval Research Laboratory, which would serve as the project center, didn't have the gamy reputation of Redstone Arsenal. It was known as a true center of research, with well-regarded scientists who had made important contributions in their fields.

So for Don Quarles, the choice was a no-brainer. He couldn't just announce it, though; tongues would wag. The last thing he needed was to get the Army upset, for its friends could cause trouble on Capitol Hill. Nor could he disclose his reason; WS-117L had to stay under wraps. But he could easily play the old game of appointing an advisory panel. His task became simpler late in May, as the National Security Council endorsed the IGY satellite—with a proviso that the peaceful purposes of the effort were to receive particular emphasis. This helped sway Quarles's panel, whose members could readily agree that Rosen's proposal was more peaceful than von Braun's. They also liked the fact that Rosen was offering a true instrumented satellite along with means for tracking. By a vote of seven to two, these advisers elected to go with the Navy.

There was only one problem: The Army had the means to allow the United States to be first into space, and the Navy did not.

Quarles's staff had reviewed the IGY proposal from a military standpoint, noting in a report that "a group of Russia's top scientists is now believed to be working on a satellite program." A presidential assistant, Nelson Rockefeller, had added comments that stressed the importance of being first: "I am impressed by the costly consequences of allowing the

Russian initiative to outrun ours through an achievement that will symbolize scientific and technological advancement to people everywhere. The stake of prestige that is involved makes this a race that we cannot afford to lose." The National Security Council had not been moved; its policy paper, NSC-5520, included a classified section and strongly emphasized the importance of protecting WS-117L.

In addition, von Braun tried to play catch-up by offering to launch a radio transmitter rather than an orbiting balloon. But it still would weigh only five pounds, which wasn't enough. In August, when the advisory panel came out in favor of the Navy, Major General Leslie Simon of Army Ordnance registered an angry protest. By fitting the Redstone with better solid-fuel upper stages, he promised to launch an eighteen-pound instrumented satellite, declaring that "the first orbital flight can be scheduled for January 1957 if an immediate approval is granted." The Navy's contractors—General Electric, Martin, Aerojet General, Thiokol Chemical—responded with their own assurances of quick action, and Simon's plea went unheeded.

The chairman of the advisory panel, Homer Stewart of Jet Propulsion Laboratory, went to Huntsville and told von Braun to keep his Redstone available, just in case. Von Braun soon added the upper stages, though not to launch a satellite. Instead, he would fly test models of nose cones, and would start with a flight to demonstrate that his multistage rocket, called Jupiter-C, would actually work.

Von Braun had it on the launch pad on September 20, 1956, ready for liftoff. It looked almost exactly like the one that would put up America's first satellite, Explorer 1, sixteen months later. Then the telephone rang; it was his boss, Major General John Medaris. "Wernher," he said, "I must put you under direct orders personally to inspect that fourth stage to make sure it is not live."

Though it could not reach orbit, this top stage broke all performance records as it flew 3,355 miles downrange, reaching an altitude of 682 miles. Even this made no difference. As the historian Walter McDougall would conclude:

> If being first was the primary consideration in U.S. satellite policy, the DoD could have overridden its advisory committees. But speed was *not* the primary consideration; in the end, assuring the strongest civilian flavor in the project was more important. The administration was advised of the propagandistic value of being first into space. Of all these critical policy areas, however, the last had the lowest priority. For there were two ways the legal path could be cleared for reconnaissance satellites. One was if the United States got away with an initial small satellite—and had no one object to it. The other way was if the Soviet Union launched first.
>
> The second solution was less desirable, but it was not worth taking every measure to prevent.[5]

Right at the start, then, America's space program featured three distinct types of activity: the IGY effort, WS-117L, and manned flight. Manned activities started with the *Collier's* scenario; when real programs materialized, they would retain that magazine's strong emphasis on hype and exaggerated expectation. The IGY project had its own blare of publicity, beginning with its announcement at a White House press conference hosted by the president's press secretary, James Hagerty. Of course, it was little more than a cover for the true space project, but Hagerty wasn't talking about that one. Indeed, he didn't even know about it.

In September 1955 the Deputy Secretary of Defense Reuben Robertson formally directed the Navy to proceed with its satellite project, giving it the name Vanguard. This name tempted fate, for the Communist Party had been asserting for decades that it stood in the vanguard of the world's workers. And in Moscow, Sergei Korolev was well along with his own plans.

Although Korolev had cherished astronautical hopes since his first year in Moscow, his first opportunity for serious involvement dated only to 1948. It grew out of the studies of an old friend, Mikhail Tikhonravov, whom Korolev had known since those student days. They had worked together in MosGIRD when it was still in the wine cellar and had both gone on to join RNII, the Scientific Research Institute of Reaction Propulsion, though Tikhonravov had kept his freedom during the purges. Korolev's biographer, Yaroslav Golovanov, notes Tikhonravov's "romantic enthusiasm and flights of fancy. He painted landscapes in oil, assembled a collection of wood-eating beetles, and studied the dynamics of flight in insects. He simply loved learning."

After the war, Tikhonravov joined a new rocket institute that would define requirements for an operational missile force. In 1947 he decided to look at the possibility of a satellite. His boss allowed him to gather a small group of fellow enthusiasts, who did the calculations on a "rocket packet": a cluster of identical rockets that would fall away during ascent as they expended their propellant. Through their analyses, the group became convinced that the rocket technology at the time already sufficed to build a packet that would place a spacecraft into orbit.

In June 1948, Tikhonravov received an invitation to give a paper at a meeting of the Academy of Artillery Sciences. Naturally he knew what he wanted to present, but the idea seemed so strange that he decided to talk with the president of the academy, General Anatoly Blagonravov. The general read his paper and told him, "We cannot include your report. Nobody would understand; they would accuse me of getting involved in things we do not need to get involved in."

But that evening, rereading the paper, Blagonravov began to think that it might not be so easy to dismiss after all. In addition, Tikhonravov's colleagues urged him not to take no for an answer, but to present stronger arguments and try again. He went back to the general, who smiled and listened carefully. "All right, we will include the report," he decided. "Be prepared—we will be embarrassed together." He knew what they would face, for after Tikhonravov gave the paper, another military officer looked past Blagonravov's head and asked, "The institute must not have much to do and decided to switch to the realm of fantasy?"

But Korolev was also there, and he approached his old friend and said, "We have some serious things to talk about." Blagonravov himself took part in the discussions that ensued, then arranged for several V-2s to fly from Kapustin Yar as sounding rockets. In June 1951 this venture broadened to include flights of dogs. The first two, Dezik and Tzigan, arced above a hundred kilometers and experienced four minutes of weightlessness; seven others soon followed.

By then Korolev was deeply involved in design work on the R-3, and he was aware that it would be quite capable of launching a satellite, by adding two upper stages. Already he and Tikhonravov had drawn up a document, "Research on a Manmade Satellite of the Earth," which had begun to address technical issues in a serious way. Then, in May 1954, came the definitive decision to proceed with the R-7 ICBM. Six days later, Korolev wrote a letter to the Council of Ministers: "The current development of a new product with an ultimate speed of nearly 7000 meters a second makes it possible for us to speak of the possibility of developing in the near future an artificial earth satellite. By reducing the weight of the payload somewhat, we will be able to achieve the ultimate speed of 8000 m/sec that a satellite needs."[6]

In July Korolev had an outline of the complete R-7 design, in fifteen volumes, and engineers were preparing the first working drawings. The design called for a missile that would carry a warhead of 5,400 kilograms for 8,600 kilometers. Also in July, Tikhonravov reported that the satellite could weigh from 1,000 to 1,400 kilograms, or as much as three tons. (Fred Singer's MOUSE called for a mere hundred pounds.) At the end of the month, in Washington, Press Secretary Hagerty announced America's program for an IGY satellite. In Copenhagen, at a conference of the International Astronautical Federation, a member of the Academy of Sciences, Leonid Sedov, replied a few days later with a statement of his own. He told journalists that his country would do the same, adding: "It is possible that our satellites will be finished before the American satellites and will be heavier."

Meanwhile, development of the R-7 was proceeding rapidly. It needed a new launch site; it couldn't use Kapustin Yar, because of the risk of crashing onto inhabited territory. Korolev joined a commission to recommend an appropriate location, the choice ultimately going to Tyuratam, a railhead on the Syr Darya River, a hundred miles east of the Aral Sea and

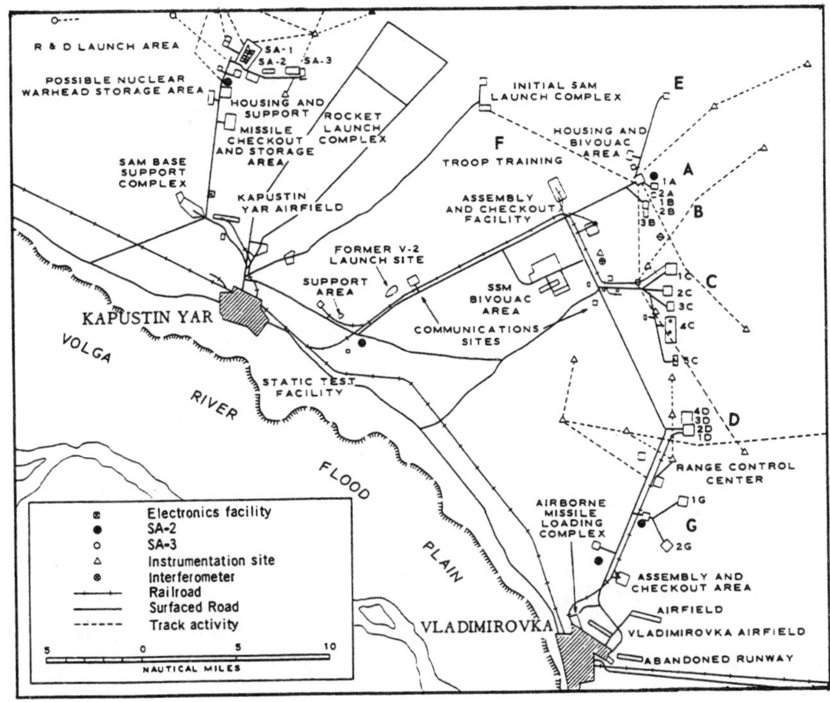

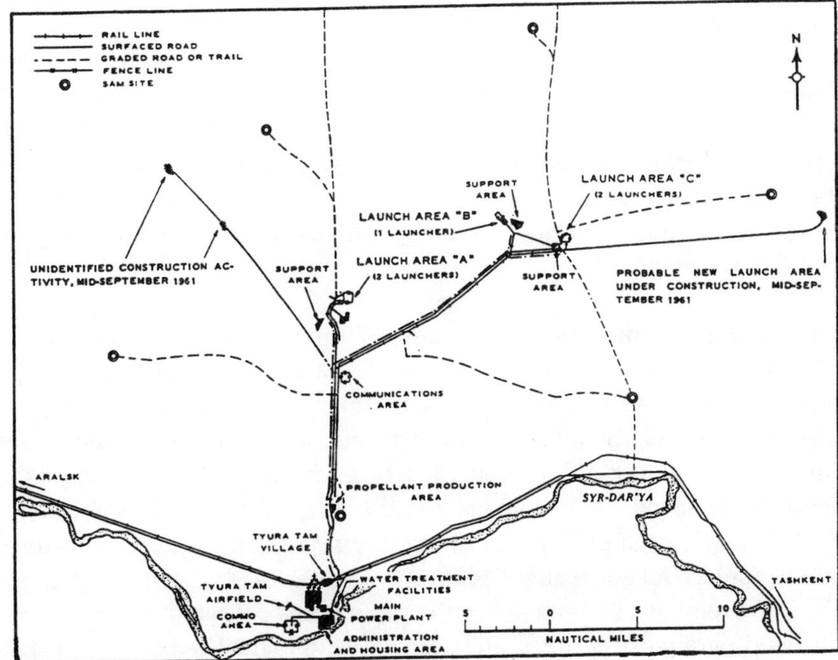

Soviet launch sites, circa 1960. *Top,* Kapustin Yar; *bottom,* Tyuratam. (Central Intelligence Agency)

close to the Kyzyl Kum Desert. Korolev had placed this site at the bottom of his list of choices, but he was overruled.

Glushko also continued with his work on the engines, performing static tests at a center near Moscow. Each engine would feed a cluster of four thrust chambers with a single set of turbopumps. Trials began in mid-1955, with single chambers. In December Glushko moved on to two-chamber assemblies; a month later, in January 1956, the four-chamber motor went on a test stand for its first successful firing. The complete R-7 would attach four boosters to a central core, and this core reached the stand in August. Single boosters followed, and the complete R-7, boosters and all, underwent static testing the following winter.

In pursuing his satellite, Korolev worked closely with the mathematician Mstislav Keldysh. Keldysh made his name in aeronautics by studying vibration in aircraft and directing postwar studies of ramjets. In 1950 his associates used his results to build an experimental ramjet that reached Mach 2.7, some 1,800 mph. He then went on to work with Korolev in studying ballistic missiles. He addressed such issues as designing a guidance system with which a large rocket could maintain stability in flight while rolling on its axis, with propellants sloshing in the tanks. To deal with such matters, he took the lead in using the first Soviet computers.

Keldysh was soft-spoken and slender, with a healthy tan and a taste for well-cut suits. He also was a full member of the Academy of Sciences. Korolev had become a corresponding member, or associate, in 1953, but Keldysh's standing gave him access to the broad scientific community, which he approached to build support for a satellite. In mid-1955 Keldysh assembled a group of colleagues in his study. Tikhonravov gave a talk on the subject and got carried away: "I know how exciting a rocket launch is, and I am convinced that if you see one even once, you will never forget it and will dream about another one."

Talk soon turned to power sources, cooling methods, and potential contributions from various institutes. "Tomorrow morning we have to send a letter to the academicians and the corresponding members," Keldysh concluded, "because we need their suggestions. Invite everyone who's needed for the creation of a magnetometer and an instrument for studying cosmic rays." Someone asked how long it would take to build the complete spacecraft. "A year and a half or two," he replied. "And then, I don't know. It's too big a project we're beginning, and right now it's difficult to foresee its consequences."[7]

In November 1955, the Academy of Sciences prepared a letter setting forth a program of space research. Copies went to the Council of Ministers and to the Communist Party's Central Committee. As a result, late in January 1956, the Academy obtained permission to set up a commission on a satellite, with Keldysh as chairman and Korolev and Tikhonravov as his deputies. The R-7 already had been slated to carry several types of nuclear warhead, designated Objects A, B, V, and G, in Cyrillic alphabetical order. The spacecraft would be Object D, the next letter in the sequence. It would enter history as Sputnik 3.

In December 1956 the first flightworthy R-7 arrived at Tyuratam for installation on the launch pad and prelaunch tests. Clearly, Korolev soon would have the means to launch his satellite. However, less than a year had elapsed since the start of work on Object D, and it wouldn't be ready for some time. This raised the prospect that the Yankees might be first in space after all. Tikhonravov then made a suggestion: "What if we make the satellite a little lighter and a little simpler? Thirty kilograms or so, or even lighter?"

Korolev quickly realized that he could build such a craft using the resources of his own organization. Early in January 1957 he wrote a report calling for two spacecraft, an early version weighing forty to fifty kilograms and then Object D, at 1,200 kilograms. He received approval, and the first satellite took the name *Prosteishy Sputnik* (Simple Satellite). It was a polished aluminum sphere with a set of antennas, holding a radio transmitter and a set of batteries. It carried no instruments, but it could transmit a clear *beep beep* that was easy to receive.

As these Soviet efforts unfolded, Americans were also stepping up the pace. The CIA already knew about Kapustin Yar; it had come to light during 1953. A Royal Air Force bomber, a Canberra, then had carried out a daring reconnaissance flight from West Germany to Iran, landing with what a CIA official would describe as "some fair pictures" of that rocket center.

Next, the Air Force's Trevor Gardner took the initiative in setting up a radar station near Samsun in Turkey, on the northern coast. It had a clear shot past the Caucasus Mountains to Kapustin Yar, and in mid-1955 it hit pay dirt. Ruben Mettler, an assistant director at Ramo-Wooldridge, recalls that this radar picked up "the top segment of a ballistic trajectory, visible over some mountain ranges." It was an R-5, and its range of 1,200 kilometers—750 miles—shocked American officials, whose best rocket, the Redstone, could fly barely one-fourth as far. Then in February 1956, another R-5 flew from that site, and a few minutes later observers at the impact area reported, "We observed Baikal." The missile had carried its first nuclear warhead, which had detonated successfully.

There was more unpleasant news in the realm of bombers. At the May Day air show in 1954, the Soviets had displayed their jet bomber, the Bison. A year later, amid preparations for the next such event, American observers saw a formation of ten of these aircraft in flight. In mid-July came the real surprise. On Aviation Day, Colonel Charles Taylor, the U.S.

air attaché in Moscow, counted no fewer than twenty-eight Bisons as they flew past a review in two groups. This bomber now was obviously in mass production, and the CIA promptly estimated that up to eight hundred Bisons would be in service by 1960.

In fact, Taylor had seen an elaborate hoax. The initial group of ten Bisons had been real enough. But they then had flown out of sight, joined eight more, and this combined formation had made the second flyby. Still, as classified estimates leaked to the press, Senator Stuart Symington demanded hearings and used them to grill Pentagon officials concerning a "bomber gap." The flap forced Ike to build more B-52s than he had planned, and to step up production of fighter aircraft in the bargain. This episode reflected the difficulty of learning what was really happening within the Kremlin, for the Soviets would build only fifty-six such bombers by 1960, rather than hundreds. Yet even when analysts discovered the Aviation Day hoax, they took little comfort. If Moscow was trying to fool the CIA, it might mean that they were putting their real effort into missiles rather than bombers.

But during 1955 the U-2 was coming along nicely. Early that spring, Richard Bissell and Kelly Johnson had flown over Nevada to look for a dry salt lake that could serve as a flight-test center. Edwards Air Force Base was the standard facility, but while it might be good enough for the Air Force, it wasn't secret enough for the CIA. Johnson found what he wanted in Groom Lake, a site ringed by mountains. A hundred miles north of Las Vegas, it adjoined the Yucca Flats nuclear test area, which made it easy for Bissell to convince Eisenhower to annex it to that site. Groom Lake thus received the security protocol of the Atomic Energy Commission.

Early in July, the first U-2 left the Skunk Works in pieces and in crates. A C-124 cargo aircraft carried it to the test area for final assembly. The finished U-2 would resemble a sailplane, with an enormous wingspan; the broad drooping wings would help it to cruise at high altitude and to achieve long range. On July 29, as Hagerty announced the IGY satellite program, this U-2, the prelude to WS-117L, made its first flight. It wasn't supposed to; the test pilot, Tony LeVier, intended merely to see how it would handle on the ground. But it wanted to fly, and at 65 mph, no faster than a car on a freeway, it climbed several feet into the air. The first true flight took place a week later, and the U-2 quickly showed that it would offer outstanding performance.

The WS-117L concept also progressed. Three companies—RCA, Martin, and Lockheed—responded to the Air Force's request for proposals of March 1955. Lockheed's approach was particularly intriguing, because it advised that the CIA would get sharper photos, and more of them, by avoiding the Rand Corporation's techniques for transmitting images to ground stations. Instead, the satellite should store its exposed film in a protected capsule, which would re-enter the atmosphere and descend to earth for physical recovery. This viewpoint dovetailed with that of Rand itself, which was preparing a secret new report, "Physical Recovery of Satellite Payloads: A Preliminary Investigation." The report came out in June 1956, the same month that Lockheed won the contract.

Lockheed's proposal also described the satellite's appearance. It would be a cylinder, nineteen feet long and five feet in diameter, with the largest part consisting of the upper stage that would boost it into orbit. This stage, carrying 7,000 pounds of liquid propellants, would use a 15,000-pound-thrust rocket engine from Bell Aerosystems that had originally been intended to propel a bomb carried by the B-58. Its name, Agena, was that of a bright star in the constellation Centaurus. This upper stage would provide considerably more lifting power than the modified Aerobee-Hi that formed the second stage of Vanguard, and would become a mainstay of the space program.

Meanwhile, the U-2 was entering service. The first unit—four planes, six pilots—was ready in May 1956 and operating from a base at Incirlik, Turkey. A month later, Bissell and Allen Dulles went to the White House to get Ike's permission to overfly the Soviet Union. On the first good flying day, with clear weather, Bissell met again with Dulles prior to the flight. Dulles asked about the U-2's intended route. Bissell replied that it was to overfly Moscow and Leningrad, and Dulles turned pale. "Do you really think that is wise?" he asked. But the mission was a complete success.

Each flight featured a well-defined set of photographic targets, but pilots could deviate from their planned course to look at targets of opportunity. Bissell in time would recall an occasion when this happened: "He was flying over Turkestan, and off in the distance he saw something that looked quite interesting and that turned out to be the Tyuratam launch site. He came back with the most beautiful photos of this place." The CIA had not even suspected the site's existence, while the Soviets later tried to hide it by referring to it as Baikonur, a location four hundred kilometers away. The Air Force soon stepped up its surveillance, building a second radar site near Trabzon in Turkey, the fabled Trebizond of medieval Byzantium, along with a listening post near Istanbul. It would pick up telemetry from missiles in flight, as they radioed performance data such as fuel consumption, thrust, burn time, speed, warhead separation point, altitude, and range.

The R-5 posed a particular threat to NATO and Europe. During the war, German troops had fired the V-2 using mobile equipment that required only a small forest clearing, and the woodlands of East Germany would hide forward bases that could cover the whole of Great Britain. This raised the interesting question of whether the British could build a missile that could threaten Moscow. It turned out that they couldn't, at least not very soon, and this meant that once again the Air Force would have to come to the rescue. Specifically, the United States would have to develop an intermediate-range ballistic missile (IRBM), based in England, with a 1,500-mile range that would reach the Kremlin.

Following approval of the Atlas, the IRBM emerged as a hot topic. Its advocates declared that it would be simpler to develop and could be ready more quickly, with some Navy officials even proposing that it might be launched from submarines. The Killian Committee report, strongly endorsing these ideas, recommended that "an intermediate-range missile be developed for strategic use. Both land basing and ship basing should be considered."

Wernher von Braun was quick to address this new opportunity. With development of the Redstone missile approaching completion, his people would soon be twiddling their thumbs, but an IRBM would offer new grist for their mill. In March 1955, only a few weeks after the Killian report, he proposed that Redstone Arsenal should develop this missile as an Air Force contractor. General Schriever turned him down flat, noting, "It would be naive to think that the Army would develop a weapon and then turn it over to the Air Force to operate."

Schriever thought it could fall out from the two-stage Titan ICBM, as a development based on its second stage. But in July he learned that the best such missile would fly only eight hundred miles, barely half the requirement. He then invited two associates, Robert Truax and Adolf "Dolph" Thiel, to outline the design of an entirely new rocket. Truax was serving as a Navy representative in Schriever's Western Development Division, while Thiel had left Huntsville to join Ramo-Wooldridge. Their proposed missile would be the Thor.

Truax and Thiel had contractor proposals available for study, along with talented people in the two organizations. In addition, with the Atlas program well under way, they had up-to-date studies on guidance, nose cones, and engines. Because easy transportability would be essential, they designed the missile to fit in a C-124. It would fly with the standard Navaho engine from North American, which Thiel would describe as "the only one available." The finished proposal reached Schriever by Labor Day, and he proceeded to run it up the chain of command for approval.

Meanwhile, Trevor Gardner again was stirring the pot. Although Atlas had top Air Force priority, he felt this wasn't enough, for it still had to compete with other federal agencies for the best scientists and engineers. In addition, Gardner knew that it would be managed in the usual Air Force way. Schriever would write that "in the Pentagon we have a dozen or more Assistant Secretaries for the Secretary of Defense, and the same number in each of the Departments. Each of them has his own little kingdom to rule over, and each one has a hundred or more bureaucrats under him, and all of them have their own little axes to grind. So when you put a systems program into the Pentagon, it spreads out in about twelve or fifteen different directions. You never see the program in the same form again, because everyone is putting his body English on it."

Gardner found his opening once again in the Killian report, which had urged the National Security Council to recognize the ICBM as "a nationally supported effort of highest priority." To him this meant that Atlas should have more than top Pentagon priority; it should receive the highest national priority, a status no previous project had obtained in peacetime. He spent the summer pushing this point and gained it in mid-September, when Eisenhower agreed.

The Air Force had a new secretary, Donald Quarles, and the president's directive authorized him to cut red tape very extensively. As Schriever would recall, "I came in with a big bitch about the whole Pentagon, and went into Trev Gardner's office, and Trev had a chart. We went down to Don Quarles's office. I went through the charts to show how it was impossible to deal with the Pentagon as it was organized. We just could not wait to get all these individual approvals."[8]

Quarles directed his deputy for program management, Hyde Gillette, to set up a committee that would recommend, ironically, how to dispense with existing committees. The resulting Gillette Procedures gave sweeping authority to Schriever, confirming that his organization would have sole responsibility for planning and directing Air Force development of ballistic missiles. These procedures chopped the number of reviewing authorities from forty-two to ten. They set up a single level of formal approval within the Air Force, called the Ballistic Missiles Committee, with a similar group to exercise ultimate authority at the level of the secretary of defense. With these arrangements, Thor would go forward at top speed.

However, Thor first needed approval from the Joint Chiefs, and here it faced competition from the Army. That service's Chief of Ordnance had picked up von Braun's proposal following Schriever's rebuff, inviting him to develop his IRBM for Army use. This missile, the Jupiter, would amount to a second Thor, which raised the issue of wasteful duplication. It was true that Jupiter offered advantages, for von Braun's people had unmatched experience, while Redstone Arsenal included an in-house group that had already shown its strength by developing inertial guidance for the Redstone missile. But within the JCS the Army's chief of staff, General Maxwell Taylor, found he could gain headway only by making common cause with his naval counterpart, Admiral Arleigh Burke. Burke wanted to send missiles to sea, but the Thor wouldn't do; with a length of sixty-five feet, it wouldn't fit into a submarine. But Jupiter, shorter and fatter, might do splendidly, and such dual use could address the matter of waste in the Pentagon. Accepting these arrangements, Defense Secretary Wilson issued a decision on November 8 that directed both Thor and Jupiter to proceed. Furthermore, both new programs were to enjoy the same high priority as Atlas.

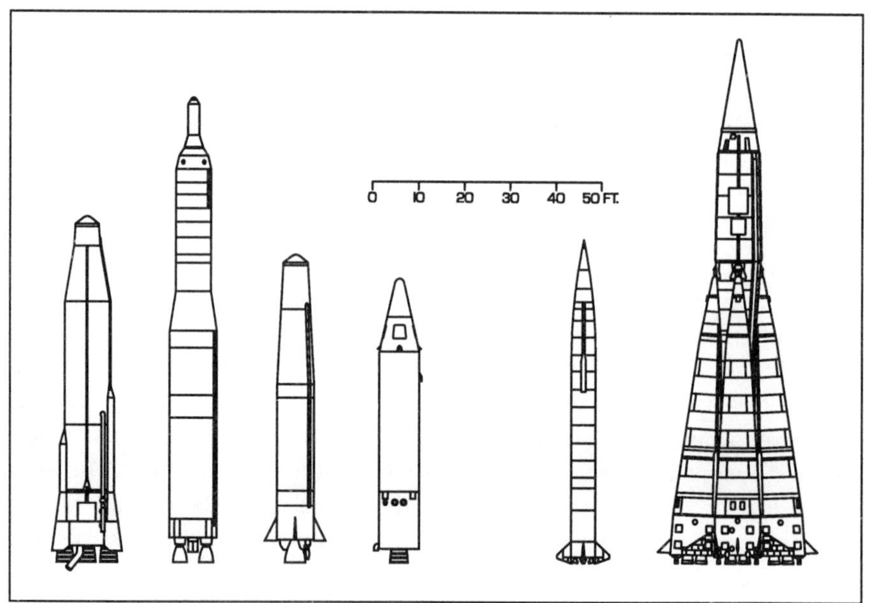

The missile race: Atlas, Titan, Thor, Jupiter; Soviet R-5 and R-7. (Dan Gauthier)

Nevertheless, at the outset the two programs faced very different prospects. The head of the Army effort, Major General John Medaris, found himself shoved into bed with the Navy. Nor would the Navy prove to be a willing mistress; the thought of storing and handling liquid-rocket propellants in a submarine would quickly prove to be more than that service cared to accept, and its admirals would drop Jupiter in a matter of months. In addition, Medaris faced a problem with engines. He would use the same one intended for Thor, but this was an Air Force engine, and it took quite a while before he could purchase the rocket motors directly from the manufacturer, North American. Until then he would have to go to Schriever, paperwork in hand, and take what he could get.

By contrast, Schriever was now in Management Heaven. He was running the nation's most important weapons programs, he had all the authority he could want, and his empire was expanding. Earlier in 1955 he had added Titan, which complemented Atlas; he was about to take on WS-117L, and now he would have Thor in the bargain. In addition, Douglas Aircraft and Bell Labs were champing at the bit for the contracts that would allow them to build this missile. The firms had worked together as prime contractors on the Army's Nike antiaircraft missiles, and their managers had a strong interest in the new IRBM.

Schriever couldn't use his authority to award the contracts directly; he had to rely on competitive bidding. But he could preselect the competitors, and he invited North American and Lockheed to prepare their own proposals. Just after Thanksgiving, a message from the Pentagon clattered from a teletype in his offices: "Program approved, proceed at maximum pace." Schriever's deputy, Colonel Charles Terhune, quickly made long-planned phone calls to executives at all three aircraft companies. He met with them and emphasized that Truax and Thiel already had defined the general features of Thor. As Terhune recalls, "The contractor would receive specified engines, guidance systems, and nose cones, and would largely have the task of integration and manufacturing. They were going to get everything to do the job; they were going to tell us how they'd do it." He described Thor as "a maximum-risk program." Two colleagues then described its essence: "There would be no time to make mistakes."

"We gave them one week to prepare proposals," Terhune adds. He then presided over a source selection board, which would pick the contractor. Douglas won, which wasn't surprising, and Terhune then flew east for the necessary approvals. It took only two days to get them, from officials that included the Air Staff and Air Force Secretary Quarles. Douglas Aircraft had its program management ready to roll, including a Nike veteran, Jack Bromberg, who would direct the new project. People would call him "Thorhead."

With 1955 not yet over, Douglas officials responded to a formal request to "present themselves, with corporate seal," by signing the contract. A mere seven weeks had elapsed since Secretary Wilson had given the go-ahead, and this pace would continue. As Schriever would put it, "We did not always have lawyers at our elbows, and every time we decided to do something, we did not have a long legal brief to contend with. I could tell any contractor, do this or do that, and they would do it, and the paperwork would follow."

The nation now was pursuing four major missile projects: Atlas, Titan, Thor, and Jupiter. In time, each would achieve renown as a space booster, but for now the focus was on nuclear weaponry. There would also be competition in this area, for the nuclear lab at Los Alamos now had a rival, Lawrence Livermore Laboratory, founded by Edward Teller. Harold Agnew, a weapons designer who succeeded Norris Bradbury as head of Los Alamos, directed an effort that competed with Livermore in designing warheads for the new missiles. Harold Brown, his counterpart at that rival center, would become Jimmy Carter's secretary of defense.

"At Los Alamos," Agnew recalls, "our feeling always was that our primary problem was Livermore and not the Soviet Union. Livermore started late; they had to play catch-up. Their early endeavors did not prove to be very successful." In the missile competitions, "we wiped 'em. Every one—Thor, Jupiter, Atlas, Titan—every warhead was Los Alamos. I don't think Harold's ever forgiven me."[9]

The four projects all depended on the same engines, which demanded extensive development and testing. This involved more than proving out the basic design; there were modifications that called for additional tests, particularly after General Medaris negotiated contracts with North American that called for boosting the thrust from 135,000 to 150,000 pounds. Turbopumps posed particular problems, but sometimes a quick fix would allow a run to proceed. Bill Ezell, a manager at the Santa Susana field lab, remembers that "we'd lose a blade on one side of the turbine, then eyeball it and knock a blade off on the other side." With its balance thus restored, after a fashion, the turbine could run again.

In addition, each individual engine used in these programs needed qualification testing, like autos off the assembly line that receive an initial drive. North American split off its rocket activities as a new corporate division, Rocketdyne. Its employment grew rapidly, and it was considered essential for each new employee to see at least one engine firing at "the Hill"—Santa Susana. But the facility didn't have enough test stands, so the Air Force granted the use of several more at Edwards Air Force Base. These had served the Bomarc, a long-range antiaircraft missile; rapid modifications made them ready for use.

The test area was out in the Mojave Desert, which was known for its snakes. The author Julian Hartt wrote that "after dark no one stepped from a pickup truck in the asphalt parking areas, or steadied himself by grasping the metal handrails along the walkways, without first using a flashlight for careful inspection—for these were the heat retainers, most attractive to deadly sidewinders seeking to warm themselves against chill night winds."[10] There was wildlife at Santa Su as well: rattlesnakes, deer, raccoons, the occasional mountain lion. Gophers were a problem until someone found how to get rid of them by pouring nearly pure hydrogen peroxide down their holes.

"We were engineers in the field," remembers Paul Castenholz, who worked on the Hill. "It got up to 105 or 110 degrees during the summer. We had some swamp coolers, water running through excelsior with a fan, but we never had air conditioning. I don't think the weather made a big difference to us. It was a very easy atmosphere; no one wore ties. The work was so interesting, you really enjoyed it. We were kind of an elite group."[11]

"It was hectic," adds Ezell. His boss, Roy Healy, would serve twice as president of the American Rocket Society, and as Ezell recalls, "I'd had two years' experience and I was Healy's number-one man. We worked six days a week; if there were problems we worked seven. We had eight engine test stands going, at Edwards and at Santa Su. We were running two shifts. I would work all day; then at night I'd be up on the Hill, helping the young guys learn how to run a test stand. And on other nights, either Roy Healy or I were always at home, so if some of those fellows got into trouble they could call one of us." Failures were par for the course: "I blew up a Redstone motor one day—and the next day a promotion came through."[12]

Time magazine printed its own impressions:

> Santa Susana is a fabulous place, a three-sq.-mi. area fenced and guarded, and crowded with up-and-down ridges dotted with rounded red rocks. A steep road winds over a pass and plunges into an amazing array of futuristic structures. There is no natural level land. Big buildings, fat tanks and weird testing equipment perch on crags or nestle in rocky crannies. New construction is being pushed with frantic urgency. The whole place swarms with hard-hatted workers. Bulldozers climb like mountain goats, pushing parts of the mountains ahead of them. A plant is in construction that will take from the air 600 tons of liquid oxygen per day.
>
> Tucked away in ravines, to reflect sound upward, are the massive steel structures where rocket motors are put through their paces. Their beams are as strong as the piers of suspension bridges, and they are "fishhooked" into the rock to keep them from being lifted by the thrust of the rockets. Seven hundred feet away are squat blockhouses with periscope windows. When a powerful motor is under test, an enormous flame licks down the precipice, sometimes bounding upward in a billow of yellow fire. A sound like the rumble of doomsday rolls among the rocks, making the flesh quiver like shaken jelly.[13]

Construction was also under way at Cape Canaveral. It had supplanted White Sands as the nation's principal missile center because it faced the Atlantic and had a clear range of ten thousand miles directly off the coast. The Pentagon had picked it for use as a missile test center in 1947; the British government had subsequently allowed the United States to build tracking stations on some islands of the Bahamas, which lay immediately downrange. Inaugural flights took place in June 1950, with successful launchings of two V-2s with WAC Corporal rockets as the second stage. This was the same combination that had reached 244 miles over White Sands the previous year. After that, the center began to grow, but in desultory fashion. For several years its most impressive test craft was the Snark, a subsonic cruise missile powered by jet engines. It tended to splash into the Atlantic, and people spoke of "Snark-infested waters."

The land was low-lying, swampy in places, with mildew and plenty of mosquitoes. They swarmed after heavy rains, which were frequent; a single acre of salt marsh could produce fifty million of them, and in bad areas, more than five hundred could land on you in a single minute. One old-timer would recall that "everyone wore long shirt sleeves and gloves, even in the summer. You couldn't wear a white shirt. The mosquitoes would be so thick they'd turn it black."

Tourists hadn't discovered the Cape; there were no motels. The nearest town of any size was Titusville, the county seat, whose 1950 population of 2,604 had shown little change since the previous census. But by the time the big push in missiles got under way, during 1954 and 1955, the Cape's basic facilities were in place. These included radar, a power plant, a telephone exchange, a firehouse, and a central control building, as well as hangars for aircraft and missiles. The area was still pretty scrubby, with wild birds and even an occasional armadillo. But growth now would come in a spurt, for Navaho, Vanguard, Jupiter, Thor, Atlas, and Titan all needed launch facilities.

Thor would lead the way and would be first off the launch pad. Its managers laid on an aggressive schedule that called for early flights, and plenty of them. They wouldn't wait for a true guidance system, even the radio-inertial variety that might not take too long to develop; they would start instead with a simple autopilot, which would guarantee no real accuracy but at least would allow Thor to fly. The managers also made a commitment to begin production of test rockets at an early date, even though they would have flaws for want of development. People expected that some would fail, but they also expected to learn valuable lessons from the flights, even if the rockets did blow up.

The schedule called for proceeding from contract signing to the first flight of a working missile in no more than a year. By the 1980s—an era when it would take almost three years merely to fix the O-rings on the boosters of the space shuttle—this would seem like mad folly. But Thorhead had a nearly complete design late in July 1956, just seven months into the program. In October the first missile, No. 101, went to the Cape aboard a C-124, followed in November by No. 102. A Thor still might fly by the end of the year.

Complex 17-B was already waiting, with a launch pad, a blockhouse, and a towering gantry painted in red and white stripes that would serve for checkout on the pad. No one at Douglas had any experience in designing blockhouses, but this had not bothered the managers. "We can't afford the time to make a blockhouse design," said Dolph Thiel, "because we want to launch somewhere between November and January. Let's look around at what blockhouse designs are available already." A colleague known as Colonel Jake replied that even this wouldn't be necessary: "There's a blockhouse down here all designed, for the Redstone. Borrow a set of Redstone plans and build it like that." Thiel decided it would be a bit crowded, but it would do.

The crew at the Cape didn't launch during 1956, but they came close. They had No. 101 on the pad on December 20, ready for a static firing that would demonstrate final readiness—and the engine failed to ignite. The fault was traced to a minor part, an electrical relay, but it was enough to spoil everyone's Christmas. But during 1957 Thor would indeed fly, followed by Jupiter, Atlas, and Vanguard. And at Tyuratam, Sergei Korolev was making similar preparations for his own R-7.

5

"The Russians Are Ahead of Us!"

The Space Race Begins

WERNHER VON BRAUN had a way with people. "He always seemed to be older than he was," remembers Paul Castenholz of Rocketdyne. "You looked up to him as having a lot of experience, a very strong vision. There was an aura about him that gave him credibility. He read quite a bit and seemed to have a flair in many areas. He liked to scuba-dive; he liked to fly. He wanted to be out there, a man that didn't care to be contained." Sam Hoffman recalls him as "a very personable fellow. He had charisma; he was a leader. He was a handsome fellow, smart—he was elite."

He also had a very commanding presence. One of his Huntsville colleagues remembers that "we had a party one evening, with about a hundred people talking and socializing. Von Braun was last to arrive. When he stepped in that room, with his wife Maria—all conversation stopped. Everyone wanted to hear what he had to say."[1]

However, none of this helped him in late November of 1956, when Defense Secretary Wilson handed down a decision on roles and missions for the three armed services in the area of long-range missiles. The Army could deploy missiles with a range up to two hundred miles, that of the Redstone, but everything beyond this limit would be the responsibility of the Air Force. General Medaris described the decision as "a crushing victory" for the rival service. Jupiter's range would be 1,500 miles; hence the Army would be out of luck.

This development meant that von Braun had struck out four times in less than two years. He had pitched Jupiter to the Air Force early in 1955, only to receive a blunt rebuff. His proposed IGY satellite, Project Orbiter, had lost out to the Navy's Vanguard. The Navy also had scuttled a plan to launch Jupiter from submarines. And now Secretary Wilson was also ruling out the Army's use of its own missile.

Wilson did not cancel the Jupiter program; he merely left it dangling. After all, von Braun had a proven track record that included the successful Redstone, whereas the Thor contractors* had only built Nike antiaircraft missiles. Development of Jupiter thus would continue; after all, Thor might falter. But if Jupiter succeeded, only the Air Force could build it and deploy it. And General Schriever already had made it clear just how little he thought of the idea.

Some people didn't mind rubbing it in. Von Braun was testing Jupiter components by flying them aboard Redstones, and during one such launch, the rocket tilted over and started right down the Cape in the direction of the Thor launch area. The range safety officer blew it up, and it crashed harmlessly into an empty field. Still, in Medaris's words, "We were accused of having a very inaccurate missile since we had found it impossible, even at that short range, to hit the Thor launch pad."

Nevertheless, Thor would fly first. As it sat in readiness, in late January 1957, it represented a true milestone, for it was the first of its generation. Everything built and flown until then, both in the United States and in the Soviet Union, had amounted to uprated V-2s. This was true even of the most advanced of the lot, Korolev's R-5. But Thor would usher in the future, by heralding an era not just of the IRBM and ICBM, but also of flight into orbit.

Civilization was beginning to come to Cape Canaveral by then, but it was coming slowly. Bob Johnson, a Douglas chief engineer, recalls that the roads were paved, but "surrounded by marshes and wild scrub." Jim Dorrenbacher, also of Douglas, remembers finding a site for a new launch pad amid a sea of palmetto and swamp: "We climbed on a bulldozer. Some old Florida guy drove us out there, with water moccasins falling off the blade. He said, 'This is where the pad's gonna be.' "

A few motels were now in the area; the Sea Missile was perhaps the first decent one. Its name was appropriate, because the sea was where so many of the missiles would end up. However, the Douglas people, who would launch the Thors, stayed in Cocoa Beach at the Clayton House, a rented estate. Dorrenbacher notes: "You didn't have a room to yourself, so you had to find a guy who didn't snore." Johnson adds that the launch crews drank bottled water because the local water contained dissolved chemicals from the aquifer: "It smelled like hell and tasted awful."

*Douglas Aircraft, Rocketdyne, Bell Labs for radio-inertial guidance; A.C. Spark Plug for true inertial; General Electric for nose cones.

Preparation for launch, Dorrenbacher remembers, "involved a lot of what-ifs. For instance, if there was a problem, you had to see if it was in the vehicle—or the test equipment. You always wondered if you were doing it right. And there were never enough people with the background and experience." Dorrenbacher adds: "You were in a situation, doing something that hadn't been done before. Without enough time, sometimes without enough assets. But a lot of conviction, a lot of belief that you could climb any hill."

The summit of that hill lay in view on January 25, as the launch countdown neared zero. "You watched the pressure gauges, the power supply, the electronics gauges—those were tense moments," says Dorrenbacher. "Everybody holding their breath and hoping that nothing wiggled." The test conductor, who would push the button to fire the missile, faced a row of controllers, each with his own console. At the appropriate moment, still watching the gauges, each in turn gave a "go" sign by making a thumb-up motion. The test conductor, said to have the "golden thumb," brought it down hard on his button.

The engine fired, and No. 101 lifted nine inches off its pad. Then it lost thrust, settled back, fell through a supporting ring, and broke in two as it exploded. People heard it in Titusville, thirty miles away. The blockhouse featured steel-reinforced concrete walls that were two feet thick, with twenty feet of sand on the side that faced the explosion. Still, in Dorrenbacher's words, "We all ducked under the table, because we weren't sure how strong the blockhouse was." Dolph Thiel was also present: "We suddenly saw a big flash; next thing I remember, Jack Bromberg and I were under the table. We looked at each other and said, 'What happened?' I felt very unhappy."

"After the fires died down, we were out in the flame pit looking for pieces," adds Rocketdyne's Bill Ezell. "I found the line that fed from a pump discharge to the gas generator. It had a check valve," which should have slammed shut when the engine started. "But it had blown apart." People laid out all the recovered pieces in a hangar and proceeded to investigate. Looking at films, someone saw a Douglas man dragging a liquid-oxygen hose through sand. This liquid is sensitive; a few grains of sand had caused an explosion. It had been minor, but had blown open the check valve and caused the failure. Ezell and his colleagues simulated this failure mode in a test at Santa Susana—and it matched exactly. They had lost a Thor, but learned to take better care in their prelaunch procedures.

Three months passed; in April it was time for No. 102. It got off the pad successfully, flew without problems for half a minute—and blew up. This was the work of the range safety officer, who thought the rocket was heading for Orlando, when in fact it was flying out to sea, just as it was supposed to. The problem stemmed from crossed wires in his own console. Thiel says, "I took my fist and was ready to hit him." Schriever adds, "He was reassigned to a new station way down in the South Atlantic about a week later."

However, in Dorrenbacher's words, "We coped with failure by not accepting it. We simply said, 'We're better than the hardware; we're gonna get it up and running.' " Win or lose, there'd be a party back at the Clayton House, where "the tension-reliever was a pitcher of Manhattans." Then it would be back to work. "We put the rocket back on the pad; if it took eighty hours a week then that's what it took."[2]

Weeks of preparation would culminate in a "hot count," a countdown preceding a static firing on the pad, followed a few days later by the actual launch. The count took four hundred minutes and featured a thousand separate events, including checks of equipment on the ground as well as aboard the Thor. If any problem showed up, the count would hold until it got fixed, and that could take a while.

They would put power into the missile and proceed with checkouts, with the guidance system causing many of the delays. During one long hold, over a difficulty that wasn't in his area, Bill Ezell went out and played nine holes of golf, only to return and find that nothing had changed. Some people took the long delays as opportunities for a catnap; they dozed in a car or lay down on a drawing board in a room that adjoined the launch center. But if the problem really was serious, the test conductor would scrub the launch. "Some counts went for thirty hours before we canceled," Ezell remembers. Then, with the fault finally repaired, the count would resume again—right from the top, at T – 400, with plenty of opportunity for more to go wrong.

Number 103 was ready for launch in mid-May of 1957. General Schriever was on hand, along with Rube Mettler of R-W and Colonel Ed Hall, the project director. In the blockhouse, some people wore red socks and red shirts for luck, a practice that dated to White Sands. But five minutes prior to launch, the rocket blew up as it sat there on the pad. People in the blockhouse broke into tears, including Hall and Mettler. So close to liftoff, and yet. . . .

This time the source of the fault turned out to be the main fuel valve. Shortly after fuel tank pressurization, which would feed the kerosene to its turbopump, the flaw had brought a buildup of pressure that blew the tank apart. As with the explosion of No. 101 in January, the basic problem was not hard to fix. But damage to the pad was another matter, and it would be August before it would be ready for another launch. And while Thor was producing the wrong kind of thunder, Wernher von Braun's Jupiter group was drawing on its matchless experience and was gaining success.

They had their first flight on March 1; the rocket arced upward into the afternoon sky and rose flawlessly for seventy-four seconds. Then it suddenly ripped apart in a major explosion. As it ascended into the upper atmosphere, the rocket exhaust had expanded until it reached the base of the missile. High temperatures in the tail then had produced the explosion. This problem called for a fiberglass flame shield along with insulating or relocating a number of ducts and electric cables. Still, it was already clear that the basic design was sound.

Late April there was another try. This time the Jupiter flew for over ninety seconds, but it went out of control and blew up again. Sloshing of propellants within the big tanks had caused the failure this time, setting up forces that the guidance system couldn't handle. In the words of General Medaris, "We made a whole bunch of big 'beer cans' out of heavy wire mesh, put floats inside so they would stay on top, and filled the whole top surface of each tank with them." This suppressed the sloshing and brought quick results.

On the last day of May, the next missile soared to a distance of 1,489 miles. It would have gone farther, but its builders had added extra test equipment, which made it heavy. Still, in the wake of von Braun's recent strikeouts, this was a home run: the first successful flight of an IRBM. Then, just to show that it was no fluke, the launch crew scored a similar success on the fourth flight, three months later.

Now the wisdom of building IRBMs became apparent, for Atlas had problems. It first flew on June 11, but lived for only twenty-two seconds before an engine lost thrust and sent it veering out of control. A second try, late in September, failed in similar fashion. Still, by then Thor had also achieved a successful flight. This happened on September 20, on the fifth flight, which reached a range of over 1,250 miles. It too had been loaded down with instruments, thus reducing the range, but it was fully satisfactory. The nation now had two IRBMs that were coming along as they should.

Nevertheless, Cape Canaveral was like a goldfish bowl. Anyone could see the firings; there was even a local ladies' luncheon club, the Missile Misses, whose members would watch from the beach. This meant that the failures and explosions were visible for all to see. Hence, it was hard for members of Congress, and the press, to believe in these programs' imminent success when eight out of eleven rockets had either blown up on the pad or had fallen from the sky in flames. And these dramatic disasters would contrast sharply with the impending achievements of Sergei Korolev, whose powerful R-7 would fly in utter secrecy when it failed.

The R-7 was far larger and more powerful than the Atlas and Thor, and its launch facility was correspondingly more elaborate. The rocket's components—its core and four strap-on boosters—came to Tyuratam by rail, for assembly in a large hangar. The complete R-7, lying horizontally,

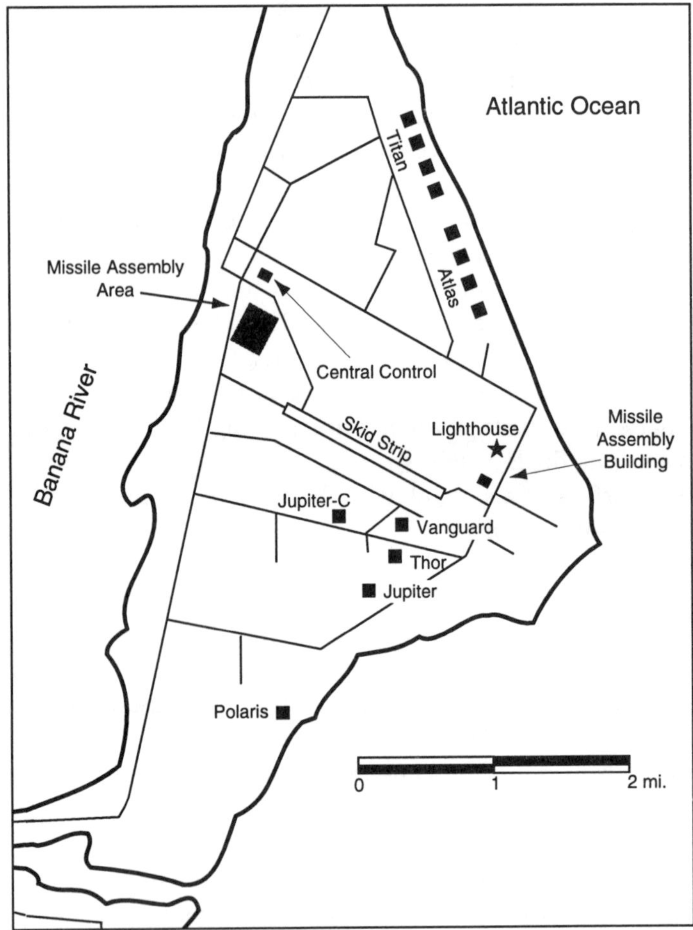

Cape Canaveral in the late 1950s. (Don Dixon)

then moved to the pad by rail as well. The analyst Nicholas Johnson writes that this pad bore "a striking resemblance to some gigantic animal trap, ready to spring up and ensnare whatever comes within its grasp." This was the *Tyulpan* launcher (the tulip), with large petal-shaped structures formed of steel girders that would hold the rocket in place while providing work platforms for checkout. During a liftoff, the petals would fold back. The rocket exhaust, with 880,000 pounds of thrust, then would strike a massive flame deflector and pour into the "stadium," an immense excavated pit next to the pad.

Korolev began preparations for the first flight in February 1957, hoping to launch the following month. He ran into delays, and opportunities to fly in March slipped away, as did dates in April. Finally, on May 15, the first R-7 took off. A propellant line ruptured, starting a fire, which produced an explosion that tore the missile apart after barely a hundred seconds of flight. This proved to be the result of a sloppy assembly job in Korolev's own production plant.

On June 9, two days before Atlas would make its first attempt, Korolev tried again. This time the missile failed to fire. Again the fault was his, for someone in the plant had installed a fuel valve upside down. This was two failures in a row, and it involved more than coincidence, for as his biographer Yaroslav Golovanov would put it, "Korolev never started failure investigation by examining his own possible faults." He much preferred to act as though the design of his rocket was flawless, and he tried to blame Glushko when he could, hoping the problem might lie in an engine. Glushko would reply in kind: "What about you and your rocket? Are you a saint?" Amid the bickering, problems sometimes went unrepaired.

This time the problems were Korolev's, but still he tried to dodge them. He had to write a report, and he prepared it as an exercise in flimflam, laboriously listing possible sources of difficulty while citing the "most probable cause" only near the end. It didn't work; his superiors were too experienced to be fooled. "What a cunning man you are," one of them said. "So much stink about what might have been caused by others, and so much perfume for your own shit."

Korolev wrote to his wife: "Things are going badly. My frame of mind is bad. I will not hide it, it is very difficult to get through our failures. There is a state of alarm and worry. It is a hot 55 degrees here"[3] (some 130 degrees Fahrenheit). Korolev's emotional involvement was evident in other ways and often strong. For example, during a two-day postponement of the first flight of a nuclear-armed R-5 in 1956, he had gotten very little sleep and his aides found he was a nervous wreck. Now a key deputy had become quite ill and his face swelled horribly. Korolev himself came down with a bad sore throat and took penicillin shots.

On July 11 the second R-7 took to the sky. As it rose and began to tilt, the cruciform arrangement of its engines placed a fiery cross in the sky. The Roman emperor Constantine, early in the fourth century, had seen just such a vision, accompanied by the words *In hoc signo vinces*. But in this sign Korolev would not conquer, for the rocket went out of control and disintegrated. This time the problem was in the guidance system, and Korolev had the pleasure of telling a colleague, "Go to Nikolai Pilyugin," the chief of guidance, "and sit there until you find the answer."

But the third try worked like a charm. On August 21 another R-7 flew successfully to a range of 4,000 miles, setting a mark as the world's first flight to intercontinental distance. The warhead, a mockup, disintegrated over Kamchatka, but that wasn't important. Wildly excited, Korolev stayed up until three in the morning, talking animatedly with staff members about the prospects for spaceflight.

Korolev had met at times with Stalin and was dealing in similar fashion with his successor, Nikita Khrushchev. As Khrushchev later wrote in his memoirs:

> Not too long after Stalin's death, Korolev came to a Politburo meeting to report on his work. I don't want to exaggerate, but I'd say we gawked at what he showed us as if we were a bunch of sheep seeing a new gate for the first time. When he showed us one of his rockets, we thought it looked like nothing but a huge cigar-shaped tube, and we didn't believe it could fly. Korolev took us on a tour of a launching pad and tried to explain to us how the rocket worked. We were like peasants in a marketplace. We walked around and around the rocket, touching it, tapping it to see if it was sturdy enough—we did everything but lick it to see how it tasted.
>
> We had absolute confidence in Comrade Korolev. We believed him when he told us that his rocket would not only fly, but that it would travel 7,000 kilometers. When he expounded or defended his ideas, you could see passion burning in his eyes, and his reports were always models of clarity. He had unlimited energy and determination, and he was a brilliant organizer.[4]

The September shot also went off successfully and reached full range as well. A government commission then gave Korolev permission to launch his initial satellite, the PS, on the next R-7. The rocket itself needed modification to reach orbit. The ground crew had to cut its weight by removing part of the guidance system along with other instruments, reprogram the engines to increase the velocity, and install a new nose cone. This nose cone enclosed the lightweight Sputnik, which weighed only eighty-four kilograms.

The launch took place late in the evening of October 4, with the rocket glowing dazzlingly in the white glare of searchlights. Golovanov gives a picture of this culminating moment:

> Korolev had arrived early, and left his car on the concrete apron. The wind was cold and piercing. He raised the collar of his heavy old overcoat. From behind him came the wooden voice of the public address system:
>
> "Attention! Time check in one minute. Prepare to fuel."
>
> The liquid oxygen was fuming like the white steam of a locomotive. The rocket was becoming shrouded. Rime was creeping up from the bottom of the oxygen tank, and soon it would be all white. Beautiful.
>
> Later, immediately before blast-off, he sat with his shoulders hunched on his usual seat at his "personal" periscope in the bunker. He was shivering slightly.
>
> "Zero minus one minute! Repeat: Zero minus one minute!"
>
> "Switch to start!"
>
> Korolev brought his face close to the periscope and felt the unpleasant sensation of the cold sweat from his face on the black rubber shield of the eye-pieces. The white oxygen cloud disappeared; the vent valves had been closed.
>
> "Auxiliary engines pressurized!"

> "Main engines pressurized!"
>
> "Launch begun!"
>
> "Ground—disengage!"
>
> Korolev clung to the periscope. The rocket was right before his eyes. He saw how the cable tower swung away after the command. Now nothing connected the rocket with the launcher.
>
> "Ignition!"
>
> "Preliminary boost!"
>
> He saw an instantaneous flash, a short flicker, before the brown cloud of dust and smoke raging in the whirlwind of the engines rapidly engulfed everything around. A blinding ball of light flared beneath it.
>
> "Main boost!"
>
> The rocket hung motionless. It'll be a few moments before it rises. It really looks as if it were pondering for a second whether to start, or not. How tedious and long are these instants of immobility!
>
> "Takeoff!"
>
> "There it goes! There it goes!" The giant white dagger is racing upward, its body looking transparent and unreal in the dazzle. Korolev's fingers tightened around the black grips of the periscope, the whole of his thick-set, heavy body stiffening. Only now did a jubilant voice penetrate his consciousness, gabbling over and over in its excitement: "All systems stable! Flight proceeding normally! Pressure in chambers normal! All systems stable!" And, at last, "Separation complete!"
>
> "Systems," thought Korolev, "but go on shouting, my dear boy, all you dear friends!" A wave of unendurable warmth and gratitude to all the people swept over him. He felt a lump in his throat. "Is this really all? Have we really done it? Of course, of course! It's time to phone to Moscow and report. But let it complete one orbit, then we'll report."[5]

In Huntsville, a few hours later, von Braun and General Medaris hosted a visit by the incoming defense secretary, Neil McElroy, who was replacing Charles Wilson. They were at a cocktail party when Medaris's public-affairs officer dashed in and interrupted a conversation: "General, it has just been announced over the radio that the Russians have put up a successful satellite!"

Von Braun's pent-up feelings suddenly gave way: "We knew they were going to do it! Vanguard will never make it. We have the hardware on the shelf. For God's sake turn us loose and let us do something. We can put up a satellite in sixty days, Mr. McElroy! Just give us the green light and sixty days!" Cautious Medaris interjected, "No, Wernher, ninety days." They presented McElroy with a formal briefing the next morning; as Medaris would put it, "We all felt like football players begging to be allowed to get off the bench." The new secretary made no commitment, but Medaris told von Braun to begin preparations as if he had a formal directive to proceed.

On the same evening of October 4, Senate Majority Leader Lyndon Johnson was presiding over a gathering at his LBJ Ranch in Texas. His guests heard the news on the radio, and after dinner they took a nighttime stroll to the nearby Pedernales River. He would write: "Now, somehow, in some new way, the sky seemed almost alien. I also remember the profound shock of realizing that it might be possible for another nation to achieve technological superiority over this great country of ours."

The press played the news to the hilt. The *New York Times* set the tone with a streamer headline that spanned the width of the front page:

SOVIET FIRES EARTH SATELLITE INTO SPACE;
IT IS CIRCLING THE GLOBE AT 18,000 MPH;
SPHERE TRACKED IN 4 CROSSINGS OVER U.S.

Newsweek ran a special section and then another one a week later. It noted: "This achievement had been reached, in a torn world, by the controlled scientists of a despotic state—a state which had already given the word 'satellite' the implications of ruthless servitude. Could the crushers of Hungary be trusted with this new kind of satellite, whose implications no man could measure?"[6]

The Democrats quickly took up the refrain. They had taken their lumps during the Korean War, when Truman had led the nation into a conflict that he could neither win nor end. Now the shoe was on the other foot, as LBJ charged that Ike had not spent enough on national defense and that his pursuit of balanced budgets had endangered the nation's security. Johnson received strong support from Senator Stuart Symington, who had been Truman's Air Force Secretary. Symington now demanded a special session of Congress and warned that "unless our defense policies are promptly changed, the Soviets will move from superiority to supremacy." Another powerful Democrat, Senator Henry Jackson, called for "a week of shame and danger."

Eisenhower, who had unmatched military experience along with photos from the U-2, tried to reassure the nation. He held a press conference a few days after Sputnik went up and said that the Soviet Union's achievement "does not raise my apprehensions, not one iota." They had "put one small ball into the air." They would soon do much more, but for the moment at least, the American public was inclined to feel the same way. In Boston, *Newsweek* found "massive indifference." Denver reported "a vague feeling that we have stepped into a new era, but people aren't discussing it the way they are football and the Asiatic flu." And on October 5, 1957, the headline in the Milwaukee *Sentinel* read: TODAY WE MAKE HISTORY. It referred to the World Series.

Yet there were excellent reasons to take the Soviet satellite seriously, reasons that went beyond its overtones of military threat. There was the simple fact that Moscow had carried off this spectacular feat, when only a few years earlier it had lagged badly in both bombers and nuclear weapons. It was easy to fear that this achievement reflected genuine merit in the Communist system. The Moscow press release announcing Sputnik played to this angle: "Artificial earth satellites will pave the way for space travel, and the present generation will witness how the freed and conscious labor of the people of the new socialist society turns even the most daring of mankind's dreams into reality."

This concern dovetailed with another: that America's freedom and democracy might be fine in times of general peace, but would lose out to the discipline of a dictatorship in the crunch. The fear was not new; it had been very much in people's minds amid the Nazi challenge. Now it showed up again, and it took the peculiar form of viewing the nation's consumer society as evidence of weakness.

"The time has come," thundered Senator Styles Bridges, a leading Republican, "to be less concerned with the depth of pile on the new broadloom rug or the height of the tailfin on the new car and to be more prepared to shed blood, sweat and tears." Lyndon Johnson scoffed at the idea that America might launch a better spacecraft: "Perhaps it will even have chrome trim and automatic windshield wipers." Bernard Baruch, an adviser to presidents, added that "if America ever crashes, it will be in a two-tone convertible." And when Leonid Sedov, a leading Soviet space scientist, met von Braun's associate Ernst Stuhlinger, he took advantage of this attitude: "America is very beautiful. The living standard is remarkably high. But it is very obvious that the average American cares only for his car, his home, and his refrigerator. He has no sense at all for the nation."[7]

In addition, there was reason to fear that in the Third World, perception of American weakness could easily become reality. Third World nations, newly independent or soon to become so, had given power to the British Empire. Yet their leaders often had little love for Western democracy; they frequently preferred one-party regimes remarkably similar to that of Khrushchev. Nor did they appreciate capitalism, which they associated with exploitation; they believed the future would favor socialism. These rulers would take heed of the United States, but only because they thought it was winning. But if they became convinced (and this wouldn't be difficult) that Moscow would win . . .

Khrushchev, for his part, had not expected so strong a response and quickly realized he had struck a nerve. He decided that from this day forward, something Soviet would be in orbit every day, to continue to impress the world with the strength of Communism. He personally presented this view to Korolev, who had been working in Tyuratam's heat and aridity since early in spring but who returned to Moscow following Sputnik's success. In addition, several of Korolev's top associates went to the

Black Sea for vacation and stayed at the country estate of Nikolai Bulganin, a government leader who was sharing power with Khrushchev himself.

The vacation did not go well. One man came down with a cold. Another, unable to stand the idleness, phoned Korolev in Moscow and received an order: "Fly back urgently. We have received the task of making a new satellite." Meeting with his associates, Korolev elaborated his instructions: "Launch a dog by the holidays." They were to place this space traveler into orbit by early November of 1957, to mark the fortieth anniversary of the Russian Revolution. The deadline was only a few weeks off.

Fortunately, everything they needed lay close to hand. Once again they would use their R-7. The dog, Laika, was also ready. Furthermore, as Korolev's deputy would recall, "back as far as 1951, we had launched dogs on high-altitude rockets. We took the pod used for those purposes, put Laika into it and placed it in a satellite-craft. We managed to do it just in time for the holiday."

This craft, Sputnik 2, weighed 1,120 pounds. So large a satellite would not raise eyebrows on the staff of Lockheed's reconnaissance project, WS-117L. But in a reversal of the usual pattern of American openness versus Soviet secrecy, WS-117L was secret whereas Sputnik 2 was out in the open. The satellite that people knew about was Vanguard; it weighed a mere 21.5 pounds and still wasn't ready to fly. Unfortunately, Sputnik 2 did not separate from its booster; its thermal control system failed, leading the craft to overheat. After no more than a day or so, Laika died for her country. She would not be forgotten; a Russian manufacturer would name a brand of cigarettes in her honor. But in America this new spacecraft, six times heavier than its predecessor, drove home the news that Moscow had the big rockets and was in space to stay. Moreover, the orbiting of a dog pointed clearly to an intention to send a man into space.

Sputnik 2 raised new concern in the press; *Life* magazine ran an editorial, "Arguing the Case for Being Panicky." This satellite also added fuel to the fire of Democrats in the Senate, who were preparing to rake Ike over the coals for doing too little. The stakes seemed to transcend ordinary politics, even ordinary issues of national defense. To LBJ, the Senate's most powerful member, space promised nothing short of magic power:

> Control of space means control of the world. From space the masters of infinity would have the power to control the earth's weather, to cause drought and flood, to change the tides and raise the levels of the sea, to divert the Gulf Stream and change temperate climates to frigid. That is the ultimate position: the position of total control over earth that lies somewhere in outer space.[8]

Was Ike truly doing too little? In fact, it was quite possible to argue that the United States was trying to do too much, for as Trevor Gardner observed, "We have presently at least nine ballistic missile programs, all competing for roughly the same kind of facilities, the same kind of brains, the same kind of engines, and the same public attention." Still, Ike would indeed do more very quickly.

Already, only days after the first Sputnik, the Army had won a reversal of Secretary Wilson's directive that had relegated Jupiter to second-class status. The project, like Thor, now would go forward at full throttle. Similarly, Sputnik 2 forced Ike's hand and resulted in swift abandonment of his reliance on Vanguard as the nation's sole unclassified satellite. Laika had barely ceased barking when Secretary McElroy directed the Army "to proceed with launching an earth satellite using a modified Jupiter C."

Von Braun would be ready early in 1958, as promised, but the Vanguard people would reach the launch pad first, with a rocket that would attempt to fly on December 6. One could hardly speak of it in the same breath as Korolev's R-7, to be sure, for its first-stage thrust, 27,000 pounds, was only three percent of the Russians'. Vanguard's first-stage engine had made only a single flight, a few weeks earlier. The second stage had never flown at all, and its builders had cut short its testing for fear that its engine would not last long when fired. The satellite was not even the 21.5-pound sphere that had been advertised; it came in at only three pounds and measured just six inches across, barely larger than a grapefruit. Yet it was weighty indeed, for it carried the hopes of the nation. Just then, it was all this country had.

The countdown reached zero and the rocket began to lift. Then, in the words of an observer, "it seemed as if the gates of hell had opened up. Brilliant stiletto flames shot out from the side of the rocket near the engine. The vehicle agonizingly hesitated for a moment, quivered again, and in front of our unbelieving, shocked eyes, began to topple. It sank like a great flaming sword down into the blast tube. It toppled slowly, breaking apart, hitting part of the test guard and ground with a tremendous roar that could be felt and heard even behind the two-foot concrete wall of the blockhouse. For a moment or two there was complete disbelief. I could see it in the faces. I could feel it myself."[9]

Reaction was swift. "How long, how long, oh God, how long will it take us to catch up with Russia's two satellites?" wailed Lyndon Johnson. Germans called it *spaetnik* (latenik). Other writers had a field day:

London Daily Herald: OH, WHAT A FLOPNIK!

London Daily Express: U.S. CALLS IT KAPUTNIK

Louisville Courier-Journal: "A shot may be heard around the world, but there are times when a dud is even louder."

Ottawa Journal: "The rocket is said by the experts not to have got off the ground because of a 'loss of thrust.' It is an arresting phrase. Loss of thrust is what the Western democracies have been suffering from."

New York Herald Tribune: "The people in Washington should damn well keep quiet until they have a grapefruit or at least something orbiting around up there."

Paris-Journal: "It seems there is a worm in the grapefruit."

Chicago American: "We have had to absorb a considerable amount of disillusionment in the past two months. All right, let's take it and get whatever benefits may derive from it, but let's not develop a taste for it."

Denver Post: "The American people are nervous, skeptical and annoyed about our conduct of scientific research and development. The people are not frightened. But they are getting pretty sore."

At the United Nations, Soviet delegates asked their American counterparts if the United States might wish to receive foreign aid under Moscow's program of technical assistance to backward nations. The *Christian Science Monitor,* summing things up, described the "bombast-pricking headlines and jokes" as "gibes at the overblown way in which public relations men and the American press built a giant anticlimax by trying to create a climax where it was not normal for a climax to come—in the midst of a delicate experiment."[10]

What, concretely, had produced this fiasco? It involved nothing more than insufficient tank pressure, which had brought low pressure at the rocket engine's injector. This permitted hot gas from the thrust chamber to shoot upward into the fuel system, burning the injector and cutting the thrust. It was not a major problem and was easy to fix. But just then, the public was in no mood to see this. At the Martin Co., the Vanguard project manager learned this personally when he tried to find someone to paint his home and met with repeated rebuffs. Finally one housepainter decided to level with him: "To tell you the truth, I don't feel much like working for anyone connected with Project Vanguard."

Amid the growing public concern, few people had clear crystal balls. With Moscow's rockets looming ominously in the distance, the novelist Ayn Rand, whose ironically titled *Atlas Shrugged* appeared in the same year, may have been the only one to argue that Communism would lead to economic collapse. Few people understood that Khrushchev was playing a cynical game, funneling disproportionate resources into a space program that he would use for propaganda, to give an impression of great overall technical strength. The president of General Dynamics would tell *Fortune* magazine: "You cannot overlook Russia's capacity to concentrate in specific areas. If the area has real military or psychological value to them, they'll put massive concentration on it, and achieve results all out of proportion to the general level of their technical ability."

Yet in at least one important respect, Soviet technology was definitely no Potemkin village. This was education, at the elementary and high-school levels. Educators had long noted the high quality of German schools; now, in the wake of the first sputniks, a number of observers would point to similar excellence in those of the Soviet Union. Two years earlier, the critic Rudolf Flesch had stirred nationwide concern by explaining *Why Johnny Can't Read.* Now John Gunther, a widely read observer of nations, came out with *Inside Russia Today:*

> The main emphasis is on science and technology, for both boys and girls. In addition to ten solid years of arithmetic and mathematics, every child is obliged to take four years of chemistry, five of physics, and six of biology. Comparison to the United States is highly pertinent. Many American high schools have no physics or chemistry courses at all. An American authority told me that the Soviet child graduating from the tenth grade (our twelfth), aged about seventeen, has a better scientific education than most American *college* graduates. The average Russian boy or girl, taking the normal course, gets more than *five times* the amount of science and mathematics that is stipulated for entrance into Massachusetts Institute of Technology.[11]

In 1956 the Soviets graduated some seventy thousand new engineers, to thirty thousand in the United States.

The Democrats had long sought to enact a program of federal aid to education, and the sputniks gave them their opening. The first such bill, the National Defense Education Act of 1958, provided funds for teaching of science, math, and foreign languages. Thus, the Soviet challenge brought more than a military response; it evoked what would become the entering wedge in an increasingly sweeping agenda of social reform. But for the moment, the main thing still was to get some hardware on the launch pad and put something into orbit.

Von Braun did it at the end of January 1958 by succeeding with the eleven-pound Explorer 1, the nation's first satellite, on his first try. *Time* put him on its cover and the nation enjoyed a moment of cheer, followed by another such moment in mid-March when Vanguard put up its own three-pound beeper. James Van Allen had built instruments for Explorer 1 as well as for von Braun's second satellite, Explorer 3, which went up late in March. With these instruments, Van Allen made a major discovery: a belt of intense radiation that surrounds the earth, trapped by its magnetic field. This discovery was all the sweeter because the Soviets had missed it.

The Soviets had a first-rate specialist in cosmic rays, the physicist Sergei Vernov, who had installed similar equipment aboard Sputnik 2. However, the Soviet Union lacked a worldwide tracking network and possessed ground stations only in the home country. When Sputnik 2 was within range of the stations, it was at low altitude, below the radiation zone. When it reached high altitude, Vernov's instrument indeed detected the trapped radiation—but by then Sputnik 2 was on the other side of the world. Even so, radio operators picked up its signals in Australia and

South America, and the data told of the radiation belt. But for reasons of secrecy, Moscow had made no arrangements to have other nations interpret the signals or to pass the data along. The radiation zone became known as the Van Allen belt, whereas with a little less secrecy it would have been named for Vernov.

But Khrushchev was not about to let anyone forget who was truly the master of space. Object D, the heavy spacecraft full of instruments, now was ready for flight. Korolev tried a launch late in April, but the rocket broke up a minute and a half after liftoff. As he prepared to try again, he learned that an onboard tape recorder was not working. It was important; Keldysh's scientists needed it to capture data when the craft was out of range of a ground station.

Korolev naturally expected to delay the launch and fix the problem. But according to Roald Sagdeev, a notable space scientist, he received a phone call from Khrushchev himself. Italy was about to hold national elections; its Communist Party was strong, and Khrushchev believed that a successful flight of the new sputnik would swing millions of votes. He ordered Korolev to launch immediately. Sputnik 3 flew to orbit, and while it had little impact in Italy, it certainly made heads turn. Its weight of 2,925 pounds was nearly three times that of the already-enormous Sputnik 2.

At a Soviet-Arab friendship meeting in the Kremlin, Khrushchev told his guests that America would need "very many satellites the size of oranges in order to catch up with the Soviet Union." But the United States could also play the weight-lifting game, for Atlas was now coming along. In addition, Ike now had a Pentagon office, the Advanced Research Projects Agency (ARPA), that would initiate new military space efforts.

Atlas mounted two 150,000-pound Rocketdyne engines as boosters, with a central engine of 60,000 pounds of thrust as the sustainer. Initial test flights used the boosters only and carried a reduced load of fuel. This meant that even when they succeeded, which wasn't often, they would fly only some 600 miles, just one-tenth of full intercontinental range. Still, this configuration made a good flight in December 1957. Other successes followed, and by the following spring the missile was beginning to show signs of reliability.

ARPA's head, Roy Johnson, had been a vice president at General Electric. In June 1958, in the wake of Sputnik 3, he went out to Convair in San Diego and met with his fellow corporate officials. "We've got to get something big up," he declared. A senior manager replied, "Well, we could put the whole Atlas in orbit." Johnson agreed to push this project in Washington, while the others agreed to keep it a secret. During the ensuing months, no more than eighty-eight people knew what was up.

Of course, before they could cook their rabbit stew, they first had to catch their rabbit, by shooting Atlas to full range. The first full-up model, with all three engines and 360,000 pounds of thrust, lifted off in mid-July and promptly went out of control. But a shot two weeks later went the full planned distance, 2,700 miles, a range that once again was reduced by a heavy instrument load. A 3,000-mile flight followed at the end of August, along with a similar success in mid-September.

The first try for full range came on September 18. It blew up eighty seconds after launch, which meant it was still too soon to remove the instruments. After all, they could disclose where the problems were. Another success restored confidence, and late in November everyone was ready to try once more to go the distance.

The test conductor, the man with the golden thumb, bore the hopeful name of Bob Shotwell. He pushed the button and Atlas leaped into the night, tracing an arc above the moon as it flew across the starry sky. It dropped its boosters; then, as it continued to accelerate, its brilliant light faded and it seemed to hang in the darkness like a new star, just below Orion. Shotwell and his crew restrained their excitement for a full seven minutes; then they erupted in shouts. They had it; the missile was soaring at 16,000 mph, bound for a spot in the South Atlantic near the island of Saint Helena, a full 6,300 miles from the Cape. In Shotwell's words: "We knew we had done it. It was going like a bullet; nothing could stop it." Later that evening the champagne flowed freely at the Starlite Motel, as some of the men hoisted their boss on their shoulders and carried him around the room.

Three weeks later came the shot to orbit, appropriately called SCORE. This Atlas carried its own tape recorder, with a message from Eisenhower, along with radio equipment that would permit it to receive and send additional messages. The secrecy surrounding the project had been maintained up to the moment of flight; even the launch crew in the blockhouse didn't know its mission. When some of them saw that the missile was deviating from its usual course, they yelled for the range safety officer to blow it up. He refused; he was part of that tight little club of eighty-eight.

"This is the President of the United States speaking" were the first words to come from orbit. "Through the marvels of scientific advance my voice is coming to you from a satellite circling in outer space. My message is a simple one. Through this unique means I convey to you and to all mankind America's wish for peace on earth and good will toward men everywhere."[12]

Now it was the turn of Americans to boast, as officials declared that they had just orbited a satellite weighing 8,600 pounds. Actually, they hadn't. The weight that counted was that of the communications equipment, some 200 pounds; the rest was simply the carrier rocket, which was nothing more than a big empty metal shell. If the Soviets had wanted to play the game that way, they could have stated as far back as Sputnik 1 that their own carrier rocket, which also flew to orbit, tipped the scales at close to 18,000 pounds. Even so, this SCORE was real. It marked an end to an initial era that had featured the makeshift Jupiter-C and the inadequate Vanguard, for it carried a message that went beyond Ike's: that America now would pursue a serious space program, with a proper array of boosters.

At that moment, during the Christmas season of 1958, the first Thor missiles were in England and would soon become operational. The first Titan ICBM, complementing Atlas, was on its pad, ready for an initial flight early in the new year that would gain its own success. Yet with these big liquid-fueled projects all running at flood tide, the Pentagon was well along with plans to leap beyond them. They all would serve brilliantly in upcoming space missions, but in their intended military role they would prove to be no more than interim weapons. Already, the missiles that would replace Atlas, Titan, Thor, and Jupiter were well along in development. These would feature solid fuel: the Navy's Polaris, the Air Force's Minuteman.

Polaris's background grew out of the nuclear submarine, the work of Captain Hyman Rickover. At the outset, in 1946, Rickover's superiors wanted to learn about atomic energy. An advance at flank speed, leading to real fighting ships powered by the atom, was not on the agenda. But it quickly became Rickover's personal agenda. He started that year with an assignment to the Oak Ridge nuclear lab. Here he would begin to learn the exotic technology of nuclear engineering.

Rickover was an Annapolis graduate, but he definitely was not a man to play the social games of the officers' club. He was already known as a plainspoken man who would set his own standards of excellence and insist that others meet them. Excellence, indeed, was his passion. He had worked during the war as head of the electrical section in the Bureau of Ships, a post that might have merely involved administering contracts and keeping track of schedules. However, that was not his way.

Rickover put together a staff of the best engineers he could find, both naval and civilian. He personally read battle reports and inspected every war-damaged ship he could visit, to learn how electrical equipment was standing up in combat. In this fashion he uncovered scores of deficiencies: circuit breakers that popped open at the firing of ships' guns, cable that would leak and carry water to switchboards, junction boxes that emitted poisonous gases inside a submarine during an electrical fire.

With his staff, he decided what changes were necessary. The staffers not only designed new types of equipment but carried out research, developing basic data on such topics as shock resistance. He picked the contractors that would build the new equipment, worked with them closely, and made sure their products would be ready on time and would meet his demanding specifications. He would follow the same methods as head of the Navy's nuclear program.

In 1946, of course, no such program existed. No naval organization was in place, ready to develop the needed power reactors, and no one anywhere had yet constructed a nuclear power plant. The wartime Manhattan Project had built atomic piles, but these were huge, low-temperature systems used to produce plutonium. A compact reactor of high power, capable of driving a ship, would be completely different, and the necessary engineering knowledge was not available. Rickover thus faced a threefold task: to set up an office with the necessary authority, to gain the technical understanding that would make his reactors feasible, and to push ahead far more rapidly than his superiors thought possible. He would have to do this while holding no higher rank than captain.

It helped that he had strong support from the chief of BuShips, Rear Admiral Earle Mills. In August 1948 Mills established a new Nuclear Power Branch within his bureau, which would become the core of Rickover's organization. Still, this naval office could do little on its own, for the Atomic Energy Act had placed prime responsibility for all forms of nuclear technology within a civilian agency, the Atomic Energy Commission. Hence the AEC would have to establish a parallel office to deal with submarine propulsion. By early 1949 this Naval Reactors Branch was also in business, and Rickover was running it as well. He thus wore two hats, naval and civilian.

These arrangements might have diluted Rickover's authority, but he turned them into a source of strength. If he met an obstacle in the Navy, he could don his AEC hat and try anew from within that agency. He soon found that the AEC would give him considerable leeway as a Navy man, while the Navy would do the same when he represented the AEC. He even drafted letters requesting help from the Navy for his AEC boss to sign and then drafted approving letters for the signature of Mills.

Within his dual office, Rickover also tried to weaken distinctions based on rank or civilian status. He knew how a naval officer could get his way by pulling rank, and he wanted all decisions to rest on technical merit. He thus sponsored a vigorous program of education for his staff, arranging for MIT to offer courses on nuclear technology while setting up similar courses at Oak Ridge. He also encouraged staffers to debate the technical issues vigorously at meetings, and to feel free to declare that the emperor had no clothes. If you worked for Rickover, you didn't get bogged down in organization charts. What mattered was what you knew and how much responsibility you could handle.

From the outset, Rickover had to steer between two shoals. He knew the first from his wartime days: It was important not to assume that the industrial contractors were fully capable. The second danger was peculiar to the nuclear community: It was top-heavy with research physicists, but their work did not deal with the engineering advances necessary for a naval reactor. So while the main contracts went to such heavy hitters as General Electric and Westinghouse, Rickover took care that much effort would go into accumulating data for engineering handbooks, which would form a foundation for reactor technology. In particular, his power plants would rely on such exotic materials as zirconium, hafnium, liquid sodium, and beryllium. He made sure that the necessary know-how would be ready, along with the production facilities.

Under usual procedures, Rickover might have had the AEC build a succession of test reactors, leading up to their use in an experimental ship. The Navy, in its own good time, then might have gotten around to writing a requirement for a nuclear-powered fighting ship. But Rickover insisted instead that there should be only a single test reactor, Mark I. So far as possible it would be the same as the shipboard version, Mark II; in his words, "Mark I equals Mark II." This approach would cut years off the program, and with the AEC building Mark II, the Navy would be very embarrassed if it arrived at the construction yard and there was no hull to receive it. With his two-hat management, then, Rickover could put the Navy on the spot. There would have to be such a vessel, provided for under the official shipbuilding program.

The vessel would be the USS *Nautilus*, and Rickover insisted that it should be a true fighting submarine, complete with torpedoes. Furthermore, to meet his schedule, he needed to obtain high-level approval during the spring of 1950. Events now lent a hand, as the Soviets detonated their first atomic bomb in August 1949. Truman responded with new initiatives in nuclear weaponry, including a commitment to build the hydrogen bomb. However, these efforts would favor the Air Force, which held the nation's striking power. Rickover's submarine offered a chance for the Navy to steer by the nuclear star. In April 1950 its General Board, which had responsibility for shipbuilding plans, decided that the *Nautilus* would proceed as Rickover wished.

In June 1952, Truman visited the Electric Boat Company shipyard in Groton, Connecticut. Off to his side lay a huge, bright-yellow steel plate that would become part of the keel for the sub. Truman gave a speech; then a crane lifted the plate and laid it before him. He stepped forward and chalked his initials on its surface, and a welder then burned them into the metal. Two and a half years later, the vessel was ready for her first sea-trials. On January 17, 1955, her skipper, Eugene Wilkinson, an engineering specialist who preferred sea duty, took her down the Thames River into Long Island Sound. A signalman blinked to the escort tug: "Underway on nuclear power."

In the history of naval warfare, this event would carry the significance of the transition from sail to steam. Submarines had played vital roles during both world wars, with German U-boats threatening to break the back of British commerce and their American counterparts strangling Japan's seaborne trade during the Pacific War. Yet they had merely amounted to auxiliary vessels, operating mostly on the surface and relying on the limited power of electric batteries when submerged, and hence spending little time at depth. Nuclear power now would remove such restrictions, allowing the sub to sail in the deep ocean for weeks at a time. Unseen and undetected, the nuclear sub now could challenge the aircraft carrier as a capital ship, capable of controlling the sea.

The *Nautilus* presented one route toward the new era of weaponry. A second route derived from new types of solid propellant, which grew out of the work of a struggling firm called Thiokol Chemical Corp. Its stock-in-trade was a liquid polysulfide polymer that took its name from the Greek for "sulfur glue," and could be cured into a solvent-resistant synthetic rubber. During the war it found limited use in sealing aircraft fuel tanks, but after 1945 this market disappeared. Indeed, business was so slow that even small orders would draw the attention of Joseph Crosby, Thiokol's president.

So when Crosby learned that California Institute of Technology was buying five- and ten-gallon lots in a steady stream, he flew out to see why. He found that rocket researchers at Jet Propulsion Laboratory were using it to advance beyond John Parsons's asphalt-based solid propellants; they were mixing Crosby's stuff with an oxidizer and adding powdered aluminum for extra energy. What was more, they were using this new propellant in ways that would make it possible to build solid-fuel rockets that were particularly large in size.

Crosby soon realized that he too could get into the rocket business, with help from the Army. Its officials could spare only $250,000 at first to help him get started, but to Crosby this was big money. In 1950, Army Ordnance awarded a contract to build a rocket with 5,000 pounds of propellant. It was successfully fired the following year in ground test, and with this experiment, missile designers had something new.

Thiokol-based fuel emerged as the first of a new class of solid propellants. These would draw on polymer chemistry to form as thick mixtures resembling ketchup. They could then cure or polymerize into resilient solids that would not crack or develop gaps along a rocket's inner wall, thus ensuring that their burning would be well controlled. The Navy was also interested in these propellants, for use in antiaircraft missiles, and in 1954 a contractor in suburban Virginia, Atlantic Research, set out to achieve further performance improvements.

Two of the company's scientists, Keith Rumbel and Charles Henderson, focused their attention on the powdered aluminum. Other investigators had shown that propellants gave the best performance with a powder mix of five percent; higher levels caused a falloff. Undiscouraged, Rumbel and Henderson decided to try mixing in really large amounts. The exhaust velocity, which determined performance, took a sharp leap upward. By early 1956 they confirmed this discovery with test firings. Their exhaust velocities, 7,400 feet per second and greater, would compare well with those of liquid fuels such as kerosene and oxygen.

The Navy's Bureau of Aeronautics had been developing a submarine-launched cruise missile, the Regulus. In 1954, with the ICBM at the forefront, two BuAer scientists, Robert Freitag and Abraham Hyatt, decided to try for a ballistic-missile program as well. They gained strong backing from the chief of BuAer, Rear Admiral James Russell. Their ideas also drew the attention of the Killian Committee, whose February 1955 report called for an IRBM with "ship basing."

Ship basing also won support from the heaviest of naval guns, Admiral Arleigh Burke. He took over as Chief of Naval Operations in August 1955; within a week he endorsed it. Burke saw clearly that uniting the ballistic missile and the nuclear sub would give the Navy an entirely new and major role: strategic striking power.

The Air Force had that power, with its bombers, and the Navy had nothing remotely comparable. The Navy was building carriers, but they were big and vulnerable to a nuclear attack; the carrier program was like putting too many eggs into only a few baskets. Moreover, carriers were costly, not only in themselves but also because they required extensive fleets of escorts if they were to go in harm's way. And while the big flattops could embark nuclear-capable aircraft, these planes would be restricted in size and hence in range. In no way would they compete with the nation-breaking force of the Strategic Air Command.

But missiles at sea would offer vast new prospects, for these weapons could take full advantage of the capabilities of the *Nautilus*. Such a force would not only be hidden, it would be mobile. Admiral Kinnaird McKee, Rickover's eventual successor, would note that "subs are designed to operate alone and unsupported in waters controlled by the enemy. A sub offers two things: certainty and uncertainty. Certainty as to what a sub will do—and a terrible uncertainty as to when and where he will do it. We deal with stealth, mobility and firepower that gives us leverage. You get leverage all out of proportion."[13]

The immediate result, in 1955, was the Navy's shotgun marriage with the Army's Jupiter. The Air Force already had its Gillette Procedures, which would chop through the red tape, and Burke proceeded similarly. Either BuAer or the Bureau of Ordnance might run the new program, and

Burke chose neither. Instead he set up a new outfit, the Special Projects Office. To direct it, he chose Rear Admiral William Raborn, a former carrier pilot and Pacific War veteran who had been deputy director of the Navy's guided missiles. Burke then joined with the Secretary of the Navy in issuing a letter to Raborn, granting him the authority of General Schriever and giving his SPO unparalleled independence:

> If Rear Admiral Raborn runs into any difficulty with which I can help, I will want to know about it at once along with his recommended course of action for me to take. If more money is needed, we will get it. If he needs more people, those people will be ordered in. If there is anything that slows this project up beyond the capacity of the Navy and the department we will immediately take it to the highest level and not work our way up through several days.[14]

"We used a philosophy of utter communication," Raborn later observed. "There was no such thing as hiding anything from anybody that had an interest in it. And there was nothing that got a person into trouble quicker than to delay reporting potential trouble. And, boy, if he waited until he had trouble, then he really had trouble." His colleague Captain Levering Smith, a propulsion specialist, adds: "Much of what was done was contrary to current government procedures and regulations. They were set aside." This meant that Raborn could expect at all times to have the information he would need to make a decision. He then could do so, rather than order further studies.

Raborn would give the seagoing Jupiter a good try, but he was well aware that its liquid fuels threatened the worst fire hazard since the days of wooden sailing vessels, when sailors had cooked over open flames. But when he ordered an initial study of a solid-fuel alternative, he found no improvement. The alternative would use conventional solid propellant, not the zippy new type from Atlantic Research, and its design came in with a thoroughly unwieldy length of 41 feet and weight of 162,000 pounds. Its submarine would displace 8,500 tons, to 3,300 tons for the *Nautilus*, yet would carry only four of the missiles. They might offer safety, but the overall package did not promise practical advantage.

Raborn expected that this solid-fuel behemoth would carry the Jupiter nose cone, which weighed 3,000 pounds with its warhead. Its projected design would take full advantage of the reductions in weight of nuclear weapons noted in the Tea Pot report and in subsequent studies. Even so, Raborn found himself in the position of Convair's Charlie Bossart back in 1951, when his Atlas predated this thermonuclear breakthrough and threatened to be too large to be militarily useful. But a second and similar breakthrough lay immediately ahead, and Edward Teller would carry it out.

During the summer of 1956, Teller made his proposal. Hydrogen bombs used a heavy layer of unenriched uranium that would receive a powerful burst of neutrons from the thermonuclear reaction and release additional energy through fission. Teller now asserted that he could reduce warhead weight anew by using weapons-grade uranium in this layer instead. The approach was obvious, at least within his specialized community, but it flew in the face of the Los Alamos approach. To Harold Agnew, "it just wasn't in line with our culture. We had been brought up in a culture of scarcity in the use of nuclear materials. Edward's idea was like putting butter on both sides of the bread. And he got one hell of a yield."

The upshot was a nose cone that weighed only 850 pounds, with the warhead itself coming to just 600. This and the work of Atlantic Research—and with work at MIT that forecast a correspondingly lightweight guidance system—combined to change Raborn's solid-fuel missile beyond recognition. It now would weigh only 28,800 pounds, about the same as a V-2 at launch. But at under thirty feet, it would stand barely three-fifths as tall. Yet it would carry a 600-kiloton warhead and hurl it 1,200 nautical miles. Best of all, this rocket's compact size would allow sixteen of them to fit in the lengthened hull of a *Skipjack*-class nuclear sub. These subs, with standard hulls, were already part of Rickover's construction program.

Raborn christened his missile Polaris, the constant star that guided men of the sea. But before he could contract for it and win a divorce from the Army, he first needed permission from Secretary Wilson. That autumn he presented a briefing, noting that his program would save $500 million when compared to the seagoing Jupiter. Wilson leaned forward with interest. "You've shown me a lot of sexy slides, young man," he said. "But that's the sexiest, that half-billion-dollar saving." In December, two weeks after issuing the decision that denied the Army the right to deploy its own Jupiter, Wilson directed Raborn to drop it as well and to proceed with Polaris.

Raborn's SPO by then was virtually a navy within the Navy, and his sweeping authority enabled him to seize the crown jewels from Rickover himself. He would need five of those *Skipjacks* at the outset, and he started with one that was already under construction. He took charge of it, cut it in two, inserted a 130-foot bay in its midsection to hold the missiles, and gave it the name of USS *George Washington*.

In July 1957, with the missile programs charging ahead, the old Navaho program finally gave up the ghost as the Air Force canceled it. Yet this North American project, which had contributed so much in the field of liquid rockets, now made one last contribution as it yielded a guidance system for the *George Washington* and her sisters. The vessels would have to know their positions as they navigated beneath the sea, and the Navaho's guidance provided this information quite capably. The system showed its merits a year later, as it steered the *Nautilus* beneath the ice of the Arctic Ocean on a cruise that took the vessel from Pearl Harbor to England, directly beneath the North Pole.

Polaris also drew strong interest from Colonel Ed Hall. He was managing the Thor for General Schriever, but he developed a passionate conviction that an Air Force counterpart of the naval missile would offer considerable advantage. At the outset, he addressed the problem of casting very large solid-fuel charges, called grains. Teller's new warhead meant that Hall's missile could be considerably smaller than Schriever's big liquids, but Hall knew he would need grains far larger than anyone had fabricated. Nor could he draw on those of Polaris; that was a two-stage missile, with grains of limited size, and Hall wanted bigger ones.

Hall awarded contracts to several solid-fuel companies, including Aerojet and Thiokol. Thiokol's Crosby responded with a will, for he had lost the Polaris contract to Aerojet and now saw a chance to recoup. He bought a large tract of land near Brigham City, Utah, a remote area where the roar of rockets had plenty of room to die away. In November 1957 his researchers successfully demonstrated a solid-fuel unit with 25,000 pounds of propellant, the largest to date.

But it would take more than big grains to create Hall's missile. He wanted it to fire on a moment's notice, and this meant he had to reinvent the guidance system. Standard systems, using the floated gyros of MIT's Stark Draper, called for much more than to simply turn on the power. You had to warm them, to bring the floated canister to accurate neutral buoyancy. You then had to "erect" them with respect to the local vertical and "align" them correctly in the horizontal plane. This took time—an hour and more. That was fine for Atlas and Titan, which needed even more than that for fueling and preparation; it was also all right for Polaris, which could hide beneath the sea until the skipper was ready. But Hall needed a guidance system that could run continually.

The solution came from North American's Autonetics Division, which had grown out of work on Navaho guidance, much as Rocketdyne had developed from that missile's rocket-engine activities. Autonetics had built the system that would steer the *George Washington*, and a key specialist, John Slater, had devised new gyros that broke decisively with Draper's floated versions. They featured a spinning ball, the size of a large marble, that rotated within a spherical housing and used a thin film of gas to keep it from touching. One such gas-bearing gyro, dating to 1952, was still in continuous operation five years later.

In August 1957, Hall took charge of a working group that set out to define what he called Weapon System Q. Two weeks later he had his design, which seemingly would be all things to all people. Hall wanted it to be inexpensive; the historian George Reed would describe it as "the first strategic weapon capable of true mass production." It would be robust, not delicate like the liquid-fueled missiles, and could hide underground in concrete-lined shafts that *Time* would describe as "inverted silos." Being robust, it could ignite within its silo and stand up to the severe stress and vibration of a launch directly from down-hole. No liquid rocket could do that. In turn, the silo would protect it from anything short of a direct nuclear hit.

Following the launch of Sputnik, Hall and Schriever went to the Pentagon to build support for this newest ICBM. In February 1958 they presented the concept to Defense Secretary McElroy and to the Air Force Secretary. In Schriever's words, "We got approval within forty-eight hours." By the end of the month the project had a new name: Minuteman.

Yet the nation needed more than new missiles; it also needed reconnaissance, and almost from the start the U-2 delivered far less than it had promised. When it entered service, Soviet radar promptly picked it up. Following the second overflight, the Foreign Ministry lodged a protest. The protests escalated, and after six such missions, all during July 1956, Ike ordered a stand-down. Subsequent flights required his personal approval, and during the next four years only eighteen took place. Richard Bissell, following the example of Curtis LeMay, transferred five U-2s to the Royal Air Force and arranged to receive their intelligence. Still, this was only a stopgap measure.

The U-2s' photos were unrivaled; one of them, shot from 55,000 feet, showed the golf balls on Ike's favorite course. And one 1956 mission returned with highly significant photos that showed far fewer heavy bombers than expected at Soviet bases. This started a process of downward revision of Moscow's estimated air power; one of Ike's military aides would declare that "very quickly we found the Bomber Gap had a tendency to recede. It was something that each year was going to occur. But in fact it did not occur." Yet the restrictions on flights of the U-2 made it all the more important to proceed with WS-117L.

However, although Schriever took over responsibility for that satellite program in February 1956, it failed to receive the high priority given to Atlas and his other missiles. When Sputnik 1 went up, WS-117L was limping along and needed a boost. The Air Force had already abandoned RCA's approach to reconnaissance, which relied on a TV camera, because it would give poor image resolution. Schriever was still pursuing the second approach to a nonrecoverable satellite, which would develop negatives onboard and scan them for transmission. But the associated spacecraft would need an Atlas to reach orbit, and wouldn't be ready any time soon. (Called Samos, the satellite didn't fly until 1961.)

Three weeks after Sputnik, Ike received a report calling for an interim reconnaissance satellite, which could be ready fairly quickly. He asked the CIA's Allen Dulles and Defense Secretary McElroy to tell him more, and the concept of "more" came forth in November from two Rand Corporation analysts, Merton Davies and Amrom Katz. Their study, "A Family of Recoverable Satellites," concluded that such a spacecraft could use the Thor as a booster. It had already flown successfully as a missile.

In January 1958 the National Security Council cleared the way for action by assigning highest-priority status to development of an operational system. Two weeks later, Ike made two decisions: to set up ARPA, and to have the CIA build a recoverable satellite, leaving the nonrecoverable effort to the Air Force. A CIA official gave his program the name Corona, which he took from a Smith-Corona typewriter on his desk. Its spacecraft was named Discoverer. From the start, Corona would share the U-2's extraordinary level of secrecy.

Accordingly, late in February, ARPA formally canceled the part of WS-117L that corresponded to Discoverer. This action involved legal contract terminations and other official notifications that were standard whenever the Pentagon dropped one of its projects. This sudden decision outraged both contractors and Air Force officials, who were appalled that ARPA, the newest and least experienced of outfits, would abandon the reconnaissance system that showed the best prospects for success. However, this outrage suited Richard Bissell's purposes.

Bissell quietly initiated Corona and prepared to run it as he had run the U-2 effort. He arranged for a few picked people, in the Air Force and among the contractors, to receive clearance to join it and carried out a covert reactivation of Discoverer under its sponsorship.

The Air Force would launch these spacecraft, but Cape Canaveral wasn't the right place. Discoverer was to fly in a polar orbit and would demand to be launched nearly due north or south. At the Cape, either direction would carry the rocket over populated areas, which wasn't safe. But the Pentagon already was building Vandenberg Air Force Base, a new center on the California coast, 150 miles northwest of Los Angeles. It offered a clear shot to the south. It also would support military launches of Atlas and Thor, which meant Discoverer would fit right in. The new base even provided security, which the Cape did not. It could keep out tourists in the manner of any defense facility, though the Southern Pacific Railroad ran through Vandenberg and presented views from passenger trains. But Bissell could schedule his flights by consulting the railroad timetables and wait until the trains were out of sight.

Lockheed, builder of the U-2 and contractor for WS-117L, would build Discoverer as well, mounting it atop its Agena second stage. Bissell worked closely with an Air Force counterpart, Brigadier General Osmond Ritland, and pushed the new project with characteristic energy. "The program was started in a marvelously informal manner," Bissell would relate. "Ritland and I worked out the division of labor between the two organizations as we went along. Decisions were made jointly. There were so few people involved and their relations were so close that decisions could be and were made quickly and cleanly. We did not have the problem of having to make compromises or of endless delays awaiting agreement."[15]

Lockheed, with a reputation for fast action, had a complete design late in July. The first Thor-Agena with its Discoverer reached the launch pad early in 1959, and the ground crew included veterans from the V-2 days who called themselves "broomlighters." They said that when a V-2 proved balky, one of them would ignite its engine using a kerosene-soaked broom. The Thor of Discoverer 1 needed no such heroics as it thundered into space at the end of February. However, the Agena fell short of orbit. A CIA analyst would write that "most people believe it landed somewhere near the South Pole."

Mid-April saw the launch of Discoverer 2, which reached orbit successfully and circled the earth for a day. It was supposed to come down near Hawaii, and a ground controller transmitted a command to have its capsule begin re-entry at a particular time. Unfortunately, he forgot to push a certain button. The director of the recovery effort, Lieutenant Colonel Charles "Moose" Mathison, then learned that it would actually come down near the Norwegian island of Spitzbergen.

Mathison telephoned a friend in Norway's air force, Major General Tufte Johnsen, and told him to watch for a small spacecraft that was likely to be descending by parachute. Johnsen phoned a mining company executive on the island and had him send out ski patrols. A three-man patrol soon returned with news: They had seen the orange parachute as the capsule drifted down near the village of Barentsburg. That wasn't good, for its residents were expatriate Russians.

Bissell sent aircraft and helicopters, while local people eagerly joined in the hunt. "There's not much to do here," said one resident, "so this search is pretty popular." However, no one ever found more than the tracks of snowshoes. Those could only belong to the Russians, for the island's Norwegians customarily used skis. The incident later inspired the novelist Alistair McLean when he wrote *Ice Station Zebra*, but for the CIA, its only consolation was that Discoverer 2 at least had not carried photographic equipment.

This was as close as the program would come to succeeding during the next sixteen months. Of ten Discoverer flights that ensued, only five reached orbit. All behaved bizarrely. One pointed its retro-rocket in the wrong direction; instead of coming down from space, it wound up in a higher orbit. On another one, the rocket did not fire at all. An electrical problem left another one tumbling uselessly, while three missions had their cameras fail during the first or second orbit. Discoverer 8 achieved a proper re-entry, but its capsule's parachute failed to deploy. It struck the ocean and sank.

Following that flight, in November, Bissell suspended further launches while he tried to take stock. "It was a most heartbreaking business," he later declared. "In the case of a recce satellite you fire the damn thing off and you've got some telemetry, and you never get it back. You've got no hardware. You never see it again, and you have to infer from telemetry what went wrong. Then you make a fix, and if it fails again you know you've inferred wrong. In the case of Corona, it went on and on."[16]

Meanwhile, Korolev was stepping up his launches of the R-7 as an ICBM, firing them once a week in a clear demonstration of military readiness. In January 1960 he fired two of them onto long-range trajectories that reached the central Pacific, and the Kremlin announced that this missile now was operational. Also during that month, a debate developed in Washington in response to a new intelligence estimate, which predicted that Moscow would possess up to 450 ICBMs in mid-1963. This would be twice America's anticipated strength, which was frightening enough, but what raised skepticism was that the estimate actually represented a substantial reduction from earlier ones. Senators Stuart Symington and Lyndon Johnson asked whether Ike perhaps was cooking the books, downgrading the perceived threat during an election year. New U-2 photos were obviously necessary, and Ike approved the flights.

Then on May Day, a surface-to-air missile downed a U-2 near Sverdlovsk. Khrushchev, who would have loved to use such aircraft himself, declared that he was shocked, *shocked* to find the Yankees spying on him. He then walked out on a planned meeting with Eisenhower in Paris. With its cover blown and now vulnerable to Soviet missiles, the U-2 would never again overfly the USSR. This meant that all Bissell could rely on were his reconnaissance satellites.

The camera problem proved fixable. The film had an acetate base, and in vacuum it tore or broke, jamming the equipment. Eastman Kodak introduced a polyester base, which did the trick. In addition, the satellites were becoming too cold, which kept the batteries from working. Dark paint absorbed sunlight and made them warmer. To make sure he wasn't overlooking anything, Bissell put more instruments aboard his spacecraft, which would report on their operation.

Discoverer 11, which flew in April, was the next to reach orbit. Its cameras worked; its batteries worked, too, but small attitude-control rockets, which were to point it in the right direction for re-entry, didn't work. Its capsule was never seen. Bissell replaced the rockets with gas jets that used compressed nitrogen and yanked out the photo equipment to make room for still more instruments.

In August, Discoverer 13 wound up as the lucky one. Moose Mathison had aircraft ready to snatch it by grabbing its parachute, but heavy clouds prevented this. No matter; it hit the water and floated, transmitting a radio signal, releasing a yellow-green dye marker, and flashing a strobe light. Three hours later, a shipborne helicopter arrived and a swimmer made the recovery. With this encouragement, Bissell allowed Discoverer 14 to fly with its camera. Its capsule too was recovered successfully, this time in midair, on August 19.

Discoverer 14's film soon arrived at the CIA's Photographic Interpretation Center, whose photo interpreters gathered in an auditorium. The director, Arthur Lundahl, spoke to them about "something new and great we've got here." His deputy then presented a map of the Soviet Union. These maps had previously featured a single narrow line to indicate the path of a U-2. This one had eight broad swaths running north to south across the USSR and Eastern Europe, covering over one-fifth of their total area. They represented the regions that this single mission had photographed, and people broke out in cheers. Some photos were fogged by electrostatic discharges, but the resolution was twenty to thirty feet, which analysts described as "good to very good." Clearly, this was a turning point.

There would be other disappointments, including a September heartbreaker when Discoverer 15's capsule was seen by aircraft as it floated in the ocean—and sank in stormy seas before anyone could rescue it. But on the next flight, Bissell began using the Agena B second stage, which was seven feet longer than its predecessor and allowed Discoverer to grow larger and more capable. With this, the CIA flew a successful three-day mission, Discoverer 18, in December. Its image quality matched the best obtained from Discoverer 14, and its fogging of the photos was greatly reduced.

President Kennedy took office in January 1961. He had campaigned on the issue of the missile gap, and his deputy defense secretary, Roswell Gilpatric, believed in it strongly. Thus, after taking office, Gilpatric and his boss, Defense Secretary Robert McNamara, went to the Air Force intelligence office on the fourth floor of the Pentagon and spent several days studying Discoverer photos.

The Air Force held the view that Moscow was building large numbers of well-camouflaged missile sites. Among the presumedly disguised bases were a Crimean War memorial and a medieval tower. But McNamara and Gilpatric preferred the view of Army intelligence that the R-7, being large and unwieldy, could move only by rail or military road. Discoverers had taken photos along the Soviet Union's railroads and principal highways—and had found no missile launchers. In February, at an off-the-record press conference, a newsman asked about the missile gap. McNamara replied that "there were no signs of a Soviet crash effort to build ICBMs." Reporters raced to their phones, newspaper articles blossomed with the word that no such gap existed, and Kennedy himself had to step in, declaring that it was too early to draw such conclusions.

But during the spring and summer, new and compelling information came both from space and from an agent in Moscow, Oleg Penkovsky. He was a colonel in the Chief Intelligence Directorate of the Soviet Army who had recently begun working for MI-6, Britain's intelligence agency. He had previously offered his services to the CIA, only to be rebuffed as a potential double agent, but this British connection soon stilled American doubts. He acquired a CIA handler and a code name, Alex (not Cardinal, as in Tom Clancy's novels), and received a request for a report on the Soviet missile program.

In May 1961 he delivered three rolls of microfilm that included technical information on the R-7, a review of its schedule and its delays, and minutes of Kremlin meetings in which officials had decided to use it for space launches—but not to deploy it as an ICBM. Not only was it big and cumbersome, its launch complexes were highly vulnerable to attack, and each missile would need up to three hours to fuel it and prepare it for flight. Soviet leaders had expected to keep its bases secret, to prevent them from being hit, but they were well aware that the Discoverer overflights had made this impossible. They were proceeding with a new ICBM, the R-16, which would use storable fuels and deploy in silos, taking only thirty minutes for preparation. But it was only in flight test, with operational use still some time off.

Then in June and July, Discoverer 25 and Discoverer 26 flew with full success. They were only the third and fourth missions to return photos having intelligence value, but together these four flights gave useful coverage of some thirteen million square miles. This included more than half of the regions best suited to ICBM deployment.

Within this vast area, photo analysts found no more than two new and previously unsuspected ICBM bases. Three others were photographed again. By comparing them with one another, and with a known testing complex at Tyuratam, the analysts came away with a clear understanding of just what an ICBM base would look like. That made it possible to eliminate a number of "suspect" launch sites and to give a clear and definitive estimate of Moscow's ICBM strength, which came out on September 21. The National Intelligence Estimate 11-8/1-61, entitled "Strength and Deployment of Soviet Long Range Ballistic Missile Forces," read in part:

> We now estimate that the present Soviet ICBM strength is in the range of 10–25 launchers from which missiles can be fired against the US, and that this force level will not increase markedly during the months immediately ahead.

> The low present and near-term ICBM force level probably results chiefly from a Soviet decision to deploy only a small force of the cumbersome, first generation ICBMs, and to press the development of a smaller, second generation system. On this basis, we estimate that the force level in mid-1963 will approximate 75–125 operational ICBM launchers.[17]

Clearly, it was not the Russians who were ahead in the missile race, but the Americans. Furthermore, Eisenhower had followed a time-honored approach, one that America had used as far back as the Civil War, to defeat the Soviet military challenge. The approach conceded the foe its initial strength, while using existing force to protect the country. Then, protected by this shield, the nation marshaled titanic industrial and technical resources, which no other nation could match or even attempt to meet, to build the weapons that would achieve supremacy.

In this manner the Union had beaten the Confederacy in the world's first industrial war. In World War I the United States became a vast granary, arsenal, and source of war loans, thereby buttressing Britain and France when their own strength failed. Germany's master of battle, Field Marshal Paul von Hindenburg, said of the Americans, "They understood war." In 1943 Stalin met with Roosevelt and offered a toast: "To American production, without which this war would have been lost."

In facing the latest challenge, Ike had brought about a revolution in armaments, taking the nation's strategic forces from the shadow of World War II to a form we would recognize today. On his inauguration, in 1953, the Air Force relied on piston-driven bombers while the Navy still operated the ships of the Pacific War. But in 1961 the Strategic Air Command had the hydrogen bomb, with over 1,500 jet bombers that could deliver it. Reconnaissance satellites, big carriers, and nuclear submarines all were at the forefront. Thor was overseas; Atlas was operational. The *George Washington*, carrying sixteen Polaris missiles, put to sea on its first combat patrol in November 1960. And less than two weeks after Kennedy entered the White House, a three-stage Minuteman flew to its full range from Cape Canaveral, on its first try. By 1967 the Navy would deploy forty-one such submarines, while a thousand Minuteman silos would dot the northern Great Plains.

Yet when NIE 11-8/1-61 came out, representing the consensus of the intelligence community, and when the Soviets proved to be weak rather than strong, there was no national celebration. No one kissed the girls in Times Square. The report was classified Top Secret, after all, and was available only to people cleared to handle U-2 and Discoverer photos. To everyone else, which was just about everybody, the space race still warned of Moscow's looming danger. And in space the Soviets were well ahead, as they explored the moon and sent the first cosmonauts into orbit.

6

A Promise of Moonglow

Space in the Wake of Sputnik

ALTHOUGH KOROLEV'S SPACE PROJECTS served civilian objectives such as politics and science, their funding came from the Ministry of Defense. He would pursue new initiatives, sending unmanned spacecraft to the moon and placing cosmonauts in orbit, by building more powerful versions of his R-7. But to develop these versions, he had to point to their military uses. Fortunately, this was not difficult. He could build reconnaissance satellites that would correspond to America's U-2. He also could introduce a new way of delivering nuclear weapons, by launching them into orbit and then using a retro-rocket to cause each bomb to drop down from space and onto its target. The warhead could then approach the United States from any direction and strike without warning.

To do these things, he needed an upper stage that would ride atop his basic R-7. In addition, just as he had accepted Tikhonravov's suggestion to launch the simple PS as Sputnik 1, he needed a similarly simple spacecraft as an initial payload for this upper stage. He could simultaneously serve the causes of science, propaganda, and space exploration, and could make Khrushchev happy, by building the payload as a spacecraft only slightly more complicated than the PS—and shooting it at the moon. As he emphasized in *Pravda* in December 1957: "Reaching the moon by a rocket, launched from the earth, is technically possible even at the present time."

Korolev then worked with the mathematician Mstislav Keldysh to prepare a formal proposal, which they outlined in a letter that they sent to the Central Committee of the Communist Party, near the end of January 1958. They would start with a simple craft that would impact onto the lunar surface, carrying a magnetometer along with cosmic-ray and meteoroid detectors. Later, a more demanding mission would fly around the unseen lunar farside, taking pictures with a television camera and transmitting them to earth.

Approval came swiftly, and Korolev told his staff, "Comrades, we've received an order from the government to deliver the Soviet coat of arms to the moon. We have two years to do this." He would start by setting up two separate projects to build his upper stage, expecting that competition would spur each group to greater effort. In addition, he wanted the stage's rocket engine to be ready in only nine months. He then could carry out his first launches as early as the fall of 1958, barely a year after Sputnik 1.

He pushed himself as well as everyone else at the pace of a forced march. The Soviet Union had not yet invented the weekend; the six-day workweek was still standard. Even so, Korolev used his Sundays for staff meetings that could be free of phone calls and interruptions for routine problems. In addition, he imposed a rule that his associates could fly only at night when traveling to Tyuratam. In the words of one of these colleagues, Boris Rauschenbach, "Korolev could not imagine wasting a working day on travel. He considered that a nearly-sleepless night in an airliner seat would give sufficient rest."

For a time, the development of the upper stage left little time for rest, indeed. It was not only new in itself; it introduced novel technical problems, for it had to survive the severe shaking and vibration of flight atop its booster and then ignite high in the sky, where no launch crew could service it. To make the problem slightly easier, the stage would connect to the R-7 by means of an open lattice. Its engine could then start before the thrust of the booster died away, with this engine's exhaust escaping through the lattice. This would avoid the problem of starting the upper stage in zero gravity. But rocket design involved plenty of other problems, which were harder to avoid.

Valentin Glushko, designer of the R-7's engines, headed one of the upper-stage design groups. Matched against him was a newcomer, Semyon Kosberg. Kosberg had little experience with rockets, having spent the last fifteen years working mostly on jet engines for aircraft. But in 1956, working in Voronezh, he built and successfully tested a rocket engine that could restart while in flight. The problems of restarting such an engine were very similar to those of starting an upper stage, and Korolev formally directed Kosberg to build the stage in February 1958, only two weeks after sending his letter to the Central Committee. This proved to be a good decision, for while Glushko's effort faltered, Kosberg, the newcomer, came through.

Meanwhile, in Los Angeles, General Schriever was smoothing the way for Thor to serve as a competing moon rocket. Initially he was concerned with nose cone development for Atlas, and the rockets available for flight test weren't adequate. The X-17 and the Army's Jupiter-C were flying small IRBM nose cones, and Jupiter would test a full-size version. But for Atlas, Schriever needed a way to fling a nose cone a full 5,000 miles or more. He could do this by placing an upper stage atop Thor, and senior managers at Aerojet recommended that he use the Vanguard second stage, which they were building. Just before Thanksgiving of 1957, Schriever gave approval to build this two-stage vehicle for nose cone tests by pulling some standard Thors off the line.

The rocket, the Thor-Able, represented the start of efforts that would mate Thor with increasingly capable upper stages, turning it into the nation's principal workhorse in space. Already, at the Rand Corporation, people were proposing to use Thor-Able to launch an early reconnaissance satellite. After all, this rocket would be like having Vanguard with the highly capable Thor as a first stage, a combination that could certainly launch much more than a grapefruit. Thor-Able would complement Thor-Agena, a separate development, with a second stage that came from Lockheed and that would carry the Discoverer satellites.

Right after Thanksgiving of 1957, a delegation from Ramo-Wooldridge visited the Martin Co. in Baltimore, which was building the complete Vanguard rockets. They wanted to buy some second stages, but found that all were reserved for the Navy. However, Aerojet was quite willing to sell a version of the second stage that featured only the engine and propellant tanks. It would need a guidance system, but R-W had an active guidance group headed by James Fletcher and they could home-brew their own by modifying the Thor autopilot. They set up a machine shop in a hangar at Los Angeles's airport and went to work on their nose-cone test vehicle.

Meanwhile, in Washington, the shock of the first two sputniks was spurring a flood of proposals for space projects. An admiral later declared, "It seemed to me that everybody in the country had come in with a proposal except Fanny Farmer Candy, and I expected them any minute." Already the head of Jet Propulsion Laboratory, William Pickering, was pushing for a quickie moon-shot effort called Project Red Socks. At R-W, the calculations of one of Fletcher's engineers showed that the Thor-Able, carrying the Vanguard third stage, could shoot a payload to the moon as well. The Rand Corporation now weighed in with its own report, stating that the Air Force could indeed carry out such a mission.

At the White House, James Killian held the post of Eisenhower's principal adviser for missiles and space. In January 1958 he came out to R-W and received a briefing on sending a rocket to the moon. He said, "Send in a proposal and we'll go ahead." Following an all-night session, the proposal was ready, and though it came to less than thirty pages, it was enough. The proposal called for more than a simple shot at the moon; its spacecraft would carry a solid-fuel rocket from the Falcon air-to-air missile and would use it as a retro-rocket to place the craft into orbit around the moon.

The Advanced Research Projects Agency would manage the work. It had a small, tightly knit crew of eighty people, including the secretaries, with a picked group of specialists drawn from the three military services. The agency's director, Roy Johnson, issued ARPA's first order to the Army, instructing von Braun to carry out a version of Project Red Socks. Order no. 2 went to the Air Force; it called on Schriever to join the lunar sweepstakes. And late in March, Ike made a public announcement, in which he said: "This is not science fiction. This is a sober, realistic presentation prepared by leading scientists."

The first American satellite had reached orbit less than two months earlier, but already the president was committing the nation to a race to the moon against the Soviets. And this moon shot would not take place, Kennedy-fashion, "before this decade is out." Rather, if all went well, the country would have a spacecraft in orbit around the moon before the summer was out.

Louis Dunn, a general manager with R-W, had prime responsibility for the mission. As he told *Time* magazine: "I got all our people together and told them that we had taken on a new job, and that in many ways it represented the biggest challenge we had ever faced. Because while we were supposed to have this ready to fire in something less than six months, we could, under no circumstances, let it interfere with the Air Force ballistic missiles program. That meant, among other things, that the forty-hour week was out the window. I also told them that it would be impossible to pay overtime, that we would have to do most of this on our own time."[1]

During those few months Dunn's people would have to build the three-stage launch vehicle, at a time when even the two-stage Thor-Able had yet to fly. They would have to design and build the lunar spacecraft. They also had to set up a worldwide network of ground stations, to track the spacecraft in flight. And they would have to do most of this with slide rules. Their only computer was a vacuum-tube IBM 704, with the power of one of today's programmable handheld calculators.

No one knew when the Russians would strike next, so speed was at a premium. In this situation, people had little time for long committee meetings. The project manager, George Mueller, stood ready to make decisions on the spot, and kept his door open to anyone with something to say. About a month before the first launch, one of his staffers came in with a suggestion. To increase the accuracy of aim, the man wanted to use radio tracking to determine the velocity of the second stage with high precision and then send a signal at the proper moment to cut off the engine. The alternative called for loading that stage with calculated amounts of propellant and allowing its engine to fire until all of it was burned.

"We made a very simple calculation," recalls George Gleghorn, who had responsibility for the second stage. The calculation showed that this technique would indeed provide better precision. "It was not studied extensively. We showed it to Mueller—and he said, 'Fine. Build it.' We found an engineer, told him to design the receiver. He designed it and built it; we flew it to the Cape and put it on the second stage."

In Moscow the pace was similar, for like Mueller, Korolev stayed close to the job. Problems developed in welding the upper-stage tankage, and production engineers soon were blaming the designers. Korolev summoned everyone to a meeting, where he looked at them with sharp eyes. "Well, what is going on?" he demanded. "Who will report? Do you understand the consequences of a delay?"

He listened to his colleagues' recriminations, and then gave a direct order to the chief welding engineer: "I give you one month. Do whatever you have to do, but the tankage should be ready." He turned to another specialist and named a subcontractor: "Go there in person and get the missing parts of that welding manipulator. If you don't succeed—blame yourself for the consequences. If you need any help, give me your plan of action. I'll sign it. That's it."[2]

A month later the welding equipment was operating properly, and the propellant tanks were fully assembled. Still, though Korolev was racing ahead at full speed, the Yankees were beating him to the launch pad. The important preliminary Thor-Able flights, which would test nose cones, began in April 1958. The first one ended barely ten seconds prior to Thor engine cutoff, as a turbopump failed and it blew up. But two flights in July went the distance, hurling their nose cones first to 6,100 and then to 6,300 miles.

These were America's first shots to intercontinental range. They demonstrated that ablative nose cones, which were considerably lighter than the copper variety, could meet the needs of Atlas. However, neither of the test nose cones was found or recovered, and this spelled misfortune to the mice that were riding inside. All three flights carried a "mouse in Able," but each met the fate of Laika. In the words of the satirist Bill Dana, who called himself Jose Jimenez, "They closed the door . . . on that little mouse . . . I don't want to talk about it!"

Even so, the successful flights of July seemed to have qualified Thor-Able for the lunar mission, and the first rocket to the moon took off from Cape Canaveral shortly after sunrise on August 17. The NBC television network accomplished another space first by providing coverage, as the newsman Herbert Kaplow added commentary. The rocket lifted off, and people in the blockhouse went outside for a better look. Then, seventy-seven seconds into the flight, they saw it disappear in a puff of smoke in the sky.

General Schriever was among the observers. He turned to Dolph Thiel, the program director at R-W, and asked, "What happened?" Thiel replied, "It blew. It blew," and a tear ran down his face. Louis Dunn joined them, also looking very depressed. Allen Donovan, the manager responsible for the moon probe, felt "completely shot, very depressed. I was just in a state of shock."

Divers recovered the wreckage, which had fallen offshore, and the main engine went back to Rocketdyne for careful study. The problem involved the turbopump; ironically, the difficulty was one that Dunn, Thiel, and Schriever knew only too well. The Army's Jupiter used the same engine, and the previous March General Medaris had halted flight tests, because its turbopump bearings had shown a tendency to break. Schriever by then was receiving a number of flight-ready Thors, and he knew that if he took time to wait for new engines, he would delay the test program. The problem only turned up about once every seven shots, so he took a calculated risk, electing to fire the Thors for the sake of the useful data they could return when they flew properly, while realizing that some would fail. The explosion in the sky had stemmed from the luck of the draw.

But in Thiel's words: "A lot of our thinking was that there would be a new one soon, so come on, fellows, let's get going on it." Their next chance came eight weeks later, just before Columbus Day. This time, they came so close to complete success that everyone could taste it.

Liftoff took place several hours after midnight. For a moment the Thor brilliantly lit up the launch area with the stark white light of its exhaust. Then it traced a bright curving streak as it rose and turned to fly downrange. All three stages fired, as members of the launch crew threw off their guarded reserve and erupted in joy. Nevertheless, "it probably wasn't more than five minutes into the flight before we knew we had a problem," recalls Richard Booton, who had responsibility for tracking. The third stage had failed to separate cleanly, hanging up momentarily and deflecting itself from the proper direction. That threw off the trajectory, giving an inaccurate flight with too low a velocity.

The spacecraft, Pioneer 1, needed a speed of 24,100 miles per hour to reach the moon. It had fallen short by less than 500 mph. Still it was soaring high and free, and while it would not approach the moon, it would attain an altitude of 70,745 miles. This was true space exploration, with instruments reporting on conditions far beyond earth orbit. And as Pioneer I rose toward its peak, it touched the fringe of interplanetary space.

Few people on the project staff cared to sleep, not while the spacecraft continued to live. Forty-three hours after launch, it re-entered the atmosphere over the South Pacific. Its signal never wavered, continuing clear and strong—until without warning, it stopped. Plunging into the upper air, it burned up like a meteor. But it had come astonishingly close, and for the moment that was enough.

Now it was America's turn to receive the plaudits of an astounded world. The Paris newspaper *La Croix* called it "the most prodigious event in history." India's Prime Minister Nehru, a frequent critic of the United States, called it "a tremendous triumph of modern science." Closer to home, Simon Ramo offered his own perspective: "What we gained this weekend was a few seconds on infinity."

This proved to be the high-water mark. Soon it would be the turn of von Braun, who hoped to fly past the moon with a spacecraft that weighed all of thirteen pounds. It was so small that a JPL manager, Dan Schneiderman, would personally carry it to the Cape by airliner, placing it next to him in a container that occupied a seat of its own. Its tiny size showed that the mission approached the limits of what von Braun could do; and the margin proved too narrow.

This mission, Pioneer 3, took off in early December. Its booster featured a Jupiter for the first stage, with a cluster of solid-fuel rockets as the upper stages, and all fired successfully. But the Jupiter climbed a little too steeply and cut off its engine a few seconds too early. Again, the resulting shortfall in velocity amounted to only some hundreds of miles per hour, but again that sufficed to make the difference. Pioneer 3 would repeat the near-success of the Air Force, reaching a peak of 63,580 miles. It made an important contribution to space science by discovering a second Van Allen belt or zone of trapped radiation, lying outside the first. But it, too, failed to come close to the moon.

Yet while the United States was missing by little more than inches, Korolev was barely managing to clear the launch pad. He had made his first attempt on September 23, 1958, but the booster disintegrated after only ninety-two seconds of flight. There were only a few days during each month when the moon was in the best position for a mission, and his second try, in October, coincided with the launch of Pioneer 1. For a brief moment it seemed that the moon race would turn into a real race, with both nations' rockets competing like Formula One cars in the Grand Prix.

Hearing of a successful U.S. launch, Korolev asked his own crew to shake off sleepiness and fatigue, drink some strong tea, and continue to work. "Don't worry that the American rocket is flying to the moon," he concluded. "We will reach the moon several hours before the Americans." But once again, 104 seconds after liftoff, his booster exploded and fell to earth like a fireworks display.

Addition of the upper stage to the basic R-7 had obviously rendered it unfit for flight. Korolev sent a salvage team onto the Kazakhstan steppes, to recover the wreckage. It had fallen over a wide area at some distance from the launch site, and one team member would recall that "at night we

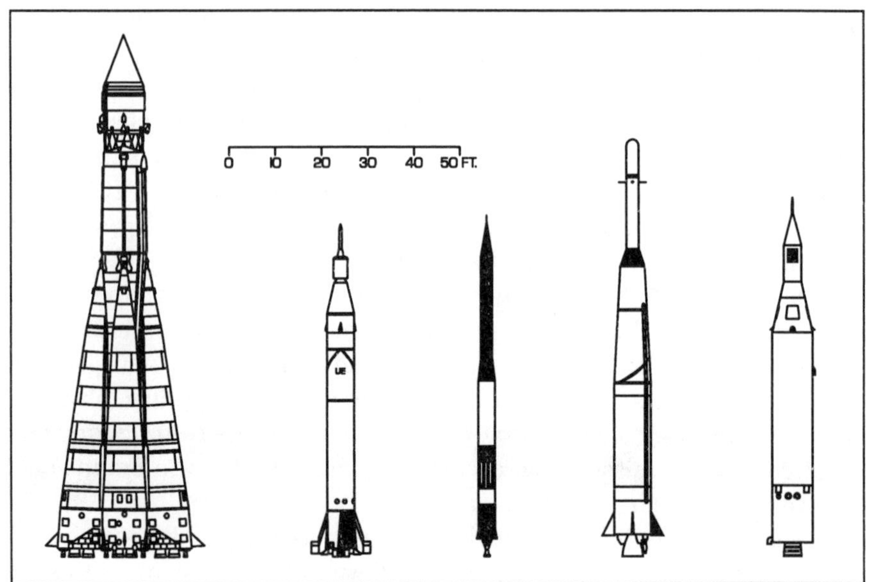

Early space boosters: Soviet Luna; U.S. Jupiter-C, Vanguard, Thor-Able, Juno II. (Dan Gauthier)

froze in tents. The dry rations were meager. Water was not always on hand; sometimes it had to be trucked in from wells several kilometers away. Anyone who has been in Kazakhstan knows how difficult it is to get your bearings," amid its flat, treeless plain. But the team's leader had brought a gun and succeeded in shooting partridge for the table. He also brought down game when a large herd crossed the steppe.

Back in Moscow, Korolev instructed his chief of ballistics, "Svet" Lavrov, to find and fix the problem. Lavrov found its source in powerful oscillations within the liquid-oxygen lines, which had torn the rockets apart. These oscillations had appeared during the flight tests of earlier missiles, and Lavrov arranged for several leading scientists, including Keldysh himself, to offer advice as consultants. After performing experiments and developing a mathematical model, Lavrov's staff concluded that they could suppress the oscillations by installing damping devices in the oxygen lines. By the end of 1958, the modified booster was ready.

His in-flight failures, coupled with America's near-successes, had put Korolev under considerable pressure. Yet he would not cut corners. One day, while visiting the shop where workers were securing the spacecraft to its mounting frame, he heard the sound of a hammer. A staff member had encountered some stuck bolts on the frame and was trying to loosen them by pounding them with a heavy wrench. An indignant Korolev said, "What on earth are you doing! Why are you pounding it? This is a spacecraft!" The worker tried to explain that he had been pounding on the mounting frame, not the payload itself, but Korolev would not listen. He cooled down only when the man pledged that he would never do it again. Still shocked that anyone could take such liberties, Korolev walked away, still muttering, "This is a spacecraft."

He came to the Tyuratam launch site just before Christmas, with Keldysh in tow. The winter weather was severe, as temperatures fell to twenty below zero. Heating pipes broke in a hostel; makeshift stoves produced so much smoke that people found it hard to breathe. Workers at the launch pad wore coats and fur-lined boots, but these provided little comfort in the gusty and freezing wind. Problems appeared in radio systems, delaying the launch.

People hoped to find relief by drinking alcohol, but a government commission was about to arrive and Korolev would not permit it. But after the commissioners left, a logistics officer approached him with a silent question in his eyes. Korolev responded with a growl, "To hell with it. Give it away!" Soon a long line of people formed at a supply shed, armed with teakettles to carry away the hooch.

The New Year holiday arrived, but Korolev returned to the pad. It was morning; the wind had gone; the weather was a little clearer. The prelaunch atmosphere was stressful the following day, but he felt he was in his element. He listened with pleasure to the familiar sounds of hissing pneumatic lines, rumbling electric motors, whining generators, and loud slaps from closing valves. "Oh, God," he thought. "Everything is in order, on schedule, and the people around me are just beautiful."

The liftoff also was beautiful, for this time the rocket, including the upper stage, performed flawlessly. Soon the probe, Luna 1, was riding its own trajectory, reaching well beyond the altitudes of the recent Pioneers, approaching the moon, then passing it closely. But as Korolev returned to Moscow, again he felt gloomy. He had not planned to fly past the moon, but to hit it with a direct impact.

However, back in the capital, the mood was upbeat. Once again everyone declared that he had scored a major success, for he had built the first rocket to achieve escape velocity. Luna 1 had indeed made a mark, even if it wasn't on the moon. It had become the first spacecraft to escape the earth's vicinity and to enter an orbit around the sun. Moreover, the 800-pound weight of Luna 1 demonstrated again the lifting power of his R-7.

Washington reacted anew with hyperbole, as a House committee declared that "inexorable changes in society and political power will follow the development of space capabilities; failure to take account of them would virtually be to choose the path of national extinction." *Time* wrote of "a race that may decide whether freedom has any future." But Eisenhower, working with his adviser James Killian, was already preparing to complement the military space projects of ARPA with a major civilian effort and a new federal agency, the National Aeronautics and Space Administration.

The agency's genesis derived from a congressional law, the National Aeronautics and Space Act of 1958. NASA's nucleus was a small organization that dated to 1915, the National Advisory Committee for Aeronautics (NACA), which occupied a niche as a specialized pocket of expertise in aerospace technology. Working on a shoestring budget, NACA operated two aeronautical labs and a propulsion-research center. It had few high-placed Washington friends and little visibility; even its research aircraft were Air Force hand-me-downs. As recently as 1954, amid a cost-cutting wave, it had received barely half its requested funds.

But NACA had a strong reputation among those who knew of it; in 1957 its chairman was Jimmy Doolittle, a research leader and top wartime general. It already had a number of pathbreaking researchers in its ranks: Richard Whitcomb and R. T. Jones, who had defined the shape of supersonic jets; and John Becker, who had built a Mach 7 wind tunnel. Other NACA investigators were making key contributions to the dawning era of missiles and space, including John Sloop, who was building the first hydrogen-fueled rocket engines. Robert Gilruth was a leader in the development of ablative heat shields for atmosphere entry. Julian Allen and Alfred Eggers had invented the blunt shape of a reentering spacecraft.

By the time the first sputniks went up, some one-third of NACA's work was related to space. Nevertheless, by itself NACA was far too small to take on the responsibility of running a major space program. But it would serve as a core, growing by taking administrative responsibility for other ongoing space and rocket programs. The existing Vanguard effort, in the Naval Research Laboratory, would be one of the first. The Army's JPL also went over to the new NASA. The senior management of the agency also had their eyes on von Braun's rocket group at Redstone Arsenal. In addition, another early initiative broke ground near Baltimore for the new Goddard Space Flight Center, which would work with unmanned spacecraft used for science.

NASA took shape during 1958, amid a surge of interest in manned space flight. Well before Sputnik, a number of specialists had initiated studies in this area, expecting that the first spaceship would be a high-performing experimental aircraft that would ride to orbit atop a big rocket. One plan called specifically for an uprated X-15 in this role. But it would take a long time to develop, and after October 1957, everyone was in a hurry. The question then became one of shooting an astronaut into space in the quickest possible way.

Von Braun, never shy in such matters, offered a concept called Project Adam, the first man in space. He wanted to put a spaceman into a sealed compartment within a nose cone and shoot it atop a Redstone, to an altitude of 150 miles. His friends backed him up, stating that Adam would "meet the U.S. Army requirement to improve the mobility and striking power of Army forces through large-scale transportation by troop-carrier missiles." But while the Redstone might fly, the argument didn't. Hugh Dryden, who had headed NACA since 1949, told the House Space Committee that "tossing a man up in the air and letting him come back is about the same technical value as the circus stunt of shooting a young lady from a cannon." Project Adam died during 1958, but the time would come, as Russia's challenge grew more pressing, when this circus stunt would become a fulcrum of hope.

The Air Force had bolder plans, which represented something of a turnabout. Prior to Sputnik, its generals had worked diligently to distinguish a nation's airspace, which was subject to its sovereign control, from outer space, where they hoped their reconnaissance satellites might fly freely. Now, having gained their point by establishing freedom of space as a matter of accepted practice, these same generals were eager to blur the distinction between air and space, so that the Air Force could lay claim to the latter. They were avidly promoting the use of the new word *aerospace*, with its suggestion that the two realms were part of an indivisible whole. And because many of these senior commanders had come up as pilots, they naturally expected that the first men in orbit would wear their service's blue uniform.

These hopes initially emphasized a concept called Man In Space Soonest (MISS), which gained the strong support of the new Vice Chief of Staff, the omnipresent Curtis LeMay. It drew on work at Convair, builder of the Atlas, which showed that this missile could place a small man-carrying capsule into a very low orbit, if all went well. The concept represented one more example of the marginal minimalism that was highly evident in those early days, but at least it held out the hope of quick results. In June 1958, General Schriever proposed a plan that would carry out this mission by April 1960, at a cost of $99 million.

NACA's own Langley Aeronautical Laboratory, near Norfolk, Virginia, represented a third center of interest in manned flight. It had taken an active role in using high-speed wind tunnels to develop specific shapes for missile nose cones, and the shaping of a recoverable capsule for an astronaut represented a natural extension of this work. The aerodynamicist Maxime Faget took the lead, seeking a form that would remain stable during atmosphere entry and minimizing the heating of the capsule while providing room for parachutes. The design that emerged featured a truncated cone with a blunt, gently curving heat shield. With modest changes, this basic shape would remain the standard in the United States until the advent of the space shuttle.

Adam, MISS, and Faget's capsule all went on to become major elements of the initial man-in-space effort, Project Mercury. The Air Force would have loved to run it, and it had the political clout to prevail over NASA, which was new and weak. But Ike himself decided that NASA would have it, and this decision says much about the true significance of manned space flight. For the plain fact, as blunt as a Mercury heat shield, was that the Air Force had no real reason to want to put astronauts into orbit.

If the cherished concept of "aerospace" had valid meaning, then Schriever might have been able to propose a manned reconnaissance satellite, a sort of orbiting U-2. In fact, even the earliest Rand Corporation studies had anticipated automated operation. As a result, Ike's decision turned out to be a mirror image of the decision of 1955, which had selected Vanguard over the Army's Orbiter to give cover to WS-117L.

The choice of Vanguard in 1955, and of NASA as manager of Mercury in 1958, both emphasized a peaceful and nonmilitary image for spaceflight. However, the 1955 decision had real merit, for it gave effective support to WS-117L, the needs of which were central. The 1958 decision had no such operational merit, for if the Air Force had no valid requirement for astronauts in orbit, NASA had even less.

But with the sputniks ascendant, few people in Washington were thinking in those terms during 1958. If Mercury carried little value in terms of what astronauts might actually do, it was enormously significant from a political standpoint. It might be hard-pressed to demonstrate American superiority in space, but at least it might offer reassurance. With space viewed as a matter of unquestioned importance, with perception in fact having become reality—public opinion gave reason enough for Eisenhower to act. He had not sought a space race and had hoped to avoid it, but events had forced his hand.

Astronauts, indeed, quickly emerged as the project's focus. Ike personally decided that they would be military test pilots with college degrees. Right at the outset, this meant that NASA's peaceful space pilots would be sheep-dipped, in much the same manner as those of the CIA's U-2s. In addition, this criterion immediately cut out some of the best, including Chuck Yeager, Bill Bridgeman, and Scott Crossfield.

Yeager had been twenty-four years old in 1947, when he broke the sound barrier. Now, less than a dozen years later, the frontier of aviation had advanced from flight in the atmosphere at Mach 1 to flight in orbit at Mach 25. Yeager was healthy enough; he was also young enough. But he'd never been to college, which left him out. Bridgeman had set speed and altitude records in the Douglas Skyrocket, living as a Malibu beach bum when he wasn't flying hot jets. But he was past forty, which made him too old. Crossfield, for his part, had been the first pilot to reach Mach 2 and was currently training to fly the X-15. But he was a civilian, which struck him out as well.

Yet no one could qualify as a Mercury astronaut by being a shrinking violet. The world of an astronaut was full of life-threatening danger. In the Navy this meant catapult-assisted takeoffs from a carrier, with a catapult that might lose power too soon. Your plane then would fall over the bows and into the sea, with an onrushing hull close behind. There also were night carrier landings, as you approached a rolling and pitching flight deck in the dark. A signalman would stand near the stern, watching you and giving a waveoff with flashlights if you were not coming in properly. You would get a good look at that fellow from about five hundred feet away, and your plane would cover the distance in three seconds. If that wasn't enough for you—well, the Navy offered a good deal on life insurance.

Air Force fighters had their own quirks. You might be roaring down a runway toward takeoff, past the point where you could abort, and suddenly see a fire-warning light. With your jet loaded with inflammable kerosene, ready to explode at any second, you would pull a cord and a charge of nitroglycerine would blast your seat upward, allowing you to save yourself with your parachute. Unless you were flying the F-104A. Its ejection seat fired *downward.*

This ejection seat cost the life of one of the country's top test pilots, Iven Kincheloe. He had reached 126,000 feet in the X-2 in 1956, an altitude that gave a clear view from Baja California to San Francisco. But he ran out of luck two years later, immediately after takeoff. He tried to roll his F-104A so he could shoot up into the sky, but it didn't work. The Air Force named a base for him in Michigan.

On the other hand, the adroit use of an ejection seat might demonstrate that you had the presence of mind that could indeed mark you for the astronaut corps. Chuck Yeager remembers an exercise in a later version of the F-104, with its seats firing upward, in which the pilots David Scott and Mike Adams had to push the throttle to full power to keep from pancaking into the ground. The engine failed; Adams ejected from the backseat, but Scott stayed with the plane and rode it in.

"It was amazing to me," Yeager writes. "Both guys made a split second decision that was absolutely correct. And both were opposite courses of action." The impact pushed the engine into Adams's cockpit space; it would have killed him if he'd still been there. But Scott would have died if he *had* ejected, because of damage to his seat. Scott went on to walk on the moon. Adams later flew the X-15, though he died when it went out of control at Mach 5 and disintegrated.

The Mercury astronauts would carry the aura of this world, but NASA took good care to exhibit them as icons of Truth, Justice, and the

American Way. This treatment started at their first press conference, in April 1959. John Glenn, who would be first in orbit, stood up and gave a little speech about God, Motherhood, and Apple Pie—and he meant it. The press ate it up. "From a nation of 175 million they stepped forward last week," gushed *Time*, "seven men cut of the same stone as Columbus, Magellan, Daniel Boone, Orville and Wilbur Wright."

Though Mercury was portrayed in the media as an exercise in science and technology, public relations was usually close at hand. The seven men of Mercury immediately acquired a NASA press agent, Lieutenant Colonel John "Shorty" Powers, who would become known as "the eighth astronaut." This was only the beginning, for the project's public-affairs officer quickly decided he needed help and approached a prominent Washington attorney, Leo DeOrsey. DeOrsey knew about show business; among his clients were a number of celebrities, including Arthur Godfrey. He offered to sell magazine rights to the astronauts' personal stories and put together a deal with *Life* magazine that came to $500,000. This was even better than fame and glory, for the Mercury astronauts were government employees, GS-12 to GS-15, with starting salaries of $8,330 to $12,770. DeOrsey closed the deal early in August. *Life* then put the astronauts on the cover of the September 14 issue, with an eighteen-page spread and a headline, "Ready To Make History." A week later, their wives received similar treatment, with a color photo on the cover that had been retouched to remove every line and wrinkle, a spread that ran to fourteen pages, and a new head, "Seven Brave Women Behind the Astronauts."

Yeager wasn't impressed. He had made his name as a pilot rather than as a symbol, and during an appearance on behalf of the Air Force he remarked in his artless way that "a monkey's gonna make the first flight" in Mercury. You had to appreciate Yeager's world of flight test to grasp the full force of his comment, for in this world, many good human pilots never got within miles of the really hot jets. In fact, he was being too kind; the first flight would take place under automatic control, untouched even by nonhuman hands. But Yeager's comment was beside the point.

Hero worship was not a sidelight to Mercury nor an incidental byproduct. It was central to the project, and the reason was simple. The astronauts were meant to put human faces on the space program, a program that was a surrogate for the arms race and that symbolized the hope that America could stand up to the Soviet threat. Human psychology could not easily grasp these large matters when they were presented in abstract and impersonal terms; it needed people, both as symbols and as central figures. Ike symbolized the nation and government; General MacArthur, during the Pacific War, had represented the hope of victory. Similarly, the men of Mercury would stand for the hope of success against this latest challenge from overseas. They would inspire a vast outpouring of pride and admiration and would play their political roles well, even if they only went along for the ride. All any of them really would have to do would be to fly into space and come back safely.

But while these pilots were fulfilling their roles as symbols of space exploration, Korolev once again was offering the real thing. He was now preparing to undertake the most demanding mission yet, the one that would accomplish the next step in his program of lunar exploration. He would seek to photograph the unseen lunar farside.

This mission demanded a spacecraft with unusually advanced capabilities. It would have to follow a trajectory of high accuracy; Luna 1's near-miss wouldn't do. It would have to orient itself while behind the moon, finding it with a sensor and pointing its camera in the proper direction. The spacecraft then would have to operate an automated photo lab that would develop the film by washing it in chemicals, a refinement with obvious value for military reconnaissance and one that even the sophisticated Discoverers lacked. In response to a radioed command from a ground station, the craft then would turn on a TV camera, scan the developed photos, and transmit their images.

To develop the orientation system, Korolev turned to another longtime colleague, Boris Rauschenbach, who had been studying similar technical issues for some time. Rauschenbach started by borrowing a thousand rubles and sending one of his engineers to a local hobby shop, to load up on secondhand electronic components. Soon his staff was building initial versions of optical sensors, attitude-control jets with compressed nitrogen, gyroscopes, and electronic controls. This spacecraft was one of the first to use solar cells for electric power. During ground tests, a crane hoisted it on a rope as bright lights played upon it, and it sparkled like a glitter-ball at a dance.

Korolev proceeded to orchestrate a one-two punch at the moon. In September 1959, Khrushchev visited the United States for a summit meeting with Eisenhower and was eager for another propaganda victory. Korolev complied with his second successful moon shot, Luna 2. Moreover, he arranged for key announcements to come, not from Moscow, but from Britain's great radio telescope at Jodrell Bank, the world's largest such instrument. Jodrell's director, Sir Bernard Lovell, had an unsurpassed reputation that would add luster to this latest achievement.

The Soviets had been conducting their space program amid considerable secrecy, hiding their failures and brandishing successes with a flourish. For his part, Lovell had worked closely with the Air Force amid its preparation for flights of Pioneers, but he had met stony silence when he had sought to work with the Russians. But as soon as it was clear to the Soviets that this latest moon rocket had fired successfully, Moscow announced that it was on its way.

Lovell was about to leave for a cricket match when a newsman got hold of him and asked what he was going to do about this moon probe. He replied that he was going to play cricket; Luna 2 would not reach the moon for over thirty hours. He returned to his office after the game—and found a telex from Moscow, giving the data he would need to track the probe. It speeded up as it entered the lunar gravity, producing a noticeable Doppler shift in its transmitter's frequency. Then, with Lovell's great telescope pointing directly at the moon, the spacecraft struck near the crater Autolycus as its transmitter cut off sharply.

Luna 2's unusual accuracy of aim would lead Korolev's biographer, Yaroslav Golovanov, to compare the feat to shooting a bird from an airplane in flight. Lovell, meanwhile, had his own assessment: "Simply astonishing, and the mind just boggles."

Then in October, on the second anniversary of Sputnik 1, Korolev unleashed Luna 3. As the probe passed over the moon's south polar region, its sensors picked up the sun. The spacecraft then turned its bottom in that direction. It remained locked on the sun as the moon moved into view of another set of sensors, located on Luna 3's top. Responding to them, the craft now turned to face the moon, locking on it and shutting off its sun sensors. It then began to take a sequence of twenty-nine photos of the farside. The film ran through its automated processor, which Korolev's staff called the "laundromat" and the "public washroom." Next, with the film developed, the spacecraft stood by to scan the images with a facsimile system, at one thousand lines per frame, and to transmit them to a ground station at Simeiz in the Crimea.

Now, soaring high over the moon's north pole, the craft began a return to earth, along a carefully planned orbit that would keep it in view of this tracking station at all times. Its transmitter lacked the power to send a clear signal at lunar distances; it would do this only when close to earth. Hence, like any tourists, the staff would have to wait awhile to see how their photos turned out. However, this time the stakes were particularly high, for here was exploration in the classic sense. Like Magellan and Columbus, Korolev and his people would be the first to view this undiscovered world and to see its farside.

Though Luna 3 was still too far for successful transmission to take place, everyone was impatient and the controllers decided to make a try by sending up the appropriate commands. Aleksandr Kashits, a participant, recalls: "We were sitting in the dark control room, staring at the screen of a monitor. Again and again we were trying to see, or rather to guess, at least a hint of a picture." However, the screen showed nothing. Nor did it show anything during a second such session, and a third.

Korolev was staying in the Black Sea resort town of Oreanda, which had a beach and a pleasant park. He tried to cope with his impatience by taking frequent walks and sometimes tried to distract himself by talking with a friend. He knew that nothing he could do would bring the photos any faster, and he kept his temper when one associate warned him, "I can assure you that there will be no pictures. The radiation of space would destroy any image."

To reduce any radio interference, authorities shut the coast road to traffic and instructed passing ships to maintain radio silence. It didn't work; a fourth session also showed nothing. Then, with hope fading, everyone gathered for a fifth attempt. This time, slowly, a lunar disk appeared on the screen. Soon the first test print came in from the Simeiz photo lab. As if in a trance, Korolev took it in his hands and said, "Well, what do we have here?" It was foggy and indistinct. "At least we know now that the farside is round also," a colleague remarked. Another man offered quick reassurance: "Don't worry. We'll add some filters and remove distortions."

They received two photos during that session. The signal from Luna 3 soon grew even stronger as it approached more closely, and the researchers wound up with seventeen usable images out of the twenty-nine. Korolev then invited Kashits to personally bring the photos to Moscow for further processing. Kashits departed several hours later, the lone passenger in an airliner, carrying the prints in a sealed envelope.

These would not provide a definitive photo atlas of the farside; such an atlas would await the much sharper photos of America's Lunar Orbiter spacecraft during 1966 to 1967. But the images were quite adequate for the first good map that noted principal craters and other features so that they could receive names. In this moment of triumph, Soviet scientists picked names with a strong international flavor. They honored their own, of course, with such choices as Tsiolkovsky, Lobachevsky, Lomonosov, and Mendeleyev. But they also selected non-Russian names: Maxwell, Edison, Giordano Bruno, Jules Verne, and Pasteur. A commission of the International Astronomical Union formally approved these designations, which then entered permanently into the lexicon of lunar geography.

But as Luna 3 rounded the moon, the United States was moving vigorously to challenge the Soviet monopoly on big rockets. The earliest moves dated to 1955, when the Air Force arranged for Rocketdyne to study the feasibility of a new engine, the E-1, with thrust of 300,000 to 400,000 pounds. Rocketdyne, soon declaring that it could do even better, proposed to reach a full million pounds. The Air Force came up with additional funds; by 1957 Rocketdyne had a full, detailed analysis of this bigger engine and was beginning to fabricate major components. The work received another boost in June 1958, when Wright-Patterson Air Force Base awarded the company a development contract. Uprated with even more thrust, this engine, the F-1, would launch astronauts toward the moon.

Wernher von Braun was also active. Secretary Wilson's decision in November 1956 took the Army out of the task of deploying long-range missiles, but it had a loophole: The decision said nothing about space—nothing about the large-scale manned missions that von Braun had long hoped to carry out. Thirty years earlier, Germany's Karl Becker had taken advantage of a similar loophole in the Treaty of Versailles, which had banned tanks and heavy artillery but had overlooked the rocket, and von Braun had been Becker's apt pupil. In April 1957 von Braun initiated studies of a clustered booster that would use four E-1 engines to achieve 1.5 million pounds of thrust. He called it the Super-Jupiter but later renamed it Saturn, the next planet out.

The pace of the project quickened during 1958, as the Advanced Research Projects Agency argued that Saturn should rely on existing engines, rather than wait for the unproven E-1. Von Braun responded to this directive with a new design that promised 1.5 million pounds by using eight Jupiter engines, an arrangement that would cut costs while speeding up the program. In mid-August, ARPA awarded a contract to initiate this rocket's development. A month later, General Medaris gave a separate contract to Rocketdyne to uprate the engines from their standard thrust of 150,000 pounds, first to 165,000 and then to 188,000. He also modified a test stand to take a complete Saturn.

Speed was of the essence, and the booster's immense tanks would not be easy to fabricate. However, Saturn would use a cluster of engines; why not a cluster of tanks? Both Redstone and Jupiter propellant tanks were available, and the designers put eight of the former around one of the latter. This was the Jupiter-C all over again, on a far greater scale, with tankage and engines lashed together in much the same way that von Braun had clustered existing small solid-fuel rockets atop his standard Redstone to launch his first satellites.

A third path toward advanced rockets involved the use of hydrogen as a high-performance fuel in upper stages. Hydrogen was not easy to liquefy, for it was much colder than liquid oxygen, and it evaporated very readily. But in important respects it was even safer to handle than gasoline, for it tended to rise into the air when released, rather than form an explosive mixture that would stay close to the ground. It also was a superb coolant. At NACA's Lewis Propulsion Laboratory, John Sloop was developing a regeneratively cooled engine of 20,000 pounds of thrust, and he would run it successfully on a test stand.

Another center of interest in hydrogen engines was the engine-building firm of Pratt & Whitney. Jet engines for aircraft were its stock-in-trade, and it was learning about hydrogen by building a turbojet that would use this fuel to power a proposed spy plane that could replace the U-2. In 1956 the company's chief engineer, Perry Pratt, decided to try to win new business in the burgeoning field of rockets. He knew his firm was a latecomer and faced such established competitors as Rocketdyne, Aerojet, and Reaction Motors. But he thought that Pratt & Whitney could make headway by working with new propellants. He brought in a rocket specialist, Branson Smith, who soon put hydrogen at the top of his list of research topics.

For over a year, Smith and Pratt tried to win Air Force interest, but they didn't get very far. Then came Sputnik—and new hope. At Convair a longtime colleague of von Braun, Krafft Ehricke, seized the moment by offering a design for a hydrogen-fueled upper stage that could ride atop his company's Atlas and carry heavy payloads to orbit. This stage, called Centaur, quickly emerged as a focus for Smith's hydrogen hopes. ARPA was supportive, as was the Air Force. In August 1958, ARPA directed that service to develop Centaur, using a hydrogen engine from Pratt & Whitney.

Hence, during the summer of 1958, the Army and Air Force were proceeding with Saturn and Centaur, while Rocketdyne had the F-1 as the prelude to a new generation of rocket engines. On July 29, Eisenhower signed the act creating NASA. The director of NACA, Hugh Dryden, was already deeply involved in planning that would shift NACA's emphasis from aeronautics to space flight. To take charge of this planning, Dryden brought to Washington Abe Silverstein, the associate director of the Lewis propulsion center. Drawing on his background in innovative engines, Silverstein went on to take over the new projects, bring von Braun's group into the NASA fold, and lead this agency into a new era of large rockets and far-reaching visions.

Sloop, who had worked closely with Silverstein, describes him as "sharp, imaginative, aggressive and decisive. He was a hard bargainer at the conference table but very warm-hearted in personal relationships. He could cast work aside like a cloak and radiate such warmth and empathy for people that those who had felt his lash in a technical discussion earlier could forget their chagrin and respond to him with equal warmth. Many damned his ways but liked the person. Sometimes he displayed a near-mania for winning the argument, especially on rare occasions when it became rather obvious that he was on the wrong side." Still, "he had an uncanny technical intuition for the right approach, and he never followed a bad argument with a bad decision."[3]

Dryden, Silverstein's boss, was a renowned aerodynamicist but had little clout in Washington. Ike now showed his commitment to a strong NASA by passing over Dryden in choosing its administrator. He allowed Dryden to have the No. 2 spot, but he followed a recommendation from James Killian in filling the top job. It went to Keith Glennan, a former member of the Atomic Energy Commission. NASA formally began operation on

October 1, and Silverstein, Dryden, and Glennan proceeded to shop for new projects and existing organizations.

Rocketdyne's F-1 was an early prize. Its Air Force contract carried funding for only a few months, for this project had been marked from the outset for transfer to NASA. Silverstein then went the Air Force one better by calling for a thrust of 1.5 million pounds, which would give this single engine the power of the eight-engine Saturn. The contract went through in January 1959, providing for full-scale development.

In turn, the F-1 held out new prospects for von Braun's pursuit of clustered boosters. He had drawn on that approach in his 1952 *Collier's* articles, in which he proposed a mega-rocket with fifty-one engines in the first stage and twenty-eight million pounds of takeoff thrust. Now, with the F-1, he could look ahead to building Nova, a booster with similar characteristics. Nova would feature a cluster of F-1s; if von Braun used twelve, then he would get a thrust of eighteen million pounds. The complete rocket would tower to a height of several hundred feet, comparable to that of the Washington Monument. It would carry astronauts to land on the moon.

This vision took on new life in March 1959, as Rocketdyne carried out an initial test firing of the F-1 in its million-pound version. This was not a true rocket engine; instead it was a thrust chamber, lacking turbopumps and fed from pressurized propellant tanks on the test stand. It had a heavy solid-wall construction, and it fired for only two-tenths of a second, just enough to demonstrate ignition and stable combustion. Still, there it was: a rocket motor foreshadowing Nova, with more thrust than Korolev's entire R-7.

NASA meanwhile was making a bid for von Braun's rocket group as well, but the Army said no, thereby forcing Glennan to settle for control of Jet Propulsion Laboratory. Anticipating a major role in space, the Army was hoping to exploit to the full the loophole in Secretary Wilson's 1956 decision. That decision still stood; Huntsville had indeed won the right to proceed with Jupiter IRBM development, but the Air Force still had full operational responsibility for all long-range missiles, and when Jupiter was ready for deployment, it would go over to the Strategic Air Command. But Saturn still might remain the Army's own.

However, NASA had better luck when it moved to take Centaur from the Air Force. It did not win the project outright—the Air Force would continue to participate. But in mid-1959, Centaur formally went over from the Advanced Research Projects Agency to NASA. At Pratt & Whitney, Centaur's engine fired successfully during an initial test in August, as a step toward a rated thrust of 15,000 pounds. This complemented the work of John Sloop at NASA-Lewis, whose 20,000-pound engine, tested repeatedly during 1959, set a standard and showed outstanding performance.

Less than two years into the space age, the nation already could look ahead to a family of boosters that could launch everything from modest-size satellites to a manned lunar mission. Thor-Able and Thor-Agena were already in service. Atlas-Agena would soon be available; Atlas-Centaur, which would lift much heavier loads, lay a few years ahead. Saturn, slated for the mid-1960s, would carry entire crews of astronauts into orbit, perhaps to a space station, while Nova promised a ticket to the moon. And while the full development of this family would require at least a decade, it was already none too soon to set goals that would reach beyond Project Mercury.

In the spring of 1959, Glennan set up a committee headed by Henry Goett, who later became head of the new Goddard center, and gave this group the task of proposing a far-reaching manned program for the 1960s. Alfred Eggers, a committee member, proposed that NASA's next goal should be a two-man mission that would loop around the moon and return to earth. A colleague warned against "setting our sights too low." Max Faget of the Mercury capsule, along with George Low, one of Silverstein's close associates, urged NASA to aim for nothing less than a manned lunar landing, as a step toward flight to Mars. Late in June, the committee heartily endorsed this goal. In the words of Goett, "A primary reason for this choice was the fact that it represented a truly end objective which was self-justifying and did not have to be supported on the basis that it led to a subsequently more useful end."

Only the president could actually make such a commitment; for the moment, Glennan was only contemplating long-range planning, and nothing more. During that summer, the gulf between plans and reality appeared ready to grow wider, as Saturn faced an imminent threat of cancellation. It was an Army project, after all, and there was a real question as to whether it could fill a valid military need. General Medaris had tried to help, telling a House committee: "I believe that the U.S. Army must make long-range plans for the transport of small combat teams by rocket. I also believe that cargo transport by rocket is economically feasible."

But there were plenty of nonbelievers, including Herbert York, who had been ARPA's chief scientist and who took the Pentagon's top post for research and development early in 1959. General Medaris was seeking a hefty increase in the Saturn budget, but York said in response, "Nothing yet suggested by the military, even after trying hard for several years, indicated any genuine need for man in space." In June York told ARPA's Roy Johnson that he would not approve new funding for Saturn. In a later message to Johnson, he wrote: "I have decided to cancel the Saturn program on the grounds that there is no military justification." Johnson could find no reason to disagree.

NASA officials protested vigorously, and in September York met with Dryden. York took the initiative in proposing that NASA should take over the Saturn effort and receive von Braun's entire rocket group in the bargain. This idea drew widespread support, but two key people disagreed: Army Secretary Wilbur Brucker—and Wernher von Braun. Brucker had fought successfully to keep Huntsville within the Army and was outraged to see the issue raised anew. But Defense Secretary McElroy outranked him and supported the transfer. Von Braun wondered if NASA could support him in the style to which he hoped to become accustomed. Glennan met with him and eased his concern. In November, Eisenhower agreed to the new arrangements.

Redstone Arsenal retained its active involvement in battlefield missiles, but the whole of von Braun's operation donned civilian dress to become NASA's Marshall Space Flight Center. Von Braun himself would now report to officials in mufti for the first time since joining the Reichswehr in 1932. NASA had already absorbed the Navy's space program, which centered on Vanguard; now the Air Force would run the only independent military space program, alongside the civilian effort of NASA. In turn, that service would handle its own space activities without going through ARPA as a middleman, and ARPA would fade in importance. Ike's decision effectively closed the loophole in Secretary Wilson's roles-and-missions decision of three years earlier, by banning the Army from space as Wilson had banned it from long-range missiles. With this development, America's space organizations assumed their permanent form.

Von Braun had focused on Saturn as a first stage and had left the issue of upper stages unresolved. Silverstein now promptly addressed this matter, putting together a panel that would offer formal recommendations to Glennan. Silverstein's experience at NACA-Lewis had made him a strong proponent of hydrogen; he had backed the work of John Sloop and had seen for himself the promise of Sloop's engines. Silverstein won over a skeptical von Braun, who had no experience with this fuel, and in mid-December he reported that Saturn's upper stages should all burn hydrogen, thus attaining the highest performance.

The pace quickened anew during 1960. A prototype of Saturn was already on a Huntsville test stand; in April it fired all eight engines, producing 1.3 million pounds of thrust—and a roar that people could hear a hundred miles away. Clearly, the empty desert was the only place for such power, and Rocketdyne found the vast spaces it needed at Edwards Air Force Base, in the Mojave. Here it erected an F-1 test stand of pharaonic proportions—250 feet high, with solid slabs of concrete supporting a flame deflector.

In addition, NASA followed up Silverstein's report by awarding contracts for Saturn's upper stages. In May, Glennan decided that Douglas

Aircraft would build the second stage, designated S-IV. It too would follow the cluster principle, using six Centaur engines for 90,000 pounds of thrust. Still, this was only an interim design, for Silverstein's panel had also recommended an entirely new hydrogen engine, with particularly high thrust. That contract went to Rocketdyne in June, with language that specified "maximum safety for manned flight." This engine, the J-2, showed anew the reach of NASA's ambition. Intended for Saturn's second stage, its 200,000 pounds of thrust would give more power than the Thor or Jupiter engine.

The same year also saw considerable activity among potential contractors, as they studied designs of manned spacecraft for lunar missions. The initial emphasis called for sending three astronauts to loop the moon, a requirement that Saturn could accommodate and that would not demand the far larger Nova. Late in July, some thirteen hundred people gathered for a NASA-industry planning session. Silverstein, who had proposed the name Mercury for the initial manned program, now had a suggestion for the new one. At the opening of the conference, Dryden announced that "the next spacecraft beyond Mercury will be called Apollo."

Eisenhower had already given Saturn his blessing. Indeed, he had authorized it to receive high priority, having directed Glennan "to accelerate the super booster program." But as Glennan prepared to present the question of Apollo, the issue of cost raised its head. Ike asked his science adviser, George Kistiakowsky, to prepare a study that would "clarify the goals, the missions and the costs" of NASA's proposed manned program. Kistiakowsky's panel worked closely with NASA officials and considered two projects: astronauts around the moon by 1970 and a manned lunar landing in 1975. The numbers that came back were appalling when compared with the cost of Mercury:

Project Mercury (Atlas booster): $350 million
Lunar orbit (Saturn): $8 billion
Lunar landing (Nova): $26 to $38 billion in addition

The Saturn effort alone would swallow much of the NASA budget, which was above $1 billion per year and rising. The full fifteen-year program was out of the question. Its potential total, $46 billion, topped the entire federal budget for so recent a year as 1951.*

Ike received this report at the White House in December, and asked for an explanation. Someone compared the lunar voyage to Columbus's discovery of America, for which Queen Isabella supposedly had pawned

*In today's economy, this would be like proposing to spend $700 billion on the space program. NASA's 1997 budget is $13.7 billion.

her diamonds. Eisenhower replied that he was "not about to hock his jewels" for Apollo. The historian John Logsdon describes the mood of the meeting as "almost sheer bewilderment—or certainly amusement—that anybody would consider such an undertaking. Somebody said, 'This won't satisfy everybody. When they finish this, they'll want to go to the planets.' There was a lot of laughter at that thought."[4]

Yet amid these visions of flight to the moon, Glennan and his Pentagon counterparts were also carrying out an unmanned program that showed both strength and vigor. It centered around five major themes and activities: military reconnaissance, weather observation, navigation, communications, and planetary science. During subsequent decades they would flourish like a sturdy tree, sending roots deep into communities of customers and of people who would rely on these services, blossoming with new branches as the communities came up with additional applications. In these areas, rather than in manned flight, the true significance of the space program would emerge.

Weather satellites became a reality during 1960, with NASA orbiting Tiros 1 (Television and Infra Red Observation Satellite) in early April. Originally an ARPA project, it grew out of an Air Force conclusion, in August 1957, that an orbiting TV camera would not provide the high resolution that was necessary for military reconnaissance. RCA, which had backed the approach, was undismayed; though its resolution wouldn't suit the military, it promised rich rewards for meteorology. NASA took over the project from ARPA in April 1959 and sent the spacecraft aloft on the first try, atop a Thor-Able.

For two and a half months it performed brilliantly, returning nearly 23,000 black-and-white images. It didn't last long enough to help during the hurricane season, and weathermen missed its photos when Hurricane Donna headed up the East Coast. Moreover, it was in a low orbit, not the geosynchronous orbit, 22,300 miles up, that would keep it stationary over the earth and offer panoramic views of an entire hemisphere. Such views would await the first Applications Technology Satellite, in 1966. But Tiros provided a vast improvement over the use of aircraft to track severe storms, and unlike the planes, it couldn't be grounded in bad weather.

Satellite navigation emerged as another initiative, growing out of the needs of Polaris. Its submarines would use inertial guidance to navigate while submerged, but their skippers had to correct for errors by updating position using known references. The means to do this weren't exactly clear at first, because it was part of a submariner's religion not to give away his position, as by surfacing. Star sights didn't work well; this approach amounted to trying to do celestial navigation through a periscope. In addition, Polaris subs would spend much time in the Norwegian Sea, where clouds often hid the stars from view.

Loran, a standard radio-navigation system, seemed promising, but it wouldn't work when a sub was well out to sea. Oceanographers also prepared undersea maps by using sonar; a skipper then could navigate from one well-marked feature to the next. But a sub could find the features only by using its own sonar, which could give away its location to nearby listeners. In addition, this method worked only within the limited areas that the Navy would map extensively.

The answer came from the physicist Frank McGuire at the Navy's Applied Physics Lab. He reasoned that just as one could determine a satellite's true location by making observations from the ground, so one could determine true location by receiving radio signals from a satellite in an accurately known orbit. He sold the idea to his boss, Richard Kershner, who pitched it successfully to Admiral Arleigh Burke, the chief of naval operations. The first such satellite, Transit 1-A, flopped in September 1959. But the following April, Transit 1-B reached orbit. It used a version of the Thor-Able that made its own contribution to rocketry: Its second stage restarted in space, thereby providing a useful new way to achieve an orbit of desired shape.

The first true communications satellites also went up during 1960. SCORE, sometime earlier, had performed a few tantalizing experiments, but the one that counted was Echo. It was an inflated balloon made of aluminized Mylar film, a hundred feet in diameter; when displayed within the interior of a naval blimp hangar, it looked like an enormous silvery beach ball. A NACA engineer, William O'Sullivan, had conceived it as a satellite that would measure density at the limit of the atmosphere by experiencing drag. John Pierce, director of communications research at Bell Labs, thought of a different use, for he knew he could bounce a radio signal off it and have the signal received on the other side of the country. When Echo 1 went up in August 1960, Pierce proceeded to do just that. And because Bell Labs was part of AT&T, this was the first satellite to serve the needs of private industry. It represented the first important move toward commercial enterprise in space, and its province, satellite communications, would become the predominant use of space for the world at large. These far-reaching prospects would flow from the leadership of Bell.

With a Ph.D. in electrical engineering from Caltech, Pierce had joined Bell Labs in 1936, as part of a picked group that included Dean Wooldridge as well as William Shockley, who went on to direct the invention of the transistor. Pierce worked on radar during the war; in 1944, visiting England, he met Rudolph Kompfner, an expatriate Austrian who had invented a highly sensitive amplifier, the traveling-wave tube. It could handle a broad range of frequencies, making it ideal for telephone links that would carry large numbers of simultaneous conversations, and

Pierce spent the next decade working with it. He not only built improved versions; he derived a theory for its operation that Kompfner described as "quite different from my own, and not only more elegant but far more powerful." In time, the traveling-wave tube would emerge as the key element in communications satellites.

Pierce first began thinking about these satellites in 1954. Talk of rockets and spaceflight was in the air; when he was invited to give a talk at an engineering meeting, he sat down and calculated the power requirements for satellite communications. Receiving a warm response, he expanded this initial work into a detailed technical paper, entitled "Orbital Radio Relays." It appeared early in 1955 in *Jet Propulsion,* the journal of the American Rocket Society.

Just then, Bell Labs was at the height of its power. Its scientists had invented the solar cell, along with new types of transistor, and were launching the age of solid-state electronics by devising ways to use silicon as a practical engineering material. The creative environment was unrivaled; the *Physical Review*, the nation's leading physics journal, was publishing more papers from Bell than from Caltech, Harvard, or Princeton. One head of research told *Fortune* magazine that it was easy to assess a staff scientist's value: "We just ask, is the field his; does he own it; did he write the book? Are other people thinking with his thoughts?" These standards fitted Pierce, for his book on his tube was viewed as definitive. In the words of Eugene Fubini, a senior Pentagon research director who counted both Pierce and Kompfner among his friends: "Rudi invented the traveling-wave tube, and John discovered it."

At Bell, a specialty of the house was to put together small research groups that could make large contributions. The transistor had come out of such a collaboration. In addition, the lab's management preferred a step-by-step approach, declining to take risky leaps into the unknown. The Echo satellite suited both these policies. Because NASA would build it, it would leave Bell free to emphasize work on the ground stations with their sensitive amplifiers, an area that played to this lab's strengths. About a dozen people handled this work, all close friends, and the total number of Bell people who worked on Echo never exceeded forty.

Still, Echo really wasn't right for practical communications. Its receiver captured only one part in a billion billion of the transmitted signal. To cope with this limitation, with what Pierce would describe as a 180-decibel loss in signal strength, Echo needed a research-quality radio telescope as a ground station, along with a maser, the most sensitive of amplifiers but one of the most delicate. The answer called for satellites that would not merely rely on passive reflection in the fashion of Echo, but would actively receive, amplify, and retransmit the signals.

Such spacecraft would rely on the traveling-wave tube, and Bell Labs was building Telstar, the first of them, though it would not reach orbit until 1962. But in October 1960 the Army launched its Courier 1-B satellite. Though it lacked Pierce's tube, its conventional electronics included five tape recorders, one for voice and four for teletype. It could transmit up to 800,000 words of text during the fourteen minutes that it was in view of a ground station. It failed after only seventeen days, but it yielded a valuable intermediate step.

Another exciting initiative emerged in 1960, as a prelude to sending unmanned missions to Venus and Mars. These spacecraft would have to operate for many months using solar power, while communicating with earth across tens of millions of miles. The first probe that did this was Pioneer 5, which flew in March. This mission was in the hands of the same group that had launched the Air Force's Pioneer 1: Dolph Thiel and George Mueller, Allen Donovan and Rube Mettler, Louis Dunn, George Gleghorn, Richard Booton, and all the others. These people, tempered by experience, would now try anew. And Sir Bernard Lovell would join in, using his immense radio telescope not only to track the spacecraft but to issue commands to it.

With a rush of fire, the rocket broke the clear morning sky, sending Pioneer 5 onto its odyssey. Weeks passed, and still its radio signals came clear and true. Student protestors started a civil-rights movement by sitting in at segregated Southern lunch counters; still the probe sailed on. Senator John Kennedy won primary elections in West Virginia and Wisconsin, setting himself on a path to the White House. The spacecraft continued on its course. Its transmitter carried only five watts of power, no more than in a Christmas-tree lightbulb; still Lovell continued to hear it clearly. At last, late in June, twenty-two million miles from home, it radioed its final message and vanished into the vast and yawning gulf of interplanetary space.

7

Afternoon in May

Kennedy Commits to the Moon

A RESTAURANT IN BOSTON, Locke-Ober's, is famous for its clam and oyster bar. During the 1950s its regulars included Jack and Robert Kennedy, who often stayed until closing time. Its maître d', who called himself Freddy, was a rocket enthusiast, and one evening he introduced them to another customer, MIT's Stark Draper. Freddy hoped that Draper might persuade them to see the merits of the space program, but both Kennedys rejected the idea out of hand. Draper later recalled that JFK, then a senator, "could not be convinced that all rockets were not a waste of money, and space navigation even worse."

Kennedy had no reason to think differently. He represented Massachusetts, where major aerospace firms were conspicuous by their absence. Hugh Sidey of *Time* magazine, an experienced observer of presidents, wrote that among the issues he faced in the White House, he "probably knew and understood least about space." But the Democrats had made considerable headway by accusing Eisenhower of indifference in this area, and Kennedy knew a winning issue when he saw it. During his presidential campaign of 1960, he took the topic of spaceflight and made it his own. He returned to it again and again in his speeches, as he charged that Republican policies had left the nation weak in the face of this challenge. And while he warned of the seriousness of the challenge, Sergei Korolev was making it more serious yet.

Since 1957, Korolev had worked to develop military boosters and spacecraft, qualifying them for service in flight tests that would seek the best possible propaganda. He had done this brilliantly with his Luna missions, which had proven the flightworthiness of his R-7 with Semyon Kosberg's upper stage. The stage's initial model had been something of a rush job, but Kosberg had built an improved version with greater thrust. With it, Korolev could prepare to launch his first reconnaissance satellites, called Zenit (zenith).

Like Discoverer, these satellites were to photograph large areas with high resolution. Each flight would cover the entire area of the United States, and while Zenit couldn't show golf balls on the presidential putting green, its photos were sharp enough to allow intelligence analysts to count the number of cars in a parking lot. In addition, Zenit resembled Discoverer by relying on a capsule that would return to earth, thereby permitting physical recovery. The program thus posed the same problem that vexed Richard Bissell: how to develop the means to execute this return reliably. Moreover, Korolev had to solve the problem before Zenit could enter operational service with the Defense Ministry, and he had to make sure that its capsules would come down in the right place. It wouldn't do to build a counterpart of Discoverer 2, which wound up in Soviet hands.

But Korolev had an important advantage, because Zenit would be much larger than Discoverer. This meant that a modified version could carry dogs—and a cosmonaut. The Discoverer capsule was so small that it could carry nothing larger than mice. But with Zenit, Korolev could put the first man into space, scoring a triumph of the first magnitude. This manned version was given the name Vostok (east).

After 1958, Korolev pursued development of a common design for both Vostok and Zenit. In the same way that the R-7 operated as an all-purpose launcher, so would this single spacecraft offer the means to carry out both manned-flight and reconnaissance missions. The magazine *Priroda* (Nature) reported in 1993, following declassification, that "if at some exhibit the two space vehicles—the widely known Vostok and the Zenit, which has never been put on display—were put side by side, the visitors, without glancing inside the compartments, most likely would not see the difference."[1]

Their most characteristic feature was a spherical re-entry capsule. Like NACA's Allen and Eggers, Korolev's designers appreciated the fact that such a craft needed a blunt rather than a needle-nose shape. At NACA, the aerodynamicist Max Faget had carried out tests at hypersonic speeds in a wind tunnel, thereby demonstrating that the blunt-faced cone of the Mercury capsule could remain stable during atmosphere entry. Korolev lacked the facilities to carry out such studies, which meant that any shape chosen for his capsule might cause it to tumble on the way in. His sphere would tumble freely, and while this would cause discomfort to a cosmonaut, it wouldn't hurt a reel of film.

The first test version of Vostok reached orbit in May 1960. It carried neither dogs nor camera and lacked thermal protection; its capsule was to burn up in the atmosphere. But its orientation system failed, and like Discoverer 5, its retro-rocket fired in the wrong direction, placing the capsule into a new and higher orbit.

Soviet sources nevertheless discussed the spacecraft in some detail, describing it as a prototype of a manned craft, with a weight of five tons. They freely admitted that they had tried to achieve re-entry but had failed, but this had caused no harm. No person was on board, only a dummy. This open disclosure caught many people by surprise, and Senator Jackson and General Medaris declared that this spacecraft indeed was carrying a man. The London *Evening Standard* ran a cartoon showing a cosmonaut sending a radio message: "Dummy to base, dummy to base—what's this about not coming back?" But Moscow was telling the truth, and it had reasons for doing so. It could score propaganda points merely by showing that it had the means to put a man in space, even without yet doing it. This in turn would add to the pressure on the Americans.

Late in July, a flight exploded and killed two dogs, named Chaika and Lisichka. The next flight of Vostok, in mid-August, demonstrated anew an uncanny parallelism between the Soviet and American programs. These nations' governments had given high-level approval to their ICBMs, R-7 and Atlas, during the same week in May 1954. Their first test flights had been less than four weeks apart, in the spring of 1957. A year and a half later, Korolev and General Schriever launched lunar missions on virtually the same day. Now, with Discoverer 13 and Discoverer 14 having achieved successful capsule recoveries during the previous week, the second Vostok flight matched this accomplishment, returning just one day after Discoverer 14. Aboard the Soviet craft rode two dogs named Belka and Strelka (Squirrel and Arrow), as an onboard TV camera sent back pictures. One American scientist grumbled, "Next thing you know, they'll have a soccer team and a Mack truck up there." Jack Kennedy warned, "The first canine passengers in space who safely returned were named Strelka and Belka, not Rover or Fido."

Encouraged, Soviet officials agreed to launch a man by the end of the year. Another test flight took place in December, but the craft burned up during re-entry and its two dogs, Pchelka and Mushka, were lost. Later in the month, another launch failed to reach orbit, but its capsule came down in Siberia and its dogs were safely rescued. Russia also opened a new front in February 1961, with the first mission to Venus. Moscow released diagrams of this craft, showing that it was considerably more advanced than Pioneer 5. It also had a more demanding mission, for it was to approach Venus and then fly past it. Its transmitter failed at 4.7 million miles, but American scientists nevertheless were ready to hail the attempt. "When I heard about the Soviets' troubles," said a member of the Pioneer team, "my split-second reaction was one of satisfaction that we did better. Then, I felt ashamed of my reaction."

Vostok still wasn't ready for a man, so March blew in with two more test flights, both carrying dogs—first Chernushka (Blackie), then Zvezdochka (Little Star). Both returned successfully. This spacecraft now had a record of four recoveries in five attempts, including complete success on the last two missions. The launch of the first cosmonaut could not be far off. By then JFK was in the White House, and he would have to deal with the consequences.

The world knows that Kennedy accepted the commitment that Ike had spurned only in December: He chose to approve Apollo and to send men to the moon. This was one of the more noteworthy turnabouts in Cold War policy making, and it had reasons. The decision stemmed in part from an ongoing broadening of federal involvement with technology. It also drew strongly on a common attitude: that the most important new technologies would not result from work in the private sector, but rather would derive from federal initiatives.

The basic federal program in technology featured public works and had begun during the nineteenth century. Lincoln built the first transcontinental railroad, not through appropriations, but by awarding generous land grants to the builders of the Central Pacific and Union Pacific. Other railways then spanned the nation after receiving similar grants. Indeed, in the world of railroads, private-sector funding was so unusual that the magnate James J. Hill became famous for relying on it, spurning federal support when he built his Great Northern.

Next came the Panama Canal, a federal project that predated Apollo by over half a century but that nonetheless showed some interesting similarities. It drew on recent military events, for during the Spanish-American War, the Navy had lacked a way to move its battleships rapidly between the oceans. There also was a challenge from overseas, for France had tried and failed to dig such a canal; the project thus became a test of American mettle. Panama and Apollo both served as a focus for the nation's best energies and skills. They both took a decade to reach completion, with the work proceeding under three presidents and both parties.

The Depression, twenty years later, saw an era of dams and water projects: the Tennessee Valley Authority, Hoover Dam, Grand Coulee. Then in the mid-1950s, Eisenhower launched the biggest public-works program of all: the interstate highway system. It demonstrated dramatically the scale of activity that Washington was now prepared to pursue. It also showed how such an initiative could change people's lives for the better.

The federal government had begun building roadways as early as 1916, and Congress passed a major highway bill in 1944. These funds built such routes as U.S. 1, along the East Coast. Others included the legendary Route 66 from Chicago to Los Angeles, Highway 61 northward from New Orleans, and the transcontinental Lincoln Highway, U.S. 30. However, these certainly were not interstates. They had narrow lanes separated only by a painted yellow stripe and were often the site of head-on collisions, particularly on curves during rainy nights. These highways also saw rear-end collisions because the cars of this era lacked turn signals; drivers relied on hand signals to suggest their intentions. Traffic could enter and leave at will on connecting driveways, and motorists often were rear-ended when they slowed suddenly to turn into a roadside stop.

The first true interstate highways stemmed from the work of state governments. State officials preferred tollways, to be built with funds from state-issued bonds, with tolls from motorists covering principal and interest. Pennsylvania led the way; its Turnpike spanned the state by 1951. However, the route ran largely through thinly settled farmland and hill country. By contrast, the New Jersey Turnpike ran through the heart of the populous Northeast. This project set the pace. The state's bond issue dated to 1950; the route opened early in 1952 and scored an immediate success. It cut the time to drive its length—New York City to Delaware—from five hours to two. It eased wear and tear on autos and drivers alike. Within months, delighted turnpike officials found that traffic and revenue both were well above expectations. This success strongly encouraged tollway advocates in other states.

In June 1952, with New Jersey's achievement evident, Ohio sold a $326-million bond issue to finance its own turnpike. Indiana officials then declared that they would build their own route to link up with Ohio's. The New York Thruway, another toll road, also entered construction. For motorists, all this activity brought a dazzling prospect. Through a combination of routes completed, abuilding, and planned, they soon could drive from New England to Chicago without a stoplight.

But even the proponents of state turnpikes knew that these routes could only be a small part of a total interstate-highway system. Because they were financed by bonds, they could be built only where traffic was heavy enough to generate the needed revenues. No one could doubt their value, for they not only saved time and spared lives but also offered great advantage to the trucking industry. Nevertheless, the limits of state action were already in sight.

At this point Eisenhower stepped in, telling advisers in 1954 that he wanted "a dramatic plan to get 50 billion dollars worth of self-liquidating highways under construction." The roads would pay for themselves through taxes paid by motorists. It certainly was no small sum; that year, the entire federal budget came to $68 billion. But the success of the toll roads attracted the necessary political support, which led to passage of the Federal-Aid Highway Act of 1956. The modern interstate-highway system stemmed from its enactment. As its freeways spread across the land, they showed vividly how this far-reaching federal program could change the way people lived and worked.

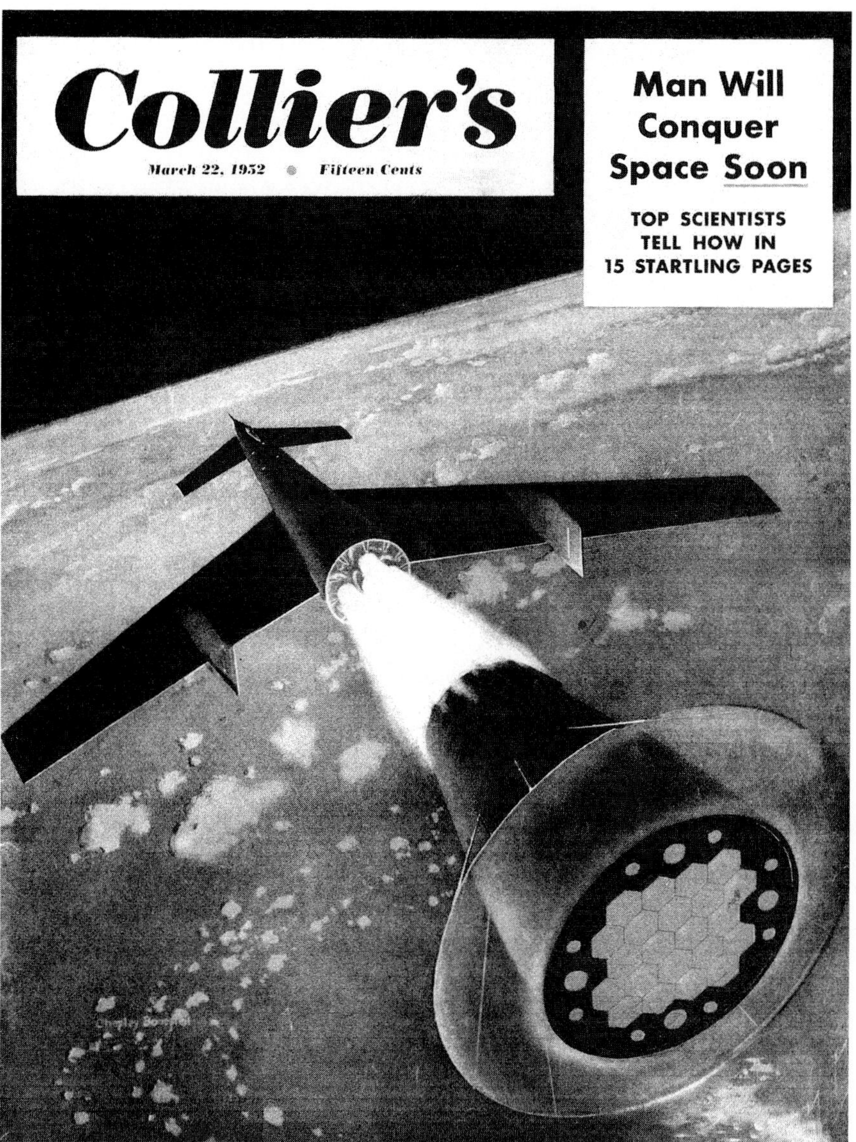

Between 1952 and 1954, *Collier's* magazine set an agenda for spaceflight that NASA follows to this day. (Courtesy Robert Kline collection)

William Bollay, America's top postwar rocket leader. (Courtesy Mrs. Jeanne Bollay)

Navaho missile, a springboard for early Air Force rocket development. (National Air & Space Museum, Smithsonian Institution, Photo No. 77-10905)

Rocket engine under test at Santa Susana. America's main line of liquid-rocket development stemmed from Bollay's work. (Rocketdyne)

General Bernard Schriever, who built the Air Force's big missiles. (U.S. Air Force)

Thor missile, the first in its era. (U.S. Air Force)

Soviet R-7 ICBM, which launched the Sputnik satellites of the Soviet Union. (Sovfoto)

Yuri Gagarin, the first man in space, *left,* with Sergei Korolev, the Soviets' leading rocket builder. (National Air & Space Museum, Smithsonian Institution, Photo No. 2B-09182)

Reconnaissance photo showing Soviet missiles in Cuba, 1962. (Central Intelligence Agency)

Satellite communications: Bell Labs's Echo 1, 1960. Radio signals, beamed from the ground, bounced off this satellite to reach a receiver thousands of miles away. (NASA)

Lunar exploration: Volcanic domes near crater Marius, photographed by Lunar Orbiter 2. (Boeing)

Rocket development: F-1 engine, which powered the Saturn V moon rocket. The F-1 developed 1.5 million pounds of thrust. (Rocketdyne)

Wernher von Braun, who built America's moon rockets. (National Air & Space Museum, Smithsonian Institution, Photo No. 2B-28636)

Liftoff of Apollo 8, the first manned mission to the moon. (NASA)

Earthrise from Apollo 11, which carried Neil Armstrong and Buzz Aldrin, the first astronauts to land on the moon. (NASA)

Buzz Aldrin on the moon. (NASA)

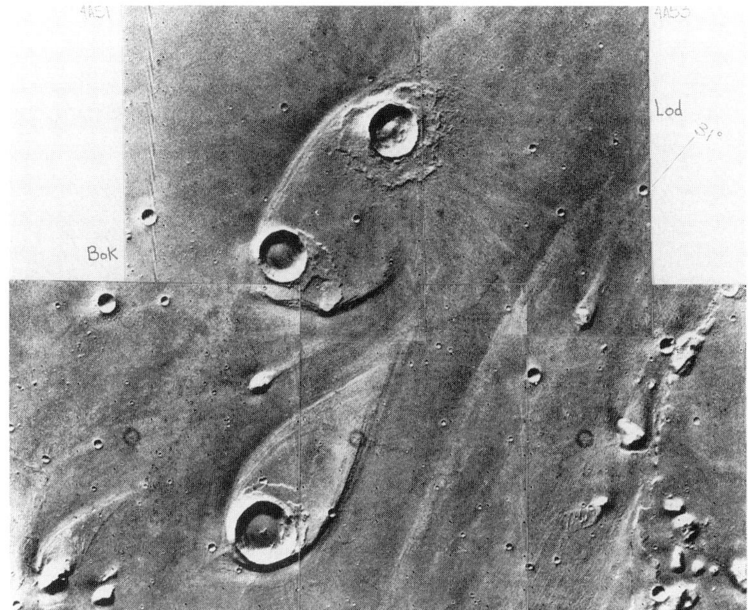

Mars: Islands formed by flows of water. (Jet Propulsion Laboratory)

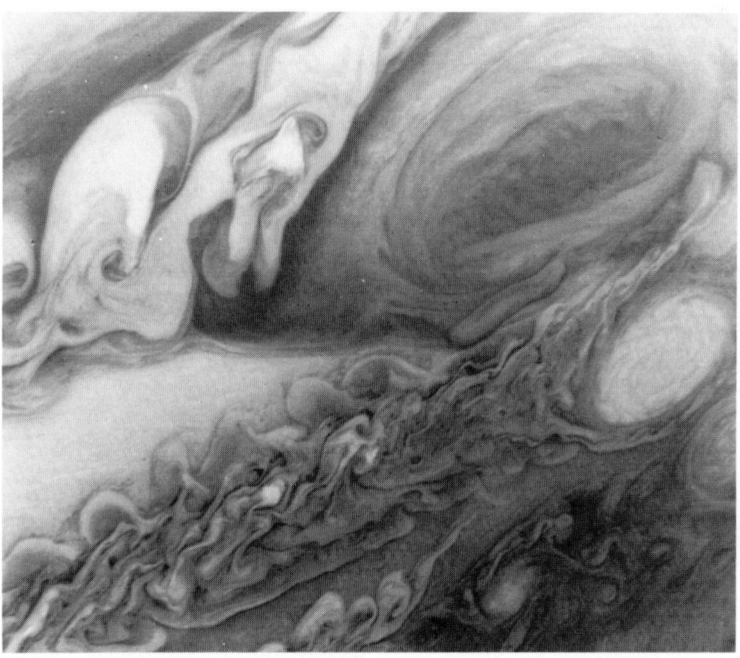

Jupiter: Great Red Spot and its vicinity. (Jet Propulsion Laboratory)

Soviet N-1 moon rocket. (RKK Energiya)

Space shuttle liftoff. (NASA)

Mir space station in 1995. (NASA)

Freeways spurred a massive move to the suburbs. Cities had long had their bedroom communities, whose residents commuted by train; commuter railroads predated the Civil War. The interstates emerged as corridors of development in their own right. Near Boston, Route 128 represented an initial portion of I-95. There were forty firms along its length in mid-1955, and over two hundred companies only three years later. Along the New York Thruway, a state report declared that construction of this highway had spurred at least $650 million in new development, including a major electronics center in Syracuse, built by General Electric. Company officials said that the Thruway would act as a pipeline, bringing in raw materials and taking finished products to the markets of the Northeast.

Like the electronics center in Syracuse, the new plants often were clean and nonpolluting, without the grimy smokestacks and noisome freight yards that had marked the industry of an earlier day. As a result, they coexisted happily with the new suburbs, and developers were quick to build more of both. Along New Jersey's Garden State Parkway, for instance, a New York syndicate set forth plans in 1959 to build 200,000 homes as well as industries and community facilities. These arrangements brought major social changes, for with jobs streaming outward from the central cities along the new routes, the cities would not continue as the nation's principal places where people worked and lived. In 1960 their outright decline still lay in the future. But by then it was already clear that the suburbs, served by freeways, would become the new centers of American life.

The success of the interstates showed dramatically what Washington could do, and it spurred hope that other federal initiatives would yield similar benefits. Another government activity helped to transform major military systems into products that would create new industries in the civilian market. These industries included jet airliners and commercial nuclear power.

The origin of America's big jets began with General Curtis LeMay, who headed the Strategic Air Command from 1948 to 1957, and who kept up a ceaseless demand for jet bombers of greater size, range, and striking power. The path to such aircraft traces to the development of a new jet engine, the J-57, by Pratt & Whitney. It had ten thousand pounds of thrust, a very high level at the time, and provided good fuel economy as well, for long range.

The firm of Boeing had built many of the nation's wartime bombers, including the B-29 that had defeated Japan. Its B-47 emerged as the most important of the early jet bombers, but LeMay wanted more. Boeing responded with the far more capable B-52, which used the J-57 engine. This company also took the lead in building tankers for aerial refueling, which could allow LeMay's aircraft to stretch their range at will. In 1949 an Air Force B-50 showed the promise of this technique by flying around the world nonstop. It could have carried an atomic bomb, and Stalin knew it.

But the early tankers used piston engines, making them an awkward fit for the new jet bombers that were much faster. In 1952 Boeing's president, William Allen, decided to spend $16 million in company funds to build a new jet, which he could offer to the Air Force as a tanker. In addition, he would try to sell the same plane as a jet airliner to Pan American World Airways and other carriers. Furthermore, he would build both versions of the plane in government-owned facilities, using government-furnished tooling and equipment.

This approach was bold, but it bore a mark of desperation, for Boeing had never built a commercially successful airliner. It had grown as a builder of warplanes, leaving the commercial market to its rivals, Lockheed and Douglas Aircraft. Donald Douglas, the founder of Douglas Aircraft and its CEO, had built most of the airliners then in use and had no need of a company-built prototype or a one-size-fits-all design. He knew that Allen's decision amounted to trying to enter the world of jet airliners while having LeMay pay the bills. Douglas would sit back, wait to see what his customers wanted, watch while Boeing painted itself into a corner by crafting a plane that would fit LeMay's needs rather than those of the airlines—and then step in with his own jetliner, a superior product that would win the market and leave Boeing to lick its wounds.

Douglas knew what he was doing. Boeing's prototype turned out to have a cabin that was too narrow for passenger comfort, along with range that was too short to avoid a stopover on transatlantic routes. By biding his time, Douglas had won a design for an airliner, the DC-8, with more powerful engines that could lift heavier fuel loads for true transatlantic range, along with a more comfortable cabin. But Allen was equal to this challenge. Willingly accepting the extra costs, he fitted his own jetliner design with a new and larger wing that would also hold more fuel. He also agreed to install the newer engines and widened his own cabin in the bargain. That did it; his airliner, the Boeing 707, went on to outsell the DC-8.

Pan Am flew its first transatlantic jet, a 707, from New York to Paris in October 1958. Other airlines soon followed with domestic service, and by 1960 the big jets were flying all over the country. The city of Seattle, Boeing's home base, gained new strength as the firm became a major builder of jetliners. Passengers flocked to the new jets, whose speed and comfort contrasted sharply with the harsh vibration and long flight times of the earlier piston airliners. In only a few years, major carriers sold the piston models to regional carriers and switched almost entirely to jets.

Moreover, although details such as range and cabin size were important to airline executives, the 707 and DC-8 both amounted to commercial variants of LeMay's tanker, the KC-135, using engines derived directly from the J-57. Similar developments took place within the electric-power industry, which had very little in common with commercial aviation. But here too the opportunity arose to turn military technology to civilian use, for the first power reactors developed from naval reactors such as that of the USS *Nautilus*.

Utility leaders initially had no interest in the atom. They were accustomed to generate their megawatts with hydro dams, or by burning coal or fuel oil under a boiler to raise steam. Atomic energy was classified at first, while by law only the government could own reactors. But during 1953 and 1954 the chairmen of the Atomic Energy Commission, Gordon Dean and Lewis Strauss, took the lead in pressing for declassification and for changes in the law. Strauss—who declared famously in 1954 that nuclear power would become too cheap to be worth metering—also built a power reactor at the AEC's Argonne National Laboratory near Chicago, to serve as a demonstration plant.

A separate action built the first true commercial plant. This initiative grew out of a partnership between Admiral Rickover, the AEC, and the utility industry, as the AEC contracted with Westinghouse to build the reactor. Duquesne Light Co., a Pittsburgh power company, built the plant at Shippingport, Pennsylvania, and agreed to operate it, while Rickover spurred the venture forward with his characteristic zeal. This plant was rated at only 60 megawatts, a relatively small size, but larger ones followed swiftly.

In 1955, to encourage other power companies to take the plunge, the AEC announced the first round of a Power Demonstration Reactor Program. It featured a package of subsidy and technical assistance to spur utilities to finance the construction of full-scale plants. Several firms took the bait, but two others—New York's Consolidated Edison and Chicago's Commonwealth Edison—announced that they would build plants entirely with private-sector funding, and with nuclear capacity up to three times that of Shippingport. They weren't about to look a gift horse in the mouth, but they wanted to gain direct experience as to the true costs of this new technology.

Shippingport entered service late in 1957, and as it built a record of operating experience, it served as a school for executives and engineering managers. Meanwhile, the reactor-building firms of Westinghouse, General Electric, and Babcock & Wilcox were aggressively promoting their products. Late in 1960, Southern California Edison launched a move toward truly large capacity as it announced that it would build a 375-megawatt installation south of Los Angeles, with help from the AEC. A few months later, San Francisco's Pacific Gas and Electric matched this with a plan to build its own 325-megawatt plant—without subsidy or federal assistance.

The real breakthrough was to come three years later, in 1964, when Jersey Central Power & Light declared that its planned Oyster Creek unit would actually produce power at less cost than a coal-fired plant. This announcement sparked a surge of activity among reactor salesmen, while coal executives tried to fight back by slashing their prices. It didn't work; during 1965, nuclear power came firmly to the forefront. For the next several years, it accounted for as much as half of the new capacity ordered by America's utility companies.

As early as 1960, there was plenty of reason to believe that nuclear power faced a bright future. It fitted the spirit of the times, in an era when people admired technology less critically than in subsequent years. It carried a sense of redemption, promising that the energy that had destroyed cities now might bring them power and light. No broad public condemnation yet existed of all things nuclear; people were willing to accept the claims of industry, and of the AEC, at face value.

In addition, nuclear power held out the hope of environmental advantages. The coal-fired plants of the day were notorious producers of fumes and soot, with few laws requiring them to clean up their smokestacks. Coal mining was known for its own forms of damage, despoiling land and polluting streams. It also had killed tens of thousands of men in mine accidents. The nation was caught up in a rapid expansion of its power industry, and if those megawatts were to come from coal, the cost in smoggy cities and ruined land and rivers would be high indeed. Rather than sticking to the coal-fired devil they knew, many executives were ready to switch to the nuclear alternative, at a time when the latter's costs and dangers were not well understood.

But as Kennedy came to the White House, these controversies lay well in the future. Most people saw government as the fountain from which new technologies would flow, with interstate highways, jet airliners, and nuclear power as prime examples. This attitude prevailed even in the field of electronics, for television had close links to radar, a military invention. Large electronics firms such as IBM, RCA, and AT&T were all private companies rather than arms of the government. But all had major Pentagon contracts, and they resembled regulated utilities rather than the Silicon Valley start-ups of the following decade.

Hence, in proposing a major new federal initiative in space, JFK knew his seeds would fall on fertile ground. People trusted government, and this feeling stemmed from personal experience. If you were a member of Kennedy's generation, you might have gone to college on the GI Bill. If you hadn't, and had gotten a job in a factory or with a big corporation, you probably were a union member, well aware that the Democrats had passed laws protecting your rights. Moreover, union wages were high enough to offer prospects that in earlier times had amounted to fantasy: a home in the suburbs and college for your children. And when you were ready to move from a city apartment to a home of your own, amid leafy lawns and good schools, the Federal Housing Administration was helping banks to offer thirty-year mortgages at five percent interest.

Kennedy was ready to offer even more. He was the first Democrat to succeed a Republican since FDR in 1933, and his advisers' minds were full of planned legislation. He would not always succeed; his bill to establish Medicare, one of the most basic reforms of the postwar era, went down to defeat in the Senate in July 1962. But JFK would not be shy in seeking new initiatives, and in his dealings with aerospace, Apollo would not be the only program of its kind. In a separate effort, during 1963, Kennedy decided to turn the Federal Aviation Administration into an aeronautical counterpart of NASA, and to have it lead the development of a commercial supersonic airliner, the SST or supersonic transport.

The SST's background involved another overseas challenge, this time from Britain and France. The British firm of de Havilland had actually built the world's first jet airliner, the Comet, and had placed it in service during 1952. During the following two years it garnered vast publicity and many sales, and it seemed that the stodgy British were leaping past the go-getting Yankees. Then these prospects literally fell apart during 1954, for the Comet proved prone to metal fatigue that caused the planes to blow up in flight. It took four years to fix the problem, and by then it was too late. The Americans had the Boeing 707 and Douglas DC-8, which were vastly superior.

After that, British aviation leaders nurtured a strong sense of having come very close and of having lost by a fluke. They were eager to try again, to challenge the United States anew, and they found an ally in Charles de Gaulle, the premier of France. His aviation industry was one of the world's strongest; its leading planebuilder, Marcel Dassault, was a pioneer in supersonic flight. And de Gaulle resented what he called "America's colonization of the sky." He wanted the world's best commercial airliners to carry the tricolor, and he was quite willing to join forces with Prime Minister Harold Macmillan to make it happen. The project that resulted was the Concorde.

The Concorde represented a challenge that Kennedy felt he had to meet. He decided that the United States would build its own SST, one that would be larger and faster than that of Europe. It was standard practice for new airliners to enter production entirely through private-sector financing, with airlines making advance payments and planebuilders raising capital through bonds or debentures. But in the early 1960s the world's airlines had their hands full with the subsonic jets, and an SST possessed technical risks that made the project unattractive on Wall Street. JFK thus decided that the government would take the lead, and he made the announcement in June 1963:

> It is my judgment that this Government should immediately commence a new program in partnership with private industry to develop at the earliest practical date the prototype of a commercially successful supersonic transport, superior to that being built in any other country in the world.[2]

The FAA would manage the SST's development while carrying up to 90 percent of the costs.

In studying the background of Apollo, one gains considerable insight by noting that during the years of Eisenhower and Kennedy, the FAA and NASA followed parallel paths of growth. The FAA grew out of the Civil Aeronautics Administration, a branch of the Commerce Department. Like NACA prior to Sputnik, the CAA's budget was low and its clout was meager, even though it had the important responsibility of ensuring the safety of commercial aviation.

In 1956 the CAA faced an event that was quite as dramatic as Sputnik, as two airliners collided in flight. The toll of 128 dead made this the worst disaster in aviation history and drove home the point that inadequate funding had made the airways unsafe. And just as Sputnik pointed to a looming threat from Soviet missiles, this 1956 accident pointed to an imminent danger that aviation soon would become even more unsafe. The new jetliners would soon enter service, and if the airways couldn't handle the fast piston-powered airliners of the day, they would face even greater trouble with the jets.

Congress responded to the crisis in its usual way, with a surge of funding, while Senator Mike Monroney, a strong supporter of aviation, did more. He introduced a bill, which passed during 1958, that transformed the CAA into the FAA, an independent agency that had a larger budget and much more influence. The bill amounted to a counterpart of the one that turned NACA into NASA. However, the responsibilities of both new agencies remained limited. The FAA would focus on aviation safety while NASA would concentrate on space activities in earth orbit.

Then came Kennedy, who greatly expanded their missions. NASA now would pursue Apollo, a far-reaching manned lunar project that Ike had rejected. The FAA, for its part, would move beyond its concern with radar and air safety to direct the SST effort as a big new program that would build the next generation of airliners. Like nuclear power and manned flight to the moon, the SST was one more federal initiative that seemed to augur bright hope during the years of JFK.

And in understanding Apollo, one must take note of this element of hope. Kennedy was youthful and as vigorous and spirited a man as ever sat in the Oval Office. He understood that by seizing the moment, by sending astronauts to the moon, he would establish a national legacy that the world would long remember. He was a serious student of history; he had written the Pulitzer Prize–winning *Profiles in Courage*. And like Winston Churchill, who was also a historian, he cherished the English language. He knew how to summon fellow citizens to large duties and challenges. He used Apollo as a motif, as during a speech at Rice University in Houston, in September 1962:

> The exploration of space will go ahead whether we join in it or not, and it is one of the great adventures of all time, and no nation which expects to be the leader of other nations can expect to stay behind in this race for space.
>
> For the eyes of the world now look into space—to the moon and to the planets beyond—and we have vowed that we shall not see it governed by a hostile flag of conquest, but by a banner of freedom and peace.
>
> We set sail on this new sea because there is new knowledge to be gained and new rights to be won, and they must be won and used for the progress of all people.
>
> But why, some say, the moon? Why choose this as our goal? And they may well ask, why climb the highest mountain? Why, thirty-five years ago, fly the Atlantic? Why does Rice play Texas?
>
> We choose to go to the moon! We choose to go to the moon in this decade, and do the other things, not because they are easy, but because they are hard. Because that goal will serve to organize and measure the best of our energies and skills. Because that challenge is one that we are willing to accept, one we are unwilling to postpone, and one we intend to win.[3]

Yet there was more to Apollo than bright hope; there was darkness as well. For JFK, like his party, carried a heavy burden. They governed in the shadow of the fall of China in 1949.

It is hard to overstate the dismay with which America faced the Communist threat of the postwar years. It was almost as if to say that our victory in the war was meaningless, that we had defeated Japan and Germany only to face the far greater power of Stalin. Truman had supported Chiang Kai-shek as a client, even though his Kuomintang regime was thoroughly corrupt and lacked public support. When Mao Zedong overthrew it, sending its remnants to seek refuge in Taiwan and proclaiming the People's Republic of China, Russia had just exploded an atomic bomb. "Let China sleep," Napoleon had written. "When she awakens, the world will be sorry." In America, these events strengthened developments that already were making the era one of the ugliest in the history of national politics.

Politics has always had its cut-and-thrust, its exaggerated appeals to fear. However, there ordinarily are limits, as when the opposition party finds ways to work with the president, rather than trying seriously to bring about impeachment. In times of great stress, however, the limits break down. That happened during Reconstruction, following the Civil War, as radical Republicans treated the defeated southern states like conquered provinces. It happened again early in the Cold War, when Republicans did not rest content with charging Truman with blunders. Instead they declared that his administration was riddled with Communists—that high officials had committed treason.

The threat from overseas left Truman determined to resist further Communist advances regardless of the cost. That is why he intervened immediately in Korea in June 1950. The nation would give no quarter to North Korea and cheered as General MacArthur advanced toward the Yalu River, the border with China itself. When China intervened, driving back the Americans, the world trembled momentarily on the brink of nuclear war, as Truman weighed a response that would have included the use of the atomic bomb.

Historians associate this era with Senator McCarthy, who recklessly destroyed the reputations of good and decent people by tarring them with the brush of Communism. Yet he was no more than an opportunist, feeding off the fears of his time. He did not create those fears; he merely exploited them for his own purpose. The Senate voted to censure him in 1954, amid an easing of tension following the death of Stalin and the end of the Korean War; he died in disgrace a few years later. But JFK knew that if the Communist threat revived, so might McCarthyism.

For his ghost lay unquietly buried. His close allies had included Congressman Richard Nixon, who had become known as a leading Red-baiter in his own right. It was Nixon who had unmasked Alger Hiss, a top official in the State Department and a close associate of Dean Acheson, the secretary of state. Nixon proved that Hiss had been a spy and a member of the Communist Party—and had lied about it. This event had provided McCarthy with his greatest victory, deepening and prolonging the virulent suspicions of the time. And Nixon had been Kennedy's opponent in the presidential election, coming within an eyelash of defeating him.

It was the proud boast of Eisenhower's Republicans that while Truman had lost Eastern Europe as well as China, they had held the line. They hadn't quite; in 1960 Castro's Cuba was Communist and was only ninety miles from Key West. Still, it was right in our backyard; we would deal with it. Nevertheless, Cuba underscored Kennedy's main challenge in foreign policy: to continue to hold the line, to deny Moscow and Beijing any further victories.

Yet under the shadow of China, JFK could not proceed with the calm confidence that had marked Eisenhower and his policies. Everyone knew that Truman's mistakes in Korea, which had trapped the nation in a conflict that we could neither win nor end, had brought about his defeat and had driven him from the White House. Everyone also knew that Ike had won the war in Europe and had ended the war in Korea, while preserving peace amid subsequent dangers in a perilous world. Living in that shadow, faced with this contrast, Kennedy's Democrats would find themselves driven to be more anti-Communist than the Republicans. In conducting foreign policy, they worked amid gnawing concern that they might prove to be weak, and they would compensate by becoming overly bold. The most important consequence was the war in Vietnam.

Ike had had his chance to intervene massively in that country, when the French faced defeat in their struggle against Ho Chi Minh in 1954. He had declined to do so and had left the French to their fate. But Vietnam was adjacent to China, in the one area of the world where further Communist advance was both most likely and most unacceptable. Nor did Kennedy and his advisers view South Vietnam in context, as a client of no great importance. They accepted the domino theory, which viewed South Vietnam as a linchpin: If it fell, the whole of Southeast Asia would soon go as well. It was like proposing that the loss of Saigon would bring about a new Greater East Asian Co-Prosperity Sphere, with China rather than Japan as the imperial power. But the Kennedy administration believed this and made it the basis for policy.

Vietnam was important in another respect: It was part of the Third World, the great arena where the Cold War might be won or lost. Here was a plethora of newly independent nations that saw hope in socialism, often disdained the West, and regarded Moscow fondly because it espoused socialism but had built no colonial empire overseas. The stakes were high even in Latin America, where any of twenty nations might become the next Cuba. They were equally high in Africa and Asia.

Accordingly, it was essential to deny Moscow propaganda victories as well as military ones. A prime topic for propaganda involved spaceflight, and it was easy to envision the chain of reasoning: Russia is ahead; this means it has the best weapons; hence, it has the best technology; hence, it has a superior society (which Third World leaders were all too inclined to believe in any case).

Waging total cold war, Kennedy would in no way concede that Moscow might concentrate resources into manned spaceflight for the sake of propaganda, while failing its citizens in a host of ways that were far more important. Time and again, during the campaign of 1960, Kennedy spoke of the Third World and emphasized that leadership in space was essential if America was not to forfeit its support:

> The people of the world respect achievement. For most of the twentieth century they admired American science and American education, which was second to none. But now they are not at all certain about which way the future

> lies. The first vehicle in outer space was called Sputnik, not Vanguard. The first country to place its national emblem on the moon was the Soviet Union, not the United States.
>
> If the Soviet Union was first in outer space, that is the most serious defeat the United States has suffered in many, many years. Because we failed to recognize the impact that being first in outer space would have, the impression began to move around the world that the Soviet Union was on the march, that it had definite goals, that it knew how to accomplish them, that it was moving and we were standing still. That is what we have to overcome, that psychological feeling in the world that the United States has reached maturity, that maybe our high noon has passed and that now we are going into the long, slow afternoon.[4]

To Kennedy, then, Apollo was many things. It was a federal technology program, in an era when people admired such programs and anticipated more. It represented a well-considered plan from one of his agencies that could address a pressing problem, and he was in the business of approving such plans. Apollo suited his sense of history, his view of the future, his spirit as a man. It also was very important in that it could deny Moscow further victories and help to hold the line in the Third World. Nor was this all, for the name Kennedy chose for his administration, the New Frontier, smacked strongly of spaceflight. And like any political leader, he knew how to tailor his message to his audience. JFK made the link explicit in a statement published in the trade magazine *Missiles and Rockets:* "This is the new age of exploration; space is our great New Frontier."

Kennedy took office in January 1961 and hit the ground running, amid a flurry of statements, decisions, and actions. From the outset he wanted the public to view him as vigorous and ready to move, and while space was not a matter of pressing urgency, he dealt with it, too. He directed his vice president, Lyndon Johnson, to take responsibility in this area, knowing that LBJ had been dealing with missiles and space since the first sputnik. He met with NASA officials and agreed to boost their budget so as to speed the development of Saturn. Still, on the whole, his early decisions tended to reinforce the caution of Ike rather than to point the nation toward new frontiers.

But the Soviets were on a roll and would not stop, even for the death of a cosmonaut. This occurred on March 23. The victim, Valentin Bondarenko, was twenty-four years old and the youngest of his country's spacefarers. As part of his training, he spent ten days isolated in an oxygen-filled pressure chamber. He was wearing a woolen training suit and had sensors attached to his skin for medical measurements. The chamber also contained an electric hot plate, with which he had prepared his meals.

With the ten days at an end, he began to remove the sensors, cleaning the places of attachment on his skin by wiping them with a pad of alcohol-soaked cotton. He tossed the cotton away without looking—and it landed on the hot plate, bursting into flame. The fire spread quickly in the oxygen-rich atmosphere of his small chamber, igniting his suit. He tried to put the fire out but failed, and it blazed out of control. Nor could he escape; a pressure difference kept his hatch sealed.

A doctor saw the flames through a porthole, but also could not open the hatch. He turned a valve to equalize the pressure, but it took time before he could reach Bondarenko, and during much of that time the cosmonaut was engulfed in flames. Even so, he was still conscious. He said, "It's my own fault; don't blame anyone else."

Medics rushed him to a hospital. "I couldn't help shuddering," the attending physician later recalled. "The body was totally denuded of skin, the head of hair; there were no eyes in the face. It was a total burn of the severest degree." Still Bondarenko continued to speak: "Too much pain—do something, please—to kill the pain." He lingered for eight hours, then finally died. He left a young widow and a five-year-old son. The minister of defense ordered that they were "to be provided with everything necessary, as befits the family of a cosmonaut." And that was that.

Then on April 12, a fine spring morning at Tyuratam, a bus pulled up to the launch pad. In it were two other cosmonauts, Yuri Gagarin and his backup, Gherman Titov, accompanied again by doctors. Gagarin got off and met Korolev, who greeted him with hugs and kisses; the two men then boarded an elevator to ride to the top of the rocket. Gagarin waved goodbye; then he disappeared behind the hatch of his spacecraft. "Yuri, you're not getting bored in there, are you?" a controller soon asked. "If there were some music, I could stand it a little better," he replied. Korolev ordered up some love songs, but it took several minutes to make them available. "That's the way musicians are," Korolev told Gagarin. "Now they're here, now they're there, but they don't do anything very fast."

Liftoff took place about fifty minutes later. "I heard the valves working," Gagarin reported following his flight. The engines then started with a roar. "The noise was approximately like the noise in an aircraft. I was prepared for much more. Then the rocket smoothly, lightly rose from its place."

"T plus seventy," said Korolev. Seventy seconds from launch.

Gagarin: I read you: 70. I feel excellent, I am continuing the flight, the g-load is increasing, all is well.
Korolev: T plus 100. How do you feel?
Gagarin: I feel fine. How about you?
Korolev: Velocity and time, all normal.

Gagarin described the forces during acceleration as "completely manageable, as in normal airplanes." He continued: "About five g's," five times the normal force of gravity. "At that g-load, I reported and communicated with the ground the whole time. It was somewhat difficult to talk, since all the muscles of my face were drawn. There was some strain."

At T + 150 the nose fairing separated, allowing Gagarin to look through an optical device.

Gagarin: I see the clouds. The landing site—it's beautiful, what beauty! How do you read me?

Korolev: We read you well, continue the flight.

"Suddenly there were no clouds," Gagarin recalled during postflight debriefing. "I saw the natural folds of the terrain, a region that was a little mountainous. I could see forests, rivers, ravines. I couldn't tell exactly where it was. I think it was the Ob or the Irtysh, but I could see it was a large river and there were islands in it." The rocket was still under thrust, but soon it shut down. He was in orbit.

"I could see the horizon, the stars, the sky," he later reported. "The sky was completely black, black. The magnitude of the stars and their brightness were a little clearer against that black background. I saw a very pretty horizon, and the curvature of the earth. The horizon is a pretty, light blue. At the very surface of the earth, a delicate light blue gradually darkens and changes into a violet hue that steadily changes to a black.

"In my flight over the sea, its surface appeared gray, and not light blue. The surface was uneven, like sand dunes in photographs. I ate and drank normally, I could eat and drink. I noticed no physiological difficulties. The feeling of weightlessness was somewhat unfamiliar compared with earth conditions. Here, you feel as if you were hanging in a horizontal position in straps."

The sunlight was very intense; he had to turn away from it or cover his eyes. Now it was twilight down below, and he was enthralled as he watched: "The change from the blue to the dark is very gradual and very lovely." But when his spacecraft flew into the earth's shadow, the transition into darkness was very abrupt.

He crossed the Pacific, curved above South America, then saw the glow of morning over the South Atlantic: "When I emerged from the shadow of the earth, the horizon looked different. There was a bright orange strip across it which again passed into a blue hue and once again into a dense black color."

Seventy-six minutes into the flight, he was over Africa and ready for re-entry, which took place under automatic control. He felt the braking rocket kick in, with "a small buzz and noise throughout the structure." The rocket cut off after forty seconds, with a sharp jolt—and the craft began to spin, turning through a complete rotation about once every twelve seconds.

"I was an entire 'corps de ballet,'" Gagarin reported. "Head, then feet, rotating rapidly. Everything was spinning around. First I see Africa, then the horizon, then the sky. I only barely managed to hide my eyes from the sun." The rotation continued, as he crossed the northern coast and saw the Sahara and Mediterranean. Here was a malfunction, as electric cabling held his capsule to the rest of the spacecraft. But ten minutes after retrofire, well into his atmosphere entry, the cable burned through and his capsule separated. Its tumbling diminished to a strong back-and-forth oscillation; then this motion came to a halt.

Gagarin had shut his viewport with a shade, but "a bright crimson light appeared along the edges of the shade. I felt the oscillations of the craft and the burning of the heat shield. It was audibly cracking; either the structure was crackling, or the heat shield was expanding as it heated. I felt that the temperature was high. Then the g-load began to steadily increase. It felt as if it was ten g's. There was a moment for about two or three seconds when the instruments began to become fuzzy. Everything seemed to go gray. I strained to see, and that helped, as if everything went back into its place."

The g-forces fell off, and he heard the whistling of air as his craft slowed through the sound barrier. At an altitude of 23,000 feet, the hatch blew off with a bang. His ejection seat fired—the heavy capsule lacked the large parachute that could have permitted a survivable landing—and he came down slowly as his own chute deployed. He saw the Volga and the city of Saratov. He thought he would land in a gully, but instead he came down in a field—"well plowed, very soft, and it hadn't even dried up yet. I did not even feel the landing. I didn't even realize that I was already standing on my feet. I saw that everything was intact. That meant I was alive and well."[5]

Yuri Gagarin, a pilot in the Soviet air force, had excellent health but little experience. He had been born in 1934 on a collective farm west of Moscow and had gone to trade school to become a skilled worker. However, his record as a student proved good enough to get him into a technical school in, of all places, Saratov. He learned to fly there, in a local aviation club, and qualified for the Air Force. He went on to an officer candidate school at Orenburg, graduating with top honors in 1957.

He married a medical technician, Valentina Ivanovna; they had two baby daughters, the younger being only a month old at the time of his flight. They lived in a modest apartment—bedroom-kitchen, living room, and bath—on the fifth floor of an apartment building reserved for military test pilots, and Gagarin commuted to his cosmonaut training in downtown Moscow. Following his flight, he showed that he knew how to say the right things: "While in outer space, I was thinking about our party and our homeland. When I was coming down, I sang the song, 'The Motherland Hears, the Motherland Knows.'"

In Washington it was Sputnik all over again. It didn't help when a news reporter, seeking a response from "Shorty" Powers, the spokesman for the astronauts, phoned him at three in the morning and woke him up. Powers replied, "If you want anything from us, you jerk, the answer is that we are all asleep."

Kennedy wasn't asleep, but he certainly could feel peeved. He called a press conference on the same afternoon. Someone asked, "Mr. President, a member of Congress said today he was tired of seeing the United States second to Russia in the space field. What is the prospect that we will catch up with Russia and perhaps surpass Russia in this field?" Kennedy replied, "However tired anybody may be, and no one is more tired than I am, it is a fact that it is going to take some time. The news will be worse before it is better. We are, I hope, going to go in other areas where we can be first, and which will bring perhaps more long-range benefits to mankind. But we are behind."

In recent weeks he had received three reports. His science adviser, Jerome Wiesner of MIT, had warned against too great an emphasis on Project Mercury, because it couldn't beat the Soviets and it carried the risk of killing an astronaut, perhaps by stranding him in orbit. A separate report from Trevor Gardner called for an opposite approach, recommending a major manned effort—under sponsorship of the Air Force, of course. The National Academy of Sciences weighed in with a third viewpoint, as a panel headed by Lloyd Berkner enthusiastically called for NASA to send men to the moon. Amid the discordant recommendations, there was a common theme: To win the space race, America could accept no less a goal than a manned lunar landing. The Soviet lead was just that great.

Two days later, on April 14, as the people of Moscow thronged Red Square to give their hero an uproarious welcome, Kennedy met with his advisers. He listened to them, muttering, "We may never catch up," and then recovered his spirit. "Now let's look at this," he said. "Is there any place we can catch them? What can we do? Can we go around the moon before them? Can we put a man on the moon before them? What about Rover and Nova?* When will Saturn be ready? Can we leapfrog?"

Dryden of NASA explained that their one hope was a crash program similar to the Manhattan Project. But such an effort might cost $40 billion, and even then it would offer no more than a fifty-fifty chance of winning.

*Rover was an experimental nuclear rocket, a project of the Atomic Energy Commission; see chapter 9.

Kennedy replied, "The cost, that's what gets me." He thought for a moment, then continued, "When we know more, I can decide if it's worth it or not. If somebody can just tell me how to catch up. Let's find somebody—anybody, I don't care if it's the janitor over there, if he knows how."

He stopped again for a moment, glancing from face to face. Then he added quietly, "There's nothing more important."

They had been meeting in the Cabinet Room of the White House. Now Kennedy thanked his men for coming over and asked Theodore Sorensen, a trusted associate, to join him in the Oval Office. They talked for about five minutes, and Sorensen came out. He said, "We are going to the moon."[6]

The president had made his decision intuitively, knowing the cost would be frightful but accepting that this challenge was one he had to face, then and there. He would need to learn much more from other colleagues, to be sure, but as the meeting broke up, he knew in his own mind what he wanted to do. But as the week progressed, he found his attention diverted by an entirely separate matter: a CIA-sponsored invasion of Cuba at that island's Bahia de Cochinos.

The invaders formed a force of Cuban exiles, émigrés whom Castro called *gusanos*, worms. Richard Bissell had planned the operation, running it in the personal style he used with Corona and the U-2 as he built a small plan for covert activity into an amphibious invasion, complete with air support. His CIA had undertaken a somewhat similar action in 1954, overthrowing a Marxist government in Guatemala. But Castro had learned from this, and had prepared Cuba to meet such an assault. Its failure would bring the fall not only of Bissell but of his boss Allen Dulles, the agency's director.

The hopes of Bissell, and Kennedy, rested on the view that Cuba was ripe for counterrevolution. Even a scratch force, gaining a beachhead, could proclaim a provisional government that would win American recognition, opening the way for a march on Havana. The CIA had few assets in Cuba (most of them were imprisoned or shot) and failed to appreciate a salient point: The people of that country would indeed fight for Castro and would rally to support him.

Castro's nation had been dirt-poor, illiterate, and sickly. He was building schools and clinics, and had expropriated the holdings of absentee landlords. He also had driven out the Mafia, which had run Havana as a wide-open city, complete with brothels where the girls were as young as fourteen. So when the CIA's Brigade 2506 tried to attack the Bay of Pigs, the defenders didn't need to hear that it is fitting to die as a patriot, or to receive pep talks about love of country. It sufficed for them to know that if they lost, their daughters once again would become whores for American tourists.

The CIA provided enough support to involve America up to the neck, with no hope of hiding it. However, JFK refused to support the invasion with the naval and air power that might have ensured success. Over a century earlier, Lord Byron had written that "who would be free themselves must strike the blow," but he probably wasn't thinking of the *Rio Escondido*. This old troop transport was unpainted and rusty, with a balky engine and a hold full of foul odors. It carried the expedition's entire stock of ammunition, gasoline, and food, and it blew up in a sheet of flame when a Cuban pilot hit it with a rocket. Still, no one could deny that America had supplied at least one side with effective weapons. The pilot's jet fighter had come from the United States.

This was no honest defeat, as in the Philippines during World War II or at the Yalu in Korea. This was humiliation. Yuri Gagarin's flight had suggested Soviet strength and American weakness, but here was the real thing. James Reston of the *New York Times* later wrote that Khrushchev "would have understood if Kennedy had left Castro alone or destroyed him; but when Kennedy was rash enough to strike at Cuba but not bold enough to finish the job, Khrushchev decided that he was dealing with an inexperienced young leader who could be intimidated and blackmailed."[7]

The space race would now have to go forward as part of a very broad response to a Cold War that had suddenly become much chillier. Kennedy had to meet with congressional leaders, to gauge reaction to Apollo on Capitol Hill. He wanted Defense Secretary McNamara to present the Pentagon's perspective, and he needed to learn more from Lyndon Johnson, his point man for space. He met with LBJ on April 19, 1961, the day the Cuban invasion collapsed, and sent him a memo the next day: "Do we have a chance of beating the Soviets by putting a laboratory in space, or by a trip around the moon, or by a rocket to land on the moon, or by a rocket to go to the moon and back with a man? Is there any other space program which promises dramatic results in which we could win?"

The technical base for such initiatives was already in good shape. Saturn was on course toward an initial test flight in October, while an F-1 thrust chamber had run on a test stand early in April. It had developed 1.64 million pounds of thrust, nearly twice the power of the complete R-7. Yet when LBJ responded to Kennedy's memo, late that month, he focused principally on political issues. He briefly mentioned "the possibility of enormous technological breakthroughs obtained from space exploration." However, he mostly dealt with the significance of space as a tool of the Cold War:

> The Soviets are ahead of the United States in world prestige attained through impressive technological accomplishments in space. This country should be realistic and recognize that other nations, regardless of their appreciation of our idealistic values, will tend to align themselves with the country which

> they believe will be the world leader—the winner in the long run. Dramatic accomplishments in space are being increasingly identified as a major indicator of world leadership.
>
> If we do not make the strong effort now, the time will soon be reached when the margin of control over space and over men's minds through space accomplishments will have swung so far on the Russian side that we will not be able to catch up, let alone assume leadership.
>
> Manned exploration of the moon, for example, is not only an achievement with great propaganda value, but it is essential as an objective whether or not we are first in its accomplishment—and we may be able to be first.[8]

Then a week later, on May 5, Project Mercury scored the first success that had political value. The astronaut Alan Shepard rode a Redstone on a fifteen-minute flight that took him to an altitude of 115 miles, making him the first American in space.

The flight was really nothing more than von Braun's Project Adam, which Dryden had dismissed three years earlier as a circus stunt. The concept had nevertheless survived to become an element of Mercury, and for the same reason that Ike had turned to the same Redstone to launch the first satellite: This rocket was available, it was ready, and it could support launches of some importance well before the Atlas would become available for serious work. People were hoping that spring that an Atlas might indeed carry an astronaut to orbit by the end of 1961. But amid the urgency of the moment, the Redstone, with its range of three hundred miles, proved to be just what everyone needed.

Television networks showed the liftoff and provided the live coverage that would quickly become standard. This meant that everyone was watching; if it had blown up, much of the world would have seen it. The drama interrupted even formal courtroom proceedings, for in Indianapolis, a judge halted a trial to allow people to watch a TV that had been part of a thief's loot. Across the country, highway traffic thinned as motorists pulled to the side of the road to listen to the radio. The immediacy of the coverage captured everyone's attention, for there was actually a man atop that rocket. The nation listened, as Shepard's voice came scratchily over his radio link:

> *Shepard:* On the periscope—what a beautiful view. Cloud cover over Florida is three to four tenths near the eastern coast—obscures up to Hatteras.
>
> *Mission Control:* He is reporting the cloud coverage just about what he saw on the weather map.
>
> *Shepard:* I just saw Andros Island, identified the reefs.

Russia might have the big rockets, but America had the TV and film crews, and in the battle for prestige, this would provide a powerful advantage. In the words of the London *Evening News*, "They have done it in the fierce light of publicity."

Three weeks later, on the afternoon of May 25, Kennedy stepped into the same light as he addressed a joint session of Congress for what he billed as a second State of the Union message. His address, carried live, dealt with a range of topics. "These are extraordinary times," he declared. "We face an extraordinary challenge." He spoke of the economy, proposing a new job-training program and an investment tax credit, but for the most part he dealt with the Cold War.

He called for a buildup of the Army and Marines. He urged the nation to build fallout shelters for protection against nuclear attack. He proposed a step-up in foreign aid, along with more money for the Voice of America as a means of countering Communist propaganda. Then, near the end of the speech, he presented NASA with its charter and with the mission that would define its role for decades into the future:

> If we are to win the battle that is going on around the world between freedom and tyranny, if we are to win the battle for men's minds, the dramatic achievements in space which occurred in recent weeks should have made clear to us all, as did the sputnik in 1957, the impact of this adventure on the minds of men everywhere who are attempting to make a determination of which road they should take.
>
> Now it is time to take longer strides—time for a great new American enterprise—time for this nation to take a clearly leading role in space achievement, which in many ways may hold the key to our future on earth.
>
> I believe that this nation should commit itself to achieving the goal, before this decade is out, of landing a man on the moon and returning him safely to earth. No single space project in this period will be more exciting, or more impressive to mankind, or more important for the long-range exploration of space; and none will be so difficult or expensive to accomplish.[9]

This choice of a goal served several requirements. It was far beyond the capability of the rockets of 1961, including the Saturn, and Moscow would gain little advantage by its present lead. It demanded rockets of an entirely new character, perhaps including Nova, and thus would shift the terms of the contest. Rather than having the United States seek to match the R-7 in the near term, the Soviet Union would have to pit its technical and industrial base against the far greater one of the United States.

At the same time, the goal of a moon landing was achievable. It did not lie somewhere off in the distant future, but rather was only a couple of presidential terms away. This horizon in time was very important, for Kennedy was unwilling to wait until 1975 to land astronauts on the moon. This had been a key element of the NASA plan that Ike had turned down only five months earlier because of cost. JFK rejected it for a different reason: He knew the public would not remain patient for as long as fifteen years.

In addition, the goal had the immense virtue of being simple and understandable. Von Braun would note that "everybody knows what the moon is, everybody knows what this decade is, and everybody can tell a live astronaut who returned from the moon from one who didn't."

Jerome Weisner, the science adviser, later told the historian John Logsdon: "If Kennedy could have opted out of a big space program without hurting the country in his judgment, he would have. Maybe a different kind of man could have said to the country, 'Look, we are going at our own pace. We are going to let the Russians be first. We don't care.' I think he became convinced that space was the symbol of the twentieth century. It was a decision he made cold bloodedly. He thought it was right for the country."[10]

Now the development of rockets divided along lines resembling those that already existed within the space program. There was a well-established division between unmanned programs that could serve practical applications and the more costly manned efforts that would serve political ends. A similar division emerged between rockets for manned flight, which were to be immensely large, and military missiles such as Minuteman and Polaris, designed to be sufficiently compact to fit within silos and submarines.

In pursuing the development of these compact missiles, Moscow was well behind. Indeed, much of its attention was still focused on ICBMs, and it was developing a new one of considerable size, the R-16. This project was of great interest to Marshal Mitrofan Nedelin, a deputy defense minister and commander of the strategic rocket force. He was preparing to fight a thermonuclear war, and like Curtis LeMay, he made no bones about it.

Nedelin had directed the test of his country's first true hydrogen bomb, on November 22, 1955. It yielded 1.6 megatons, over a hundred times more than the bomb at Hiroshima. That evening he hosted a banquet and invited Andrei Sakharov, the weapon's designer, to offer the first toast. Sakharov said, "May all our devices explode as successfully as today's, but always over test sites and never over cities."

This toast displeased Nedelin, who replied with a parable: "An old man is praying before an icon. 'Guide me, harden my spirit. Guide me, harden me.' His wife heard him praying and said, 'You just pray to get hard, old man. I can guide it in myself.'" Nedelin looked at the group and continued, "Let's drink to getting hard."[11]

Sakharov was shaken, for this general, who was very powerful, had responded with lewdness and blasphemy to views that even Khrushchev shared. Five years later, that lack of respect for the power of modern weapons would cost Nedelin his life.

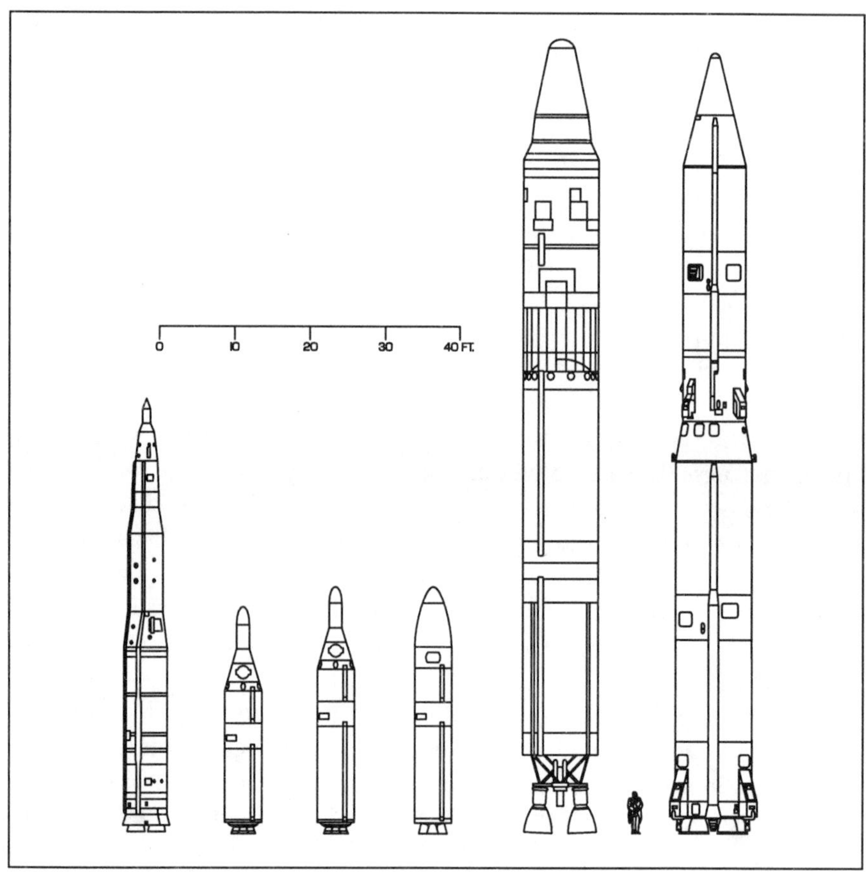

Second-generation missiles: U.S. Minuteman, Polaris (three versions), Titan II; Soviet R-16. The versions of Polaris, *from left:* 1,200-nautical-mile range, deployed in 1960; 1,500 n. mi., 1962; 2,500 n. mi., 1964. (Dan Gauthier)

He came to Tyuratam in October 1960 for the first test flight of the R-16, which was to replace the R-7. It stood a hundred feet tall and could hurl a ten-megaton bomb 6,500 miles. It was not so huge as the R-7, but was still among the largest missiles that would reach operational deployment. It represented the work of a protégé of Korolev, Mikhail Yangel, who was running a separate engineering center.

Yangel had a strong interest in storable propellants, including nitric acid and hydrazine, which he used in his R-16. Korolev had scorned nitric acid as the "devil's venom," fearing to use it because it was extremely toxic. Indeed, it produced very severe burns when it touched the skin. Yangel had no such qualms, for he expected to use it safely by adhering to strict precautions. He designed the R-16 as a two-stage rocket, with the devil's venom in both stages.

In mid-October, a few days before the launch, Khrushchev had boasted of his missile force in a speech at the United Nations. He had plenty of medium-range missiles such as the R-5, but few if any operational ICBMs, and this test of the R-16 might go a long way toward making reality fit his rhetoric. It therefore was of unusual importance, and with Khrushchev putting great pressure on Nedelin, Nedelin went to Tyuratam to show that he was on top of things. His presence did not reassure the launch crew; it made them nervous, and more likely to make mistakes.

Technicians pumped propellant into the rocket, while Nedelin and a number of others sat near it in chairs. Problems developed as the countdown proceeded, and it became necessary to make repairs. Safety rules ordinarily would have called for draining the fuel and oxidizer, to reduce the fire hazard, but Nedelin was impatient. He personally recommended that the launch crew leave the missile fully fueled and proceed with the work. He did not change this procedure even when some crew members began to use soldering irons, which might ignite any leaking propellant.

A controller sent a signal to the rocket. Because of a bad connection in the wiring, it reached the second stage—and ignited the engine. The flame of its exhaust pierced the first stage like an immense blowtorch, setting it afire. Instantly an enormous fireball engulfed the pad, as the rocket broke apart and spilled more hydrazine and nitric acid to feed the conflagration. Not all of the acid burned up; much of it vaporized, forming a poisonous and corrosive cloud that seared the lungs. In the words of a survivor,

> A thick plume of flame suddenly erupted, covering everything around it. People were running toward a special overhang. But in their path was a strip of freshly laid asphalt. It immediately burst into flames, trapping many in the hot, sticky mess. Later, outlines of men and noncombustible items like metal coins, bunches of keys, pins and belt buckles could still be seen at this site. The most terrible fate fell upon those who were on the upper levels of the servicing gantry. People were enveloped by the fire and burst into flames like intense flares. Later, scorched corpses hung everywhere on the barbed wire that surrounded the site.[12]

One of the corpses was Nedelin's, for the conflagration engulfed him as he sat near the rocket. Some men working high on the gantry had attached ropes for safety; now they dangled from those lines, as their bodies burned. Many of Yangel's top associates also died, including his deputy. Over a hundred people perished, making this the worst disaster in the history of rocketry.

Yangel had gone into a protected room for a cigarette; this saved his life. Few others were as fortunate. Marshal Nedelin was buried in the

Kremlin Wall, with full honors. Most of the others went into a mass grave at Tyuratam.

The disaster caused an upsurge in concern for safety. Gagarin's orbital flight, which had been set for late December of 1960, now was put off to the spring, to permit additional tests with dogs. The R-16 launch program resumed in early February, with success; this put the missile on schedule for deployment in 1962. Then, following the Bay of Pigs, Khrushchev met with Kennedy in Vienna. He came away convinced anew that he could win by intimidation, and threatened to allow East Germany to seize West Berlin.

Kennedy's Apollo decision, and the other measures he announced in the same speech, represented only part of his response. In late June he requested more funds for the Pentagon, a call-up of reserves, more combat troops, a tripling of the draft calls, and the reconditioning of aircraft and ships that were in mothballs. Khrushchev made his move in August—as he built the Berlin Wall. He also carried out tests of nuclear bombs with extremely high power, including one of 50 megatons. But he did not take West Berlin; instead, having sealed it off with his wall, he allowed the crisis to die down. Still, Kennedy felt he had to do more, as he turned his eyes to Southeast Asia. General Lyman Lemnitzer, chairman of the Joint Chiefs, predicted that if South Vietnam were to fall, "we would lose Asia all the way to Singapore." In December, Kennedy initiated a buildup that would take his commitment there from eight hundred advisers to sixteen thousand troops in the field.

By then, Khrushchev was running short on strategic options. His long-range bomber force numbered only fifty-eight Bison jet bombers—and seventy-six Tu-95s, which were turboprop-powered, slow, and highly vulnerable to American air defenses. Khrushchev was beginning to deploy the R-16 missile, but he had only a total of twenty ICBMs. He could attack the United States with some 220 nuclear weapons, but many would fail to get through.

And American power was awesome: It had 122 operational ICBMs as well as 1,381 jet bombers, some two-fifths being B-52s. It also had sixty Thors in England, forty-five Jupiters in Italy and Turkey, plus forty-eight Polaris missiles aboard three submarines on station, with more in reserve. The United States had some 4,000 megatons ready to deliver. The Joint Chiefs estimated they would kill up to half a billion people, eighty times more than in the Holocaust.

Khrushchev nonetheless could do one more thing to redress this imbalance. Working in secrecy, he would install forty medium-range missiles in Cuba, to greatly augment the number of weapons he could aim directly at the United States. The result was the Cuban missile crisis, in October 1962. For a few days the world came closer to nuclear war than ever before or since, as the grim logic of rocket and missile development, pursued relentlessly since 1945, appeared ready to bring about its intended conclusion.

How close did we actually get to World War III? One answers this question by noting the state of the crisis on Saturday, October 27, nearly two weeks after Kennedy learned of this imminent threat.

He had blockaded Cuba, which was itself an act of war, but the missiles were already there and crews were working feverishly to make them operational. Secretary McNamara was ready to launch air strikes; Curtis LeMay was now the Air Force's chief of staff and would carry them out with pleasure. But McNamara advised JFK that he could not guarantee that those strikes would take out each and every missile. If any survived, commanders probably would fire them immediately in reprisal, on the principle of "use it or lose it," before America could strike again.

McNamara had also built a powerful force that would invade Cuba when Kennedy gave the word. Pentagon analysts were estimating forty thousand to fifty thousand American casualties.

Diplomatic moves were approaching a standoff. Khrushchev had sent a personal message on Friday, offering to withdraw the missiles in exchange for a promise that the United States would not attack Cuba. This brought momentary hope of a settlement. But then a new message came in, apparently drafted by a committee, that stiffened the terms. Moscow now insisted that Washington must withdraw Jupiter missiles emplaced in Turkey, a NATO ally. This condition was unacceptable; it would amount to selling out Turkey to protect the United States. Such an act would badly damage the NATO alliance, which rested upon mutual commitments of military support.

McGeorge Bundy, the national security adviser, suggested that JFK ignore the second letter and reply to the first, picking the terms he liked best. This ploy smacked of desperation—of responding to an offer Khrushchev had not made. Still, it held out one last chance for diplomacy, by avoiding for the moment a rejection of Moscow's terms that would heighten the risk of war.

The Joint Chiefs, including LeMay, made their own recommendations: an air strike on Monday, followed by the invasion. Kennedy demurred: "It isn't the first step that concerns me, but both sides escalating to the fourth and fifth step—and we don't go to the sixth because there is no one around to do so. We must remind ourselves we are embarking on a very hazardous course."

Kennedy then approved a response to Khrushchev's note, sent it, and announced publicly that he was accepting Moscow's conditions. His brother Robert then told the Soviet ambassador that time was very short. Only a few hours were left; Khrushchev must respond the next day.

In Florida, American forces were ready to move. In Cuba, the Soviet force held twenty nuclear warheads for their missiles. They also held nine tactical nuclear weapons, and the local field commander had the authority to use them. Kennedy didn't know about them. His advisers did know that Khrushchev himself had taken the lead in defining his country's war-fighting doctrine.

This doctrine held that if Moscow went to war, it would fight an all-out world war. Such a course might begin with a local conflict that escalated, but further escalation to the nuclear level was "inevitable," unleashing "simultaneous" strikes against cities as well as military targets, along with the "maximum" use of nuclear weapons from the very beginning. The war would involve the onetime use of the Soviet Union's entire accumulation of strategic striking power.

It still was Saturday. In the evening, McNamara returned to the Pentagon. He wondered how many more sunsets he would see.

Sunday morning came, along with an announcement from Radio Moscow. This might well be the last chance for peace. If Khrushchev rejected the American terms, then Kennedy was prepared to respond as early as Monday with an initial air strike against Soviet surface-to-air missile sites in Cuba. This could be the first of his steps toward nuclear war. The world might be within twenty-four hours of a countdown to World War III. The Soviet radio announcer read Khrushchev's message:

> In order to eliminate as rapidly as possible the conflict which endangers the cause of peace, the Soviet Government, in addition to earlier instructions on the discontinuation of further work on weapons construction sites, has given a new order to dismantle the arms which you describe as offensive, and to crate and return them to the Soviet Union.[13]

The crisis was over.

That evening, the Kennedys talked about it, and John remembered the end of the Civil War. He said, "Maybe this is the night I should go to the theater." He and Robert laughed. Then Robert added, "If you go, I want to go with you."

8

High-Water Mark

The Manned Moon Race

AS JOHN GLENN RODE IN ORBIT in February 1962, he saw the sun, intense and clear, not yellow but bluish white. It sank toward the bright horizon. As it approached, a black shadow moved across the earth and left its surface in darkness. Now the horizon stood out more sharply, as it changed from nearly white to a brilliant rainbow that extended to either side of the sun for hundreds of miles, along the curvature of the earth.

Here was the sun itself, flattened and distinctly yellow just before it sank from view. Above it were bands of iridescent orange and molten gold, marked with silhouettes of distant clouds—then reddish brown, light blue, a darker blue, and a magnificent royal purple that blended into the black of space. The colors were sharply defined, glowing, alive with light. As Glenn's spacecraft continued into shadow, the sun set and the band of color narrowed. Yet a rim of light persisted for several minutes, marvelously blue, until this twilight also faded to black.

Then, just as the sun was rising, he saw fireflies. All around him were swarms of small particles that reflected the light, until he entered full sunlight and lost them in the glare. But he saw them again repeatedly during subsequent sunrises, showers of them that looked like luminescent snowflakes. These were bits of ice that had dislodged from the capsule's cold surface, but Glenn didn't know this at first. To him they were a mystery of space.

A moment of danger followed, as ground controllers saw a signal indicating that the heat shield might have separated prematurely. If hot gases leaked behind the shield during re-entry, they would burn Glenn alive. A retro-rocket lay against the shield, secured with straps, and the controllers told Glenn not to jettison this rocket pack following retrofire, but instead to leave it in place. The straps then would help to hold the heat shield tightly in position. But the signal proved false. There was nothing wrong with the shield, and it protected Glenn as he plunged through the atmosphere toward a safe splashdown.

John Glenn grew up amid the small towns of Ohio, which equipped him with strong religious faith and a sincere commitment to family and country. He had come up as a combat pilot in the Marines, flying nearly sixty missions in the Marshall Islands during World War II, then topping this record in Korea. This background qualified him for training as a test pilot; in 1957, as project officer on the new F8U Crusader, he set a record by flying it from Los Angeles to New York in under three and a half hours. During his career he married his hometown sweetheart, won five Distinguished Flying Crosses, and rose to lieutenant colonel.

His orbital flight still couldn't match the Russians' best, for the previous August Gherman Titov, Gagarin's backup, had spent a full twenty-four hours in space, well ahead of Glenn's three orbits and five hours. But his mission put America solidly into the race, and the country pulled out all the stops as it hailed him. President Kennedy, who had welcomed Alan Shepard at the White House the previous May, now flew to Cape Canaveral to greet Glenn on his own ground. "This is a new ocean," said JFK, himself a Navy man, "and I believe that the United States must sail on it."

In Washington, Glenn addressed a joint session of Congress, an honor given to few heads of state. Then it was off to Manhattan, where four million people gathered to roar and cheer. His motorcade drove up Broadway through what looked like a late-winter snowstorm but was actually a deluge of paper flung in welcome from office windows. This was the mother of all ticker-tape parades, with nearly 3,500 tons of the stuff, burying the previous record of 3,249 tons for General Douglas MacArthur in 1951.

Glenn's fellow astronauts accompanied him, along with their families, and everyone stayed at the Waldorf-Astoria, in suites that each featured two lavish bedrooms and a luxurious living room. They returned to Broadway to see a current hit, Frank Loesser's *How to Succeed in Business without Really Trying;* it turned into a command performance, as people in the audience gave up choice seats, while the play's company delayed starting until Glenn's party arrived. He was feted at the United Nations. He met the lord mayor of Perth, Australia, whose city had turned on its lights as Glenn orbited overhead, and who traveled to New York just to greet him briefly. Then it was back to Ohio, for further welcoming from the people he knew best.

His capsule was marked for the Smithsonian, for display with Lindbergh's *Spirit of St. Louis.* In Utah, a move got under way to add another *n* to the partially completed Glen Canyon Dam. Nikita Khrushchev sent greetings, as did over thirty other heads of state. A message came from the artist Pablo Picasso: "I am as proud of him as if he were my brother."

Through it all, Glenn showed a natural savoir faire, modesty, and coolness. He took in stride the oleaginous officials of Washington, who cheered when he said, "I still get a real hard-to-define feeling when the flag goes by." He put up with the constant attention of Lyndon Johnson. Riding in motorcades, he flashed a winning grin and turned his head to the crowds like a campaigning Kennedy. Pundits measured him for high political office; he would go on to become a senator from his home state.

This flight stood out as a high point during the years that followed JFK's decision to put a man on the moon. Moreover, Glenn's accomplishment took place within a NASA that already was expanding rapidly. During the second half of 1961, NASA committed itself to supporting Apollo by building four major new centers and facilities, all in Texas and the Deep South. The locations served a requirement for ice-free transport by water of the immense boosters, which would be too large to travel by rail or highway. They also reflected a Kennedy policy of funneling federal dollars into the South, a region that was strongly Democratic and that lagged economically behind the rest of the country. In addition, this policy suited the needs of southern congressional barons who controlled key committees in the House and Senate.

In August, NASA kicked off its buildup of facilities by announcing that it would erect a moonport on 125 square miles of land on Merritt Island, north of Cape Canaveral. The Cape itself had plenty of launch pads for existing missiles, including ICBM Row, where massive gantries stood close to the ocean, only a few hundred yards apart. But Apollo needed more. Its Merritt Island base would center on the Vehicle Assembly Building, the largest in the world, with 130 million cubic feet of enclosed space. Its enormous bays would provide room for the simultaneous assembly and checkout of four moon rockets.

NASA also needed a plant where industrial contractors could assemble the rockets' first stages. The government owned a good one, the Michoud facility near New Orleans, which fronted on the Mississippi River. It had been built during the war for the production of Liberty ships and had its own record for size, with over 1.8 million square feet under roof, the equivalent of a thousand suburban homes.

In addition, complete rocket stages would require static test, at a site close to Michoud that offered plenty of buffer space. Huntsville's Marshall Space Flight Center wouldn't do; following initial tests of the eight-engine Saturn, the complaints from the townspeople had been nearly as loud as the rockets themselves. Von Braun found what he needed on the Pearl River in Mississippi, only thirty-five miles from Michoud. The land was flat pine forest, thinly populated and easily acquired through eminent domain. In addition, the river provided deep-water access.

A fourth NASA initiative gave Apollo an institutional base within a new Manned Spacecraft Center. The project staff of the Mercury program, called the Space Task Group (STG), had operated within NASA's Langley Research Center near Norfolk, Virginia. This was the agency's oldest such center, dating back to the early days of the former NACA. But it lacked room for expansion, whereas the STG now was to serve as a nucleus that would grow to manage the whole of Apollo. It needed a center all its own, preferably a large one, and the firm of Humble Oil was willing to donate a large tract of pastureland in Texas, twenty miles from Houston. This suited the needs not only of NASA but of Texas's delegation in Washington, which included Vice President Johnson and House Speaker Sam Rayburn. In September 1961 the agency's new administrator, James Webb, declared that here he would build "the command center for the manned lunar landing and follow-on manned space flight missions."

Amid this flurry of activity, von Braun was finding himself in the unusual position of scaling back his plans at precisely the moment when the dream of a moon landing had become national policy. He had expected to accomplish this mission using an eight-engine version of Nova, generating twelve million pounds of thrust. But this plan was out; the Nova couldn't be developed during that decade, which was what Kennedy insisted on. The choice of the Michoud plant in September made this clear; its roof was too low to accommodate the enormous Nova first stage, even when set on its side.

There was no clear understanding of how to fill the gap between the too-small Saturn, which could not carry out a lunar landing, and the too-big Nova. Both within NASA and at Huntsville, planners expected that some rocket of intermediate size would emerge, perhaps with four rather than eight F-1 engines in the first stage. Two such rockets would be necessary for a lunar landing; each of them would carry half of a moonship, with astronauts assembling the halves in orbit. But throughout the second half of 1961, the question of how to design this booster was up for grabs.

Nevertheless, other major program elements were receiving attention, including the lunar spacecraft and the moon rocket's second stage, the S-II. In pursuing them, North American Aviation now came to the fore. The company had not only used its Navaho experience to build a strong technical base, but had wrested primacy in Air Force bombers from Boeing. In 1957 it had won the contract to build the XB-70, a Mach 3 aircraft that could be described as a Navaho with a flight crew. In addition, it had built the X-15, whose pilots would repeatedly attain the status of astronauts by flying it to altitudes as great as sixty-seven miles. At the center of these accomplishments stood the president of the company's space division, Harrison "Stormy" Storms.

Like William Bollay, Storms had grown up in the Chicago area and had followed much the same course: Northwestern University, Caltech and Theodore von Karman, then on to North American. But while Bollay became known as a technical director, Storms came up as a designer of aircraft. During World War II, he made a solid contribution by arranging to install a fuel tank within the fuselage of the P-51 Mustang. This was more than a routine engineering job; it turned out to be a key to victory. The Luftwaffe had regained air superiority over Germany because Allied fighters lacked the range to escort their bombers, but with the new tanks, the P-51 could accompany them all the way to Berlin. When Hermann Goering saw that the bombers raiding his city now had fighter escort, he told his staff, "The war is over."

Storms went on to enhance his reputation with a succession of path-breaking aircraft that rivaled those of Lockheed's Kelly Johnson. He directed the aerodynamic design of the B-45, the first production jet bomber. He had overall design responsibility for the F-100, the first fighter to break the sound barrier in level flight. He showed similar leadership on the X-15 and XB-70.

His resume would show active involvement with a total of forty-eight aviation and spacecraft projects in the course of his career, from the piston-driven AT-6 trainer to the space shuttle. This meant that during the exciting decades of rapid advance, he could learn fast and contribute effectively, as new programs followed one another in quick succession. In addition, his temperament suited these opportunities. People said that "while other men fiddle, Harrison storms." He surrounded himself with picked associates who called themselves, proudly, the Storm Troopers.

The design of the moon rocket's first stage was still up in the air when von Braun sent out notices to industry groups, inviting them to compete in proposing designs for the second stage. It was a very important stage, for it would thrust the elements of the lunar mission into earth orbit. The Storm Troopers were eager to try for this prize, but they had only about ten weeks in which to prepare the proposal, which was to fill a stack of documents two feet tall.

"We spent long nights at it," recalls Rocketdyne's Bill Ezell, who worked on the project. "Coffee was the fuel that drove us. We all gathered in a big room; guys who'd had offices were now out on the floor, helping with the writing. You'd start at seven in the morning and run to midnight." People would make calculations and prepare drawing-board designs, drafting text on quadrille-ruled paper and sending it on to a typist. "There'd be periods of writing," says Ezell. "Periods of reviewing, critiquing each other's work; then more writing."[1]

Saving weight was of utmost importance, and these people found a new way to do it. The S-II would amount to a set of tanks for liquid hydrogen and oxygen, fitted with five J-2 engines to deliver a million pounds of thrust. To seal in these supercold liquids, the tanks would need insulation. Standard practice called for putting it on the tanks' interior, but Storms's designers found they could make the stage lighter by putting the insulation on the outside. The tanks' aluminum metal, exposed to the frigid cold of their contents, would gain strength. The S-II then would need less aluminum, and less weight, to carry its load. Indeed, it would come forth as the merest eggshell of a rocket, weighing over a million pounds when fully fueled but light enough to ride a specialized truck when empty. This innovative approach won the competition and, in September, the contract.

The award underlined North American's position as a leader in missiles and space. The firm formerly had measured output in pounds of airframe metal; its new measure was engineering man-hours. Indeed, research and development was accounting for over 70 percent of its sales. Within the Apollo effort, it had contracts for the principal engines, the F-1 and J-2; now it would build the S-II stage as well. Yet Storms thought he could win even more, as he set out to garner a contract to build the manned lunar spacecraft.

The moment of decision came just after Thanksgiving, as a selection board matched North American's bid against those of four competing companies. The Martin Co. came in on top, and when word of this leaked out, its employees heard an announcement on the public-address system that they had won. The next day they learned the truth: The prize had gone to North American. Martin's bid had been slightly stronger, but the board cited North American as having the best technical qualifications, noting such projects as Navaho and the X-15. The board therefore recommended this firm as the best alternative. In addition, NASA's leaders liked it because they had worked closely with it on the X-15. Still, even Storms was stunned at this latest success and said repeatedly, "I just can't believe it." The company's NASA contracts now topped those of its two nearest competitors combined.

Meanwhile, von Braun was defining the shape of the moon rocket's first stage. In September 1961 he had invited industry representatives to a meeting in the Huntsville municipal auditorium. He asked them to prepare proposed designs for an "Advanced Saturn," which would be smaller than Nova but still capable enough for the lunar mission. "The lunar landing is our present focal point, our big step into manned space travel," he told the gathering. "But it will not end there. The Advanced Saturn does not care what its payload is. It is just a big truck to increase this country's capacity to carry cargo into space."

The people in the audience included George Stoner, a top man at Boeing's Space Division in Seattle. Back home, huddling with his senior managers, he remembered those words. "Von Braun says the Advanced Saturn is a truck," he remarked. "It is; it's a space logistics system. Let's find out what are the heaviest pieces any other truck or boat or train has to haul. Let's see if we could build a Saturn that would lift that big a piece out of the earth's gravity."[2]

Stoner found that typical cargo compartments on ships carry a hundred thousand pounds. It was about the same for railroad cars and for the heaviest air freight, about seventy thousand pounds for large trucks. An Advanced Saturn with five F-1 engines would lift such a load. It would have 7.5 million pounds of thrust and would toss ninety thousand pounds at the moon. Furthermore, it could be developed in the 1960s decade. Stoner's group came up with a preliminary design and decided to recommend this five-engine version with vigor.

The work fitted in with the views of Milton Rosen, who had built Viking in the long ago and who now was NASA's director of launch vehicles. As early as March 1961, two months before JFK's speech to Congress, Rosen had recommended strongly that this first stage have five engines rather than four. He won the support of William Mrazek, one of the top rocket men at Huntsville, and together they convinced von Braun. As a key argument, they noted that it would be easy to add the extra engine. The original four F-1s were to mount at the ends of two heavy crossbeams in the base of the rocket; the fifth one could readily go in the center, at those beams' junction.

Eleven days before Christmas, von Braun issued his judgment. The sound of decision was the rattle of a teletype in the office of Lysle Wood, Stoner's boss at Boeing. The message stated that this company had been selected as the contractor to build twenty-four Advanced Saturn boosters. Soon von Braun was on the phone, confirming the news: The rocket would follow Stoner's proposal as a five-engine design that could throw a fully loaded railroad car at the moon. The full three-stage moon rocket, including this first stage, would be the Saturn V, and it would carry the hopes of President Kennedy.

NASA's managers generally agreed that while it would be nice to go to the moon with a single launch, only Nova could do this. The smaller Saturn V would demand two launches: one for the manned lunar spacecraft and the other for the rocket stage that would shoot it out of earth orbit toward the moon. The two vehicles would have to rendezvous in space and link up, and it was clear that this would call for pilot skills far beyond those of the Mercury astronauts. No monkey could fly this mission; it would demand the care and attention of Air Force in-flight refueling, wherein a bomber finds a tanker in the empty sky and receives its load of jet fuel.

But in the Space Task Group, and at the firm of McDonnell Aircraft that built the Mercury capsules, a number of people had been studying advanced designs that could improve on Mercury while training astronauts for Apollo. In December these plans reached fruition, as NASA announced that the next step beyond Mercury would feature a two-man craft, Gemini. It would develop experience in all the skills of a lunar mission: crews in orbit, rendezvous, docking or linkup, spacewalks, flight durations of up to fourteen days, and ground control of multiple spacecraft. In particular, a Gemini craft would dock with an orbiting Agena stage and ride it to a high orbit. This would simulate the basic operation of the forthcoming flight to the moon.

During the same eventful December, Douglas Aircraft received a contract for the third stage of the Saturn V, the S-IVB. Douglas was already building the Saturn's second stage, the S-IV, with its six Centaur engines. The new design would replace the Centaurs with a single J-2. It was a measure of the leap in size, from Saturn to Saturn V, that the second stage of the former, even when fitted with this new engine of 200,000 pounds of thrust, would serve only as the third stage of the latter.

Now NASA completed the outline of a buildup that in only six months had matched the Pentagon's surge into missiles during 1954 to 1956. The earlier effort had produced three new institutions—Ramo-Wooldridge, the Western Development Division, and Admiral Raborn's Special Projects Office—and six programs: Atlas, Titan, Thor, Jupiter, Polaris with its *George Washington*-class submarines, and WS-117L with its Agena upper stage.

For NASA, the present count was four and five. Its new institutions included the Merritt Island moonport, the Manned Spacecraft Center in Houston, the Mississippi Test Facility for static tests of complete stages, and Michoud for their assembly. The new projects, all of which were under contract by year's end, included Gemini, the Apollo lunar spacecraft, and the three stages of the Saturn V, each a demanding effort in its own right. The agency could also cherish its inheritance from ARPA and the Army: the Marshall Space Flight Center and its Saturn program, the Jet Propulsion Laboratory, and the F-1 and J-2 engines. It certainly was a giant leap—perhaps for mankind but certainly for NASA.

Still, as of early 1962 it mostly existed as paper rather than hardware. The big rocket motors were roaring on test stands, for the F-1 now had its turbopumps and the first J-2 flamed to life at the end of January. In addition, the eight-engine Saturn had made a spectacular debut in October, for while it flew barely two hundred miles downrange from the Cape, its 1.5 million pounds of thrust put the United States in the lead in rocket power for the first time. However, launching the new Apollo facilities was not always so tidy.

The Michoud plant gave a case in point. It took its name from the wealthy Antoine Michoud of New Orleans, a junk dealer and recluse who had owned the property. As if in keeping with its checkered past, the government had allowed it to lie dormant since 1954 and it was a mess. It was rat-infested and grimy from a recent flood; the aviation writer Robert Serling would note its "cobwebs, dust, dirt, cracked floors, rusting rafters, and corroded overhead cranes." Yet all the forty-three acres of floor space needed was a good cleanup. The Mississippi Test Facility was worse.

The MTF had Mississippi mud, soggy cypress groves, poisonous snakes, and clouds of mosquitoes. Local attractions included Devil's Swamp and Dead Tiger Creek while among the wildlife were some wild hogs descended from local farm stock. The main test stand required sixteen hundred pilings for a foundation, each a hundred feet long, and it didn't help when heavy rains came one winter, submerging rainfall records that had been on the books for thirty years. The test stand would tower to a height of over four hundred feet, making it the tallest building in the state of Mississippi, but it took until 1966 to make it operational. Yet not all was bleak, for snakebite kits cost only $1.25.

Even so, for all of Apollo's fast pace and gigantism, in 1962 it still remained a program in which one man could make a large difference. This was John Houbolt, a manager at NASA-Langley who had completed a successful inquiry into why Lockheed's Electra airliner had been losing wings in flight. In 1959, with post-Mercury planning on the agenda, Houbolt set up a committee to look at problems relating to rendezvous. This was a characteristic approach; in a realm where no one had any kind of hands-on experience with manned missions, it was more important to pick good people than to search fruitlessly for experienced specialists.

Houbolt's initial work focused on linkups with a space station, but he broadened it to include lunar missions, and his group became a magnet for anyone at Langley with an interest in rendezvous techniques. Then in May 1960, a colleague introduced him to a new approach to Apollo: lunar-orbit rendezvous. It contrasted sharply with the methods that were receiving the most serious attention and seemed to have enormous advantages.

Standard approaches featured a large manned moonship that would enter lunar orbit, set down on its surface, and return to earth. A Nova booster could hurl it there directly; smaller boosters, including the eventual Saturn V, would start by shooting it into earth orbit, to meet up with a large rocket stage that required a separate launch. However, following this earth-orbit rendezvous, the lunar craft would keep all its astronauts together for the rest of the mission.

Lunar-orbit rendezvous called for the spacecraft to enter an orbit around the moon, but not to descend to the surface. Instead it would serve as a mother ship, sending forth a small landing craft that would carry two astronauts down and then back, while a third continued to ride in the lunar orbit. Hence this lander would have to rendezvous with its mother, in a tricky operation that would take place a quarter of a million miles from earth.

But because only part of the moonship would fly to and from the lunar surface, its overall propellant requirements would be greatly reduced. It then could be smaller, and much lighter in weight. With this weight saving, there would be no need to launch two moon rockets from Cape Canaveral to carry out a single lunar mission. Just one of these rockets would do it all.

Houbolt understood from the start that he was on to something important. He later wrote: "I can still remember the 'back of the envelope' type of calculations I made to check that the scheme resulted in a very substantial saving in earth boost requirements. Almost spontaneously, it became clear that lunar orbit rendezvous offered a chain reaction simplification on all back effects: development, testing, manufacturing, erection, countdown, flight operations, etc. All would be simplified. The thought struck my mind, 'This is fantastic. If there is any idea we have to push, it is this one!' I vowed to dedicate myself to the task."[3]

Houbolt set to work with missionary zeal, spurring his staff to new designs and calculations while spreading the word by giving briefings and presentations to all who would listen. He encountered considerable skepticism, and some of it was of his own making. In his enthusiasm for a lunar lander that would be tiny indeed, he backed such concepts as a simple rocket-powered platform that a space-suited astronaut would ride while sitting in the open, like a sailor in a life raft. But he also faced a strong presumption that rendezvous should take place in earth orbit, for then if a linkup failed, the astronauts could come down through a simple re-entry. By contrast, a failure of lunar-orbit rendezvous would leave the spacefarers stranded with no way home, to die in a lunar craft that would become their coffin.

Houbolt's initial goal was to win the support of Langley's Space Task Group, which was doing the planning for Apollo. Its director, Robert Gilruth, would soon head the Manned Spacecraft Center. But Gilruth opposed all rendezvous concepts because they would distract from the simplicity of direct ascent. As he put it, "I am concerned that rendezvous schemes may be used as a crutch to achieve early planned dates for launch vehicle availability, and to avoid the difficulty of developing a reliable Nova-class launch vehicle."

When Nova retreated out of reach, Gilruth turned his attention to earth-orbit rendezvous. But as the STG studied it, its disadvantages took on new significance, while the merits of Houbolt's ideas became much clearer. There really was no good way to win assurance that after carrying out two Saturn V launches to assemble the expedition in earth orbit, the big moonship could land safely on the lunar surface while retaining enough fuel to return to earth. Max Faget, a leading STG designer, said that, by contrast, the best thing about Houbolt's approach "was that it allowed us to build a separate vehicle for landing." By the end of 1961, Gilruth, Faget, and the rest of the STG were solidly in Houbolt's corner.

Further support came from NASA headquarters, where a new associate administrator, Robert Seamans, demonstrated a keen interest in rendezvous. At RCA he had directed studies of its military uses, to destroy enemy satellites. In NASA, he learned of Houbolt's work during the fall of 1960 and gave him strong encouragement. Then in July 1961, Seamans named a technical assistant, Nicholas Golovin, to head a panel that would recommend a rendezvous method for Apollo. His committee's report, issued in November, had good words to say about Houbolt's ideas. However, it endorsed the conventional wisdom by coming out for earth-orbit linkup.

Houbolt wasn't satisfied. Seamans might have one of the top posts in NASA, but Houbolt had already bypassed the chain of command to write to him personally. Now he did this again, writing a letter in which he described himself as "a voice in the wilderness, appalled at the thinking of individuals and committees." He asked, "Do we want to get to the moon or not?" Seamans replied to Houbolt with his own letter, writing that Houbolt's issues "deserve serious consideration." He also praised Houbolt's spirit and initiative: "It would be extremely harmful to our organization and to the country if our qualified staff were unduly limited by restrictive guidelines."

This was not an endorsement, but it meant that Seamans would see to it that Houbolt's approach received detailed study in the industry. In addition, though Gilruth had come around to supporting lunar-orbit rendezvous, von Braun was still intensely promoting the earth-orbit technique. It was essential that they should agree, and the new studies brought von Braun around as well. At a key meeting in June 1962, he announced that he had changed his mind. The Marshall Space Flight Center, which he directed, now would join forces with Gilruth's Manned Spacecraft Center to support a lunar-orbit linkup as the key to success with Apollo.

There still were a few highly placed holdouts; the most prominent was Jerome Wiesner, Kennedy's science adviser. He persuaded Golovin to leave Seamans and work for him; the two of them proceeded to launch a strong attack on NASA's decision. As late as September, when JFK visited Marshall and heard von Braun describe the lunar-rendezvous approach, he said, "I understand Dr. Wiesner doesn't agree with this. Where is Jerry?" Jerry stepped forward; right in front of news reporters, he debated the issue with von Braun until the president told them to move on to other topics.

Wiesner's opposition held up a final NASA commitment for several months, but the agency's administrator, James Webb, stood firm. He had come up as Harry Truman's budget director, and he knew he had Kennedy's confidence. Wiesner's challenge faded during October, and Webb then made clear his commitment by awarding a contract for the lunar landing craft. The choice went to Grumman Aircraft, whose senior managers had gotten a head start by contributing to Houbolt's work as early as 1960. With this contract award, in November 1962, the last major element of Apollo fell into place.

Another key decision, made during 1963, introduced risk but helped the program to speed up considerably. The decision introduced a commitment to "all-up testing," so that from its first flight, the Saturn V would exercise all its stages. Standard practice would have called for an initial flight with only the first stage, followed by subsequent launches that would introduce the second and third stages in turn. The eight-engine Saturn was following this course; four test flights, between 1961 and 1963, used dummy upper stages, and this booster would not fly with a live second stage until January 1964. But while all-up testing would put all of the Saturn V's eggs into one basket, it would greatly compress the test schedule—if it worked.

The concept sprang from the experience of George Mueller, who had directed the Air Force's Pioneer moon probes back when the world was young. Now he had similar oversight for Apollo, for Webb had tapped him as NASA's Associate Administrator for Manned Space Flight. Mueller had introduced all-up testing in the Minuteman program; on its first launch, early in 1961, the missile had fired all three stages and had flown successfully to full range, cutting a year off the program. He did the same with the Titan II, a two-stage ICBM that used liquid fuel; it scored a similar success a year later.

On taking office in September 1963, Mueller found himself under considerable pressure to cut costs. He responded with what Webb called "a very bold move," for two months later, he directed that the first Saturn V flight would comprise all live stages and carry a complete Apollo spacecraft. Von Braun's staff were incredulous; within their cautious, step-by-step world, such things simply were not done. But after only a week von Braun issued his reply: "There is no fundamental reason why we cannot fly 'all-up' on the first flight." This approach would strongly complement Houbolt's lunar-orbit rendezvous, the two together leading to total success within the decade.

Yet while NASA was planning for rendezvous, Korolev was showing that he already knew how to do it. The orbital flights of Yuri Gagarin and Gherman Titov, during 1961, had completed the qualification of his recoverable spacecraft; it now was ready for use both as the Zenit reconnaissance satellite and as the manned Vostok craft. Zenit flights began in December 1961; over the next two decades, some four hundred of them would reach orbit. In addition, Korolev followed up Titov's day-long flight by placing two Vostoks into space at the same time.

The mission began on August 11, 1962, as the cosmonaut Andrian Nikolaev soared into orbit from Tyuratam. The following day, Pavel Popovich did the same. His orbit matched very closely that of his wingman, having the same inclination of sixty-five degrees to the equator, and differing by only three to four kilometers in altitude. The launch crew also got the timing right, allowing the two craft to approach within less than five kilometers. Then, to top it off, Nikolaev remained in space for nearly four days before returning.

It was a virtuoso demonstration of precision, for with both craft flying in orbit at a speed of eight kilometers per second, a launch delay of only a few seconds would have put this near-meeting out of reach. The Vostok vehicles would have needed onboard radar and maneuvering rockets to perform the minor flight-path adjustments needed for a true rendezvous. But an essential prerequisite involved a demonstrated ability to fire the booster at a specific moment and to guide it into a predetermined orbit. This mission showed that Korolev's R-7 was up to the mark.

Then in June 1963 Korolev repeated this dual flight—with a woman, Valentina Tereshkova, as one of the two cosmonauts. General Nikolai Kamanin, the head of cosmonaut training, had ordered this demonstration of a Soviet commitment to gender equality, but there was less to it than met the eye. Other cosmonauts had come up as jet pilots; Tereshkova's own background featured nothing more than skydiving. She received a commission as a lieutenant in the air force and acquired extensive training in two-seat MiG aircraft, but she never soloed.

But her political credentials were impeccable. She had been born in 1937 on a collective farm near the city of Yaroslavl. Her father, a tractor driver, died fighting the Germans. The family then moved to the city, where her mother worked in a factory. Tereshkova graduated from high school and then got her own job in a textile mill. At work there was a chapter of the Young Communist League; she became an active member, filling the position of secretary. Following Gagarin's flight, she wrote a letter expressing hope that she too might fly into space. This letter drew the attention of a selection board headed by General Kamanin, which selected her along with four other women as quota queens who would receive astronaut training.

During her mission, Tereshkova showed that her training had been thorough indeed. She repeatedly radioed statements: "Warm greetings from space to the glorious Leninist Young Communist League that reared me. Everything that is good in me I owe to our Communist Party." She stayed up three days, with Khrushchev noting pointedly that this was longer than all American flights put together, and then came down by using her autopilot to execute the re-entry.

She returned to Moscow in time to address a meeting in the Kremlin of the International Congress of Women. She then went off on a world tour that took her to India and Pakistan as well as the United States. On returning home, she became the sweetheart of her fellow cosmonaut Nikolaev. Their wedding that November was a high point of the Moscow social season—and the bride was two months pregnant.

In America, her flight supplied a new spin on the fear of Russian superiority, as writers and political leaders praised the equality of Soviet womanhood. But as her husband said a year later, "Nowadays we keep our women here on earth. We love our women very much; we spare them as much as possible." Tereshkova was the only one who actually flew; no other Soviet woman would reach orbit until 1982. Her marriage failed, and while she continued to receive high honors, people who knew her said she was still a dull factory girl, who was putting on airs.

Even so, in the realm of spaceflight, the twin-Vostok missions of 1962 and 1963 contrasted sharply with the limited performance of Mercury. Its capsule left some room for further improvement, and in May 1963 the astronaut Gordon Cooper, nursing his reserves of oxygen and attitude-control propellant, managed to stay up for twenty-two orbits, nearly a day and a half. But a month later, during Tereshkova's flight, her colleague Valery Bykovsky kept his own craft aloft for five days. It was the same old story: NASA pushing its limits while falling short of what Korolev could do with ease.

Moreover, there was room for growth in the Vostok craft as well. Korolev was readying a new version that could carry two cosmonauts instead of only one. He named it Voskhod (sunrise). He also was about to introduce a new and powerful variant of his R-7, with a highly capable upper stage. He had been launching the Vostoks with a version that featured Kosberg's upper stage of 1958, improved to give more thrust. But Korolev had also designed an entirely new ICBM, the R-9, as a two-stage rocket. Its second-stage engine went into a new upper stage that fit atop the R-7, to increase its payload capacity from five thousand to seven thousand kilograms. This uprated R-7 now was ready for service. It would become the most widely used booster in the Soviet inventory.

Still, Khrushchev could not rest content. There could be no final victories in the race for propaganda, for this was not like putting up a specific set of spacecraft to perform a well-defined service, such as satellite communications. There was the eternal need to do more, to reach beyond the achievements of yesterday. For while Gordon Cooper's 34-hour flight was the best America could do in 1963, less than a year later the first

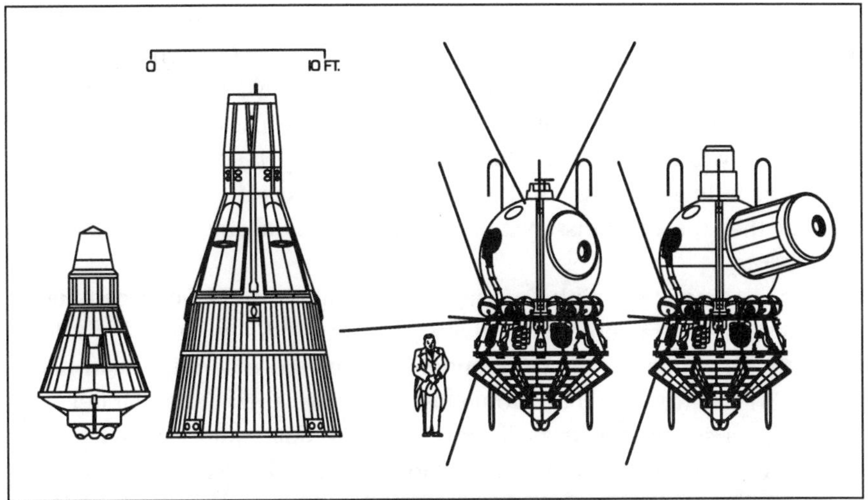

Early manned spacecraft: U.S. Mercury and Gemini; Soviet Vostok and Voskhod, the latter with an airlock for use in a spacewalk. Figure indicates scale. (Dan Gauthier)

Gemini was in orbit. This mission carried no astronauts, but very soon it would be ready to host its own two-man crews.

Accordingly, Khrushchev made a decision. If the Yankees were about to put two men into orbit in 1965, he would trump them by launching three men during 1964. The order, delivered to Korolev, instructed him to fit three cosmonauts into a space built for two, for only the Voskhod was ready to carry out this mission. The third man would add weight, but that would not be a problem now that Korolev had his powerful new R-7. The problem would be to manage the tight squeeze of this ménage à trois.

Vostok's principal designer, Konstantin Feoktistov, had opposed the conversion of Vostok to Voskhod, but now was directing the work. Korolev had won him over by promising to include him in a flight crew. Feoktistov now proposed to fit three men into his orbiting telephone booth by throwing out much of the equipment that was vital for safety. In the words of Korolev's deputy, Vasily Mishin, "Fitting a crew of three people, and in spacesuits, in the cabin of the Voskhod was impossible. So—down with the spacesuits! And the cosmonauts went up without them. It was also impossible to make three hatches for ejection. So—down with the ejection devices. Was it risky? Of course it was. It was as if there was, sort of, a three-seater craft and, at the same time, there wasn't. In fact, it was a circus act, for three people couldn't do any useful work in space. They were cramped just sitting! Not to mention that it was dangerous to fly."[4]

Russians tell a story of an architect who designed a ten-story apartment building and left out the elevators. As punishment, he was sentenced to live on the top floor. Similarly, Feoktistov was named to join the three-man crew. This of course was a high honor, not a punishment. Still it concentrated his mind wonderfully, as if he knew he was to be hanged in a fortnight, and he brought it off successfully. The mission flew in October 1964, and while it stayed up for only a single day, it convinced the world that Russia indeed possessed a three-man spacecraft. Once again it appeared that the capitalist system was powerless against the new Soviet society.

Then came 1965, with Gemini set for a major series of manned flights. Spacewalks—extra-vehicular activity (EVA), in NASA's parlance—would be high on the agenda, as an essential step toward extending astronauts' capabilities. After all, they would have to leave their lunar lander, protected by a space suit, to walk on the moon. Spacewalks would also be essential in rescuing a stranded astronaut, fixing damaged spacecraft, installing new instruments within an existing one, or carrying out large-scale construction in orbit. And there was undeniable visual appeal in the spectacle of a single man, alone and unafraid amid the vastness, with all of Planet Earth in the background.

Korolev also liked spacewalks, and set out to beat the Americans by launching Voskhod 2. The cosmonaut Alexei Leonov would be the first to walk in space, while his buddy Pavel Belyayev remained inside. Leonov went out through an airlock, mugged for a camera, and then tried to get back in. He couldn't; the pressure in his space suit was too high. This made it rigid, like an inflated balloon, preventing him from bending at the waist to get his feet through the hatch.

He knew very well that he might die, and he came close to panicking as his pulse and breathing rate shot sharply upward. Then, taking stock of the situation, he turned a valve and reduced his suit's air pressure. It worked; now he could bend slightly, and he managed to catch the toes of his boots on the sill of the hatch. This gave him leverage, and he soon was safely inside. But as the analyst James Oberg notes, "He was soaked in sweat, his eyes were stinging from the salt in his perspiration and he was panting so hard he had fogged his visor."

Soon it was time for re-entry. The autopilot failed, and the cosmonauts had to return using manual control. Belyayev waited through one more orbit and then fired the retro-rocket, hoping to land in the flat steppes of Kazakhstan. Instead they wound up amid deep snow in the Urals, in a thick and mountainous birch forest, over a thousand miles from the recovery crews.

They had only their space suits to keep them warm, and nothing but the chilly Voskhod capsule to camp in. They soon found that they needed it, for when they built a fire, it attracted hungry wolves. The spacefarers abandoned the fire and fled to the safety of their craft, where they spent a highly uncomfortable night as wolves continued to howl. But their radio beacon was working, and they soon heard the welcome sound of an airplane. At dawn a ski patrol arrived, and both of them schussed out to a waiting helicopter.

Once again the Russians had successfully stormed the heavens. Yet this was to be their program's high-water mark; as events would demonstrate, Moscow would fail to keep pace with the surging American effort, unable to match its challenge. The Gemini missions now were at hand, and with them, America would take the lead.

The Apollo launch vehicles were not yet ready, but Gemini showed from the start that even with boosters derived from the ICBM programs of the 1950s, one could still do a great deal. The program's two-man spacecraft would ride to orbit atop the Titan II, a powerful variant of the original Titan of earlier years. It towered over a hundred feet tall, and like Moscow's R-16, which it closely resembled, the Titan II was about as large as a missile could be and still achieve deployment as a practical weapon. In turn, its size reflected its military mission: to deliver a nine-megaton warhead, one of the nation's largest and heaviest.

Like the R-16, Titan II was a two-stage rocket burning storable propellants. It didn't rely on nitric acid, the devil's venom, but instead used the higher-performing nitrogen tetroxide as its oxidizer, with hydrazine for the fuel. It was quite versatile, for in Air Force service it complemented the small and numerous Minuteman by adopting its key features: downhole emplacement in silos, along with quick-response launch. Between 1963 and 1965, as the Pentagon consolidated its strategic-missile deployments within an extensive force of the Polaris and Minuteman, the Air Force brought its Thors and Jupiters home from Europe while retiring its Atlas and Titan I squadrons. But fifty-four Titan IIs would remain on alert until 1987.

NASA also liked this rocket. Its first-stage thrust of 430,000 pounds represented a welcome improvement over the 360,000 of Atlas. It also had 100,000 pounds of thrust in the second stage, greatly enhancing its ability to lift heavy loads to orbit. Its development had not gone smoothly; there was a worrisome problem called "pogo," in which the rocket oscillated strongly in flight in the fashion of a pogo stick. No astronaut could ride such a booster, and the problem took a year of intensive effort to fix. But the finished missile was simple. In the words of Alvin Feldman, a program manager at Aerojet General, "We got rid of all the garbage. With Titan I, we had to fiddle with the lox right down to the countdown, but this time there was nothing we could worry about for the last two days before the flight. You preload, and that's the end of it."

The Gemini spacecraft had the external shape of the Mercury capsule, but it was considerably larger. Its two crew members sat side by side, looking out through windows set forward in what closely resembled a jet fighter cockpit. Like jet pilots, they each had an ejection seat. The spacecraft also went Voskhod one better by featuring a highly capable maneuvering rocket. Its weight of over eight thousand pounds compared favorably with the less than three thousand of Mercury. It therefore needed extra thrust to reach orbit. But a good second stage can markedly boost a rocket's power, and Titan II had one of the best.

The first important mission, Gemini 4, flew in June 1965 and featured its own spacewalk. The astronaut Edward White did not float freely in the void; he remained attached to his craft using an umbilical cord, a flexible hose that delivered oxygen and connected to the belly of his pressure suit. He also carried a maneuvering jet, with two small tanks of compressed gas. By squeezing on a trigger, he could release small bursts and produce thrust with which to move around. He spent twenty minutes floating freely, and enjoyed it thoroughly. When the command pilot James McDivitt told him to come back in, White replied, "Aren't you going to hold my hand?" When McDivitt insisted, White said, "It's the saddest moment of my life."

The mission stayed up four days, approaching the Soviet record of five. The next flight, Gemini 5, flew to orbit in August and remained there for eight days. This brought the record for duration to the United States, which would hold it until 1970. And that mission's astronauts, Gordon Cooper and Charles Conrad, were up there for a reason: to demonstrate that they could fly long enough to reach the moon and return. Medical exams found them to be in fine shape, and the program's chief surgeon, Charles Berry, summed things up: "We've qualified men to go to the moon."

The next several launches emphasized rendezvous, with Agena as the target. That upper stage was a veteran of over 140 flights since 1959, but its engine had been modified for use in Gemini. As a result, it blew up when it ignited at altitude, and the Gemini 6 astronauts, Wally Schirra and Tom Stafford, were left on the ground with no place to go. However, the program had a good deal of flexibility, and NASA officials learned very quickly that within little more than a week they could launch Gemini 7, prepare Gemini 6 for flight, and then send it to rendezvous with its twin. When President Lyndon Johnson announced that NASA would do just this, the die was cast.

Gemini 7 took to the sky early in December 1965, on a fourteen-day mission. Sanitary facilities weren't all that great, and for the crew members, Frank Borman and Jim Lovell, the experience amounted to sitting side by side in a men's room for two weeks, with no opportunity to take a shower. Eight days into their mission, it was the turn of Gemini 6. The main engines of its Titan II ignited—and shut down. An electrical plug had dropped from its socket in the booster, and a malfunction-detection system had sensed this, shutting down the engines and scrubbing the launch.

Schirra and Stafford were lying on their backs in the spacecraft, atop a fully fueled rocket that closely resembled Marshal Mitrofan Nedelin's R-16 and that might blow up in similar fashion. Each of them held a D-ring that would trigger an ejection seat when pulled, and no one would have blamed them if they had used it. But a launch restraint held the booster to its pad during that critical moment, preventing it from rising. If it had lifted off even slightly, it would have come crashing back down, breaking open its tanks amid an immense fireball. However, Schirra, the command pilot, had felt no upward motion, and he kept his cool, saving the mission. Here was a test pilot's presence of mind, but also his knowledge that an ejection seat itself carried the danger of serious injury. In his moment of peril, Schirra rode his rocket, knowing that he would eject only if the alternative was death by fire.

Three days later, Gemini 6 reached orbit uneventfully, and its astronauts proceeded to fly to rendezvous. This involved nothing so simple as pointing their craft toward the target and firing a rocket; the dynamics of rendezvous were considerably more intricate. If they tried to close a gap by speeding up, for instance, they would thrust into a higher and slower orbit and would actually lag behind. Instead, the thing to do was to slow down slightly, so as to drop into a lower orbit where they would fly faster. The maneuver made for an Alice-in-Wonderland world where you would speed up by slowing down, but Schirra and Stafford had an onboard computer programmed with the equations governing flight in orbit. An onboard radar, which could lock onto the target from 250 miles away, would determine its distance and relative motion, feeding the data to the computer, which would calculate the needed maneuvers.

Schirra saw the other spacecraft sixty miles away, reflecting sunlight and looking like a bright star. Both craft were in darkness as he made the final approach. They were only a few hundred feet apart when Gemini 7 flashed again into sunlight, and with the astronauts' eyes adapted to darkness, it was almost too bright to look at. Moments later came Stafford's calm voice: "We are 120 feet apart and sitting." At the mission control center in Houston, people stood up and cheered.

The two craft proceeded to fly in formation, approaching as close as a foot as crew members looked through each other's windows. A day earlier, Borman had radioed, "We're getting to the stage where we're starting to itch a little bit. We're just getting a little crummy." A plastic urine-storage bag had burst in his hand, with its amber-colored fluid breaking up into globules that floated in the cabin while he and Lovell tried to grab them. But their spirits were high. When Schirra said, "You've sure got big beards," Borman replied, "For once we're in style." And when the Naval

Academy graduates Schirra and Stafford held up a sign, "BEAT ARMY," Borman, a West Point man, replied with a card of his own: "BEAT NAVY."

There still remained the task of docking with an Agena in orbit and riding it to high altitude. In March 1966 the astronauts Neil Armstrong and David Scott made a good try in Gemini 8, as they nosed their craft into a docking port. But before they could initiate maneuvers with the combined spacecraft, they began to spin uncontrollably. When Scott separated the two craft, the whirling increased to a full revolution per second. The astronauts held their heads still, knowing that they might become too dizzy to regain control. "We have serious problems here," Scott radioed. "We're tumbling end over end, and we've disengaged from the Agena."

Both men knew how to stay calm. Scott, after all, had ridden his F-104 safely to a crash landing when his buddy in the rear seat had chosen to punch out; Armstrong had made his name in the X-15, flying it above 200,000 feet and reaching a speed of nearly 4,000 mph. The problem with Gemini 8 turned out to be a stuck attitude-control thruster, but it took awhile to find. Armstrong said, "All that we've got left is the re-entry control system." By cutting off the main attitude control and using this re-entry control as a backup, they managed to stabilize the spacecraft. But mission rules were firm: Any premature use of the re-entry control meant immediate termination of the flight. They came down near Okinawa, and when Armstrong was asked about his state of mind during the emergency, he replied, "I suspect you could categorize it as anxiety."

No such problem recurred during Gemini 9, for the excellent reason that the Agena's Atlas booster went out of control and plopped it into the Atlantic. Tom Stafford and Eugene Cernan flew to orbit anyway, and Cernan carried out the first EVA since Ed White had walked in space, a year earlier. Like all astronauts, Cernan was strong and vigorous, but EVA imposed exertions that proved too much for him. "He's doing four or five times more work than we anticipated," Stafford radioed. Breathing heavily and perspiring, Cernan saturated his space suit with more moisture than it could handle. His helmet faceplate fogged over heavily, blinding him, and his suit overheated. He had planned to spend nearly three hours outside the spacecraft, but after two hours he'd had enough. His heart had reached a peak of 180 beats per minute, and when he came back in, he was panting with exhaustion.

What had happened? He had had to fight his space suit with every move, working against its internal pressure, which caused it to resist being bent or flexed. Nor could he easily rest; small movements kept pushing him this way and that, requiring him to work still more to counteract these motions. This had produced little trouble during Ed White's brief EVA, but an excursion of several hours was another matter altogether.

Much the same happened on Gemini 11, another rendezvous mission. The astronaut Richard Gordon made his way forward onto the Agena, straddling it like a horse. "Ride 'em, cowboy!" called his partner, Pete Conrad. Already, though, Gordon was feeling pooped. As he continued his spacewalk, he developed fifty percent more heat than his space suit could carry away. His face streaming with sweat and his eyes stinging, he came back in after little more than a half-hour.

These experiences made it clear that NASA had badly underestimated the difficulty of EVA, and they called for a new approach. The astronaut Buzz Aldrin, who was slated for Gemini 12, took advantage of a new method for simulating EVA by wearing his pressure suit in a big tank of water, which allowed him to float as if in weightlessness. He learned to pace himself, and when he flew to orbit in November, his spacecraft mounted restraints and handholds that made it easy for him to stay where he wanted. His EVA lasted over two hours, and he completed a list of assigned tasks without becoming overheated or tired.

Rendezvous operations also achieved full success. In July, John Young and Michael Collins docked with their Agena and held the craft steady, then fired its engine for a boost that took them to 475 miles. "That was really something," said Young. "When that baby lights, there's no doubt about it." Eight weeks later, Gemini 11 did even better, reaching 850 miles.

When its Agena ignited, Pete Conrad radioed, "It's going, it's really going. Whoop-de-do!" A half-orbit later, approaching peak altitude, Conrad couldn't contain his excitement as he talked with controllers in Australia: "It's fantastic. You wouldn't believe it. I've got India in the left window, Borneo under our nose, and you're right at the right window. The world is round!"

Astronauts in low orbit had often said that flight in space was not so different from flight in a jet plane. The earth showed its curvature, but it still filled the field of view. An astronaut might fly twenty times faster than a jet pilot, but he also would be twenty times higher, so the world below would slide past at a familiar pace. But for Conrad and Gordon it was different. At the peak, their view from horizon to horizon spanned nearly five thousand miles.

The earth did not merely exhibit curvature; it was as round and sharply defined as a beach ball. From 460 miles above the Nile, a single photo took in Sinai, Israel, and Jordan. Soaring above the Indian Ocean, another snap of the camera captured all of India, from its southern tip to the distant Himalayas. High over Australia, one more shot covered one-third of that continent.

With ten manned flights in the course of twenty months, the Gemini program produced a cadre of trained astronauts for Apollo and made manned missions routine. Its achievements easily surpassed the best ones of Russia, while yielding practical experience of enormous value in such critical areas as rendezvous and EVA. Nor was the program costly; its total budget, from start to finish, came to $1.15 billion. Gemini proved versatile, flexible, and reliable, and if its capabilities had addressed ongoing needs—if it had been more than a prelude to Apollo—then its spacecraft would be flying to this day.

Meanwhile, preparations for Apollo progressed apace. One center of activity was the immense Vehicle Assembly Building (VAB) at Cape Canaveral, a cube measuring 674 feet long, 513 feet wide, and 526 feet high. Its huge slablike sides would catch the full force of a hurricane, so its architect, Max Urbahn, anchored it to bedrock. The rest of the facility showed a similarly outsize scale.

Within the VAB, launch crews would assemble the Saturn V atop a platform carrying a 380-foot support tower. A crawler, a vehicle weighing 3,000 tons, then would move beneath the platform and lift it to ride atop a flat surface the size of a baseball infield. Eight tractor treads would begin to clank, each forty feet long and ten feet high, and the entire array—crawler, platform, tower, rocket, all together coming to 9,300 tons—would make its way through a door large enough for a 45-story office building. The crawler would head for a launch pad, three miles away, and in this fashion the mission would set out for the moon at the speed of those treads—one mile per hour.

There was further activity southeast of Los Angeles, where North American Aviation and Douglas Aircraft were building the second and third stages of the Saturn V. Their plants were only a few miles apart, in the adjacent towns of Seal Beach and Huntington Beach, which made it easy for their engineers to meet for lunch—perhaps at the Huntington pier, a place beloved by surfers. The third stage, the S-IVB, was nearly twenty-two feet across, giving it the diameter of a Boeing 747 fuselage. This raised interesting questions as to how to transport it, which Douglas solved by turning to a local entrepreneur, John Conroy.

Conroy owned a number of surplus Stratocruiser airliners, piston-powered and dating to the 1940s. He fitted one with a bulbous and distended fuselage, christened it the Pregnant Guppy, and offered it to NASA for the transport of large rocket stages. He was operating on a shoestring; to fly to Huntsville for a presentation to von Braun, he had to borrow aviation gasoline from a friend in Oklahoma. But he was persuasive, and won the contract he sought. For the S-IVB, he crafted the even more bloated Super Guppy, which looked like a blimp with wings. His company, Aero Spacelines, went on to launch a new industry: transport by air of outsize and bulky cargoes.

The S-II, the second stage, was bulkier still: thirty-three feet across. Riding on a specialized truck, it required all four lanes of Seal Beach

Boulevard, as the ultimate wide load. That street led to a nearby Navy harbor, where the stage went aboard a barge that would set out for the Panama Canal, on a sea voyage that would end at the Mississippi Test Facility. The first stage, built at Michoud, also traveled by barge and reached MTF in similar fashion.

That test center stood amid muck-sodden country that was dense with foliage, humid, and rank with mustiness. Alligators roamed here; the sun sank at dusk with terrifying suddenness, leaving a primeval and fearsome darkness. Yet now there were lights amid the dark, the stark and brilliant illuminations of the test area. When a Saturn V stage was in place for a night firing, its bright flame would cast a glow across the land. During the brief minutes of its firing it would hold back the night. And in that state, one could cherish the dream that somehow there would be other lights, brighter and stronger, to drive shadows from the hearts of men.

These preparations, combined with the success of Gemini, presented a challenge that Moscow could not ignore. Indeed, the Soviet Union was responding with its own manned lunar program, and as with so much else in that country's space activities, it was growing out of military requirements. It also was growing out of the work of a new leader in rocket development, Vladimir Chelomei.

Chelomei started as a designer of submarine-launched cruise missiles, and he scored a real coup by hiring Sergei Khrushchev, the son of Nikita, as a guidance-systems engineer. This gave him access to the top level of Soviet government. Hoping to enter the realm of new ICBM projects, Chelomei cultivated an association with Valentin Glushko, who had designed the engines for the R-7. Glushko was turning sharply toward the use of storable propellants and was preparing to build engines for both stages of Mikhail Yangel's new R-16. Chelomei agreed: Yes, he would gladly use such engines in his own designs.

His chance came in 1960, when he received permission to take over a major aircraft-development center in Fili, a district of Moscow. The planebuilder Vladimir Myasishchev had run it. He had built jet bombers with enough range to reach the United States, but Khrushchev was moving away from bombers and toward reliance on ballistic missiles. Myasishchev's bombers had not pleased him, and he was not willing to give that planebuilder a major role in the missile program. By contrast, Chelomei stood in Khrushchev's good graces. And Chelomei had a far-reaching plan for new rockets, including an ICBM to carry a thirty-megaton warhead.

Khrushchev liked big bombs; a 50-megaton version, tested in November 1961, had the power to turn the entire New York metropolitan area into a Hiroshima. But these weapons were heavy, for they relied on a large casing of uranium to receive neutrons from a thermonuclear core and to produce much of the yield through fission reactions. No existing ICBM could carry such weight, not even the R-16. The Kremlin authorized the development not only of Chelomei's missile, but also of a new engine.

Glushko received the assignment to build this new engine, the RD-253. Its thrust was 150 metric tons, or 330,000 pounds, and it would burn storable propellants. Chelomei's rocket, the UR-500, would use six of them for a thrust of two million pounds, somewhat more than the 1.5 million of von Braun's eight-engine Saturn. Significantly, while the builders of Saturn had failed to find a military use for it, the UR-500 had one from the outset. But U.S. strategic doctrine had no need for thirty-megaton bombs, let alone for an ICBM to carry them.

It was clear from the start that the UR-500 would make a superb space booster. In 1961 Khrushchev directed Chelomei to use it for this purpose, and to develop a spacecraft that could carry a man on a looping flight around the moon. This was the same circumlunar mission that NASA had proposed to Eisenhower the previous December. In this new role, the UR-500 gained the new name of Proton.

Korolev by then was completing preliminary designs for his own heavy-lift boosters, and he faced the question of engines. He could have had Glushko's for the asking, but they used the devil's venom and he wouldn't touch them. Nor could he arrange for Glushko to develop new versions that were more to his liking, for Glushko also had his hands full with a separate project, a 1.3-million-pound engine that would also use storable propellants. But Korolev had an ace in the hole. He had been cultivating a new manufacturer of rocket engines: Nikolai Kuznetsov, who had a background as a builder of aircraft turbojets.

Korolev had turned to Kuznetsov earlier and had helped him get a start in rocketry by directing him to develop the NK-9, an engine that Korolev sought to use in his R-9 ICBM, which had competed with Yangel's R-16. Korolev now arranged for Kuznetsov to develop uprated counterparts of the NK-9, designated NK-15 and NK-15V, for the first and second stages of his proposed booster. He would use the NK-9 as well, for the third stage.

On its face, this plan might have involved the sort of straightforward decision that the Martin Co. had made in designing the Titan II, where a preference for storable propellants had led that firm to pick engines from Aerojet rather than Rocketdyne. In fact, the decisions in Moscow touched off a full-blown feud between Korolev and Glushko.

The roots of the dispute traced to an earlier disagreement over engines for that R-9 ICBM. Glushko tried to build them but ran into serious problems. Korolev went to the Kremlin and insisted that he be allowed to drop Glushko's engine in favor of Kuznetsov's new NK-9. This episode recalled the competition of 1958 over an engine for the R-7 upper stage.

Glushko had lost that one, to Semyon Kosberg, another newcomer whom Korolev had also assisted. But this time, Glushko would not accept Korolev's clear penchant for turning inexperienced rocketbuilders into major rivals. He had clout in Moscow, and appealed to a commission. It ruled against Kuznetsov, forcing Korolev to use engines from Glushko in the R-9 after all.

The engines had used kerosene and liquid oxygen, the propellants Korolev preferred. For his proposed new booster, the question of using storable propellants intensified the quarrel. Korolev loathed storables; he remembered the death of Nedelin. By contrast, Glushko saw them as a key to the future, promising simplicity in design. In an ICBM, they would permit deployment in silos along with rapid launch. For space boosters, storables meant reliability and fewer launch failures.

After Korolev rejected Glushko's RD-253, Glushko returned the favor by declaring he would have no part of Korolev's big new rocket, including engines for its manned spacecraft. If Korolev was so fond of that neophyte Kuznetsov, so willing to build him up at Glushko's expense, he could go to Kuznetsov for all his engines and have done with the matter. Korolev was well aware that Glushko's engineering center had the most experience. But he would not go there for his engines, for the devil in the venom now was Glushko himself.

The dispute became personal. Glushko belittled Korolev's work, saying that with a good engine, even a broomstick would fly. The fight set back Korolev's design efforts while eroding the political support he needed. Khrushchev himself intervened, trying to patch things up between the two longstanding partners, but even he couldn't succeed. With their longtime association in tatters, Korolev indeed turned to other sources, including Kuznetsov, for his additional engines.

Korolev's new launch vehicle, the N-1, gained an acceptable design in mid-1962 amid a careful review by a commission headed by Mstislav Keldysh, who now was president of the Academy of Sciences. The design called for a true moon rocket, a counterpart to the Saturn V. It, too, was to put cosmonauts on the moon, and while Korolev at first expected to use two or three launches to assemble a lunar spaceship in earth orbit, in time he too adopted the technique of lunar-orbit rendezvous. A single N-1 then would launch a large two-man spacecraft to orbit the moon, sending a small lander to the surface with a single cosmonaut, for subsequent return to this mother ship.

Within the Soviet Union's landlocked expanses, Korolev had no opportunity to follow Saturn V practice by building the rocket's stages far from the launch center and shipping them in barges. Instead he set up an assembly plant right at Tyuratam, to build the entire N-1. Workers would receive rocket engines from Kuznetsov in the city of Kuibyshev, assembling the three stages on site, then joining them horizontally. The complete N-1 would take form lying on its side in a big cradle.

Vladimir Barmin, Korolev's longtime chief of launch-complex design, followed his experience with the R-7 in arranging to fire the N-1. As with that ICBM, the moon rocket would ride on railroad tracks to its launch pad, still in its cradle. A huge hydraulic erector would set it upright, swinging this immense booster, 345 feet tall, through a ninety-degree arc. Technicians then would fuel it and prepare it for flight.

Just then, in mid-1962, NASA's F-1 engine was well along in development, having run at full thrust and for full flight duration in May. This achievement reflected America's early commitment that dated to 1958, and gave this country a four-year lead. Korolev had no such engine in view; he would have to get along with Kuznetsov's NK-15, which had 340,000 pounds of thrust to the F-1's 1.5 million. In his eventual design, Korolev used thirty of them, for a total thrust of over ten million pounds, greatly exceeding the 7.5 million of the Saturn V.

Here was a genuine Nova-class rocket, coming as close as the world would see to the 51-engine behemoth that von Braun had discussed in *Collier's* a decade earlier. Yet it was less capable than the Saturn V, for it would carry only one cosmonaut to the lunar surface, compared to two for Apollo. Moreover, while the Saturn V was slightly heavier, it needed considerably less thrust. The reason was that the N-1 had extra engines, in the expectation that some would fail in flight.

Indeed, the entire N-1 effort was guided by an anticipation that the enormous boosters would fail during test launches. This was not fatalism; it again reflected the R-7 experience, for Korolev's designers had learned lessons from that rocket's fiery crashes, and had gone on to build a reliable vehicle. The N-1 program had no true counterpart of NASA's Mississippi Test Facility. Though Korolev had facilities for the static firing of complete N-1 upper stages, there was no center where its critical first stage could undergo such static tests. Though Kuznetsov would test his engines with care, their multitudinous arrays would thrust only during flight. Korolev knew that he could expect some spectacular fireballs, and he was prepared to accept them.

Meanwhile, Chelomei's Proton would go first. The initial flight to orbit, in July 1965, lofted a 27,000-pound payload and demonstrated anew that storable propellants brought simplicity and reliability, even when six of the engines had to work at once. The Proton then repeated this success in November with a similar flight. America's Saturn had orbited heavier loads, as test versions of the Apollo spacecraft. But these two missions set a weight record for Moscow.

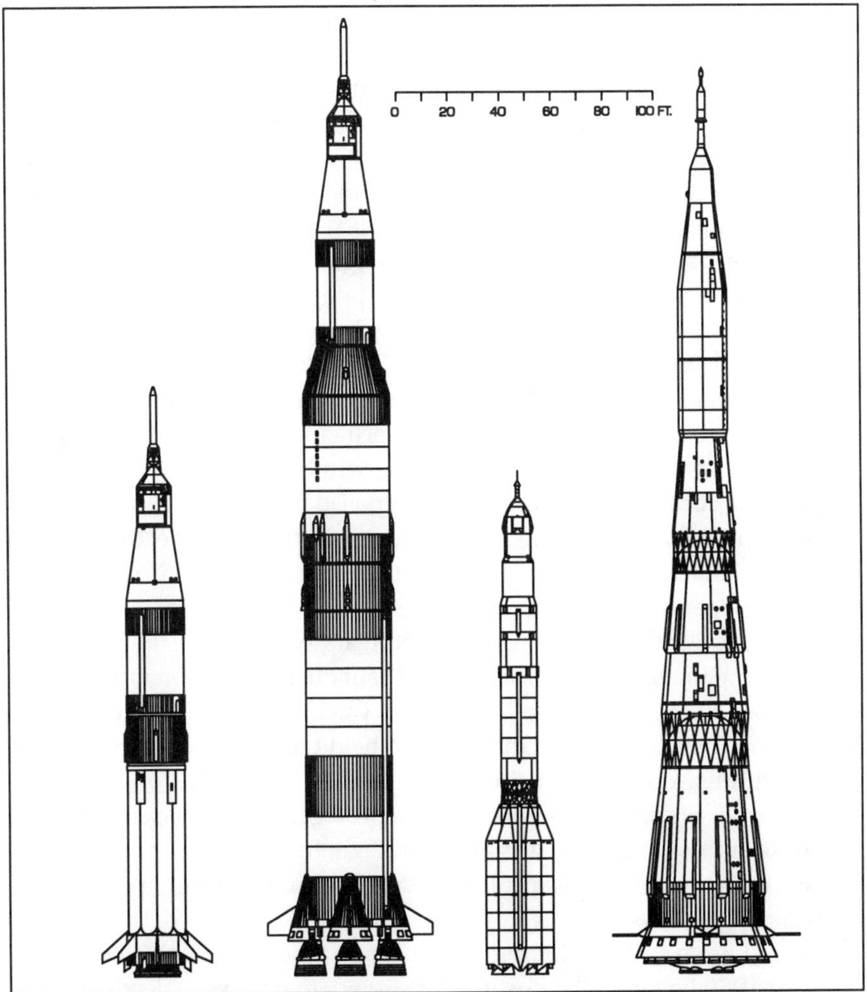

The moon race: U.S. Saturn I-B and Saturn V; Soviet Proton with Zond and N-1. (Dan Gauthier)

Chelomei was developing the manned spacecraft that would ride atop a Proton for its circumlunar flight. But his engineering center lacked experience with such craft, which gave Korolev a new opportunity. His own group's experience in this area was unmatched, for it had built both Vostok and Voskhod. In addition, Korolev now was developing a new manned spacecraft, Soyuz (union). It was to carry out rendezvous and docking, matching the achievements of NASA's Gemini.

Korolev's opportunity was particularly bright because Khrushchev, Chelomei's patron, had fallen from power in October 1964. By contrast,

Korolev was still in the good graces of his own mentor, Dmitri Ustinov, who now was heading the space program. In September 1965, Korolev moved boldly to take the Proton's circumlunar spacecraft from Chelomei by replacing it with a highly modified Soyuz called Zond (probe). He confronted Chelomei personally in a series of acrimonious exchanges, aggressively attacking his designs and pointing out shortcomings. Korolev won the contest and thereby found himself responsible for the circumlunar mission as well as for the moon landing.

Korolev was still Chief Designer. Despite his quarrel with Glushko, and in the face of challenges from Yangel and Chelomei, he now commanded a more extensive empire than von Braun's. As the director of NASA's Marshall Space Flight Center, von Braun was responsible for all Saturn-class boosters and also for their principal engines. These engines came from Rocketdyne, but its president, Sam Hoffman, looked to Huntsville for his contracts. Von Braun's domain also included Michoud and the Mississippi Test Facility. But Apollo's lunar craft fell inside the domain of Robert Gilruth's Manned Spacecraft Center in Houston, while Cape Canaveral operated as a separate and co-equal NASA center.

Korolev, like stout Cortez of the poet Keats, could see farther horizons. His purview included Soyuz and its R-7 launcher, in the version with the new upper stage. The Proton belonged to Chelomei and Glushko, but Korolev owned its Zond spacecraft. He also held the entire N-1 moon rocket along with its moonship, including the lander. Most of his engines were Kuznetsov's, whom he had nurtured as a protégé. The N-1 assembly plant, a counterpart of Michoud, was also Korolev's. In addition, Tyuratam's N-1 launch complex and transport facilities belonged to Barmin, a close associate since the days of the V-2.

Still, Korolev would not remain for long at the helm of this empire. In January 1966 he entered a hospital for surgery to remove a bleeding polyp in his lower intestine. His high status ensured that he would receive excellent care, and his surgeon, Boris Petrovsky, was sufficiently renowned to hold cabinet rank, as minister of health. But Korolev hemorrhaged on the operating table, bleeding severely. Petrovsky tried to stop the flow of blood by performing surgery on the abdomen—and found a massive cancerous tumor that had escaped detection. The operation now would take much longer. Petrovsky brought in a highly experienced colleague to help him remove the cancerous tissue. Together these two surgeons completed the operation, controlling the bleeding and suturing the incisions.

However, all major surgeries place considerable strain on a patient, and Korolev was not in good shape. He was overweight, and his heart was weak. The operation, though apparently routine at the outset, took eight hours. Before Korolev could regain consciousness, his heart gave out. He was fifty-nine years old.

His duties went to Vasily Mishin, his deputy and a colleague since 1945, but this was not the same as replacing him. Everyone knew a mighty oak had fallen, and Korolev's ashes were accorded the high honor of interment in the Kremlin Wall. It is also fitting to salute this man with words written to mark the passing of Abraham Lincoln:

> And when he fell in whirlwind, he went down
> As when a lofty cedar, green with boughs
> Goes down with a great shout upon the hills
> And leaves a lonesome place against the sky.

In that sky, Protons and Saturns were flying successfully during tests, though not yet with people aboard. But this was about to change, and again the United States would take the initiative. The eight-engine Saturn of earlier years now had given way to its definitive version, the Saturn I-B. Each of its engines now delivered a full 200,000 pounds of thrust, fifty percent more than their counterparts of a decade earlier, for a total thrust at liftoff of 1.6 million pounds. Its second stage, the S-IVB, carried an Apollo spacecraft, though without the lunar lander. It was to carry this craft to orbit early in 1967, and its three-man crew was ready. The commander, Gus Grissom, was a veteran of both Mercury and Gemini. Alongside him were Ed White, who had spacewalked during the Gemini 4 mission, and Roger Chaffee, a rookie.

They were lying on their backs in their spacecraft early in the evening of January 27, 1967, conducting a lengthy training exercise. Chaffee said in a rather casual tone, "Fire, I smell fire." A moment later came White's voice, more insistent: "Fire in the cockpit." In the blockhouse, disbelieving ground controllers watched on closed-circuit television as flames licked furiously and smoke filled the cabin. Someone yelled hysterically, "There's a fire in the spacecraft!"

In fifteen seconds the internal pressure climbed to nearly three atmospheres, and the craft's walls ruptured. Flames swirled across the cabin and then burned down, forming dense smoke along with carbon monoxide. The astronauts all sustained major burns, but those were not life-threatening. They died of asphyxiation.

What had happened? The Apollo spacecraft had a life-support system that would fill the cabin with pure oxygen when in orbit, at a pressure of five pounds per square inch, one-third of an atmosphere. This would actually provide more oxygen than exists in sea-level air. But in space this would pose no fire hazard, because a fire could start only if a flame's hot combustion gases rose through buoyancy, permitting fresh oxygen to reach the flame. In weightlessness, there would be no such buoyancy; the burned gases would stay close to the fire and would smother it before it could spread.

During Mercury and Gemini, NASA had followed the practice of using the cabin oxygen system in ground tests, filling the spacecraft with this gas at a pressure slightly greater than atmospheric. The custom had carried over to Apollo; at the moment of the fire, the astronauts were immersed in pure oxygen at 16.7 pounds per square inch, over five times the sea-level concentration. The test pilot Scott Crossfield would say that this was "like they were in a thermite bomb."

A quick-release hatch might have saved them, and the Mercury capsules had had one. Ironically, it was Grissom's own flight, in mid-1961, that had led NASA to drop such hatches in subsequent spacecraft. Grissom's capsule had come down by parachute in the Atlantic—and he had accidentally triggered the explosive bolts that blew his hatch open, prematurely. His craft filled with water and sank, while Grissom himself, treading water in a clumsy space suit, had to swim to safety and came close to drowning.

The eventual Apollo modifications were straightforward: a hatch with emergency release, a major effort at fireproofing the cabin, and, most important, a two-gas system that would dilute the oxygen with nitrogen during ground tests. Still, as von Braun put it, "We had a blind spot." The man who now was on the spot was Stormy Storms, president of North American Aviation's space division.

Storms was responsible for the Saturn V's second stage, the S-II, as well as for the Apollo spacecraft, and NASA viewed his performance on the S-II as less than stellar. It set records for light weight; to keep its weight down, the stage's builders had avoided using rivets, designing it instead for welding. It demanded some 3,000 feet of welded seams, surgically clean and flawless, and many of them accurate to 1/75 of an inch. Moreover, the designers had specified an aluminum alloy that was not favored for welding. This choice represented another weight-saving measure, for the alloy would gain strength when exposed to the stage's supercold liquefied-gas propellants, thus permitting less metal to be used in the structure.

This was no task for a traditional welder, with sparks flying from a torch near his helmeted head. It demanded automated equipment. But assembly of the first S-II stages ran into delays, and it didn't help when two of them were destroyed during tests. NASA officials became very unhappy, and Storms had to accept a shake-up of his project management before the difficulties came under control. Now, with his Apollo spacecraft having shown similar troubles, NASA insisted that he would have to go—or else North American might lose its contract.

His boss, Lee Atwood, viewed this as a gross injustice. NASA, not Storms, had made the decision to fill the spacecraft with high-pressure oxygen. Indeed, that agency had taken on full authority for activities at the Cape, and had not even bothered to tell Storms what its people were doing. But someone's head had to roll, and Stormy's was convenient. He had shown bad karma; Napoleon had fired generals for being unlucky, and NASA was ready to do the same. After all, its top officials had to feel comfortable with a man holding Stormy's position; they didn't want to wait for more bad news from him. Nor did this agency ever admit that it was at fault for the fire, with Storms being merely a scapegoat. Storms nevertheless stayed on at North American, demoted to a vice president. He now could only recommend; his new management alone had the authority to act.

The new shake-up replaced Storms with William Bergen, who had been president of the Martin Co. With Bergen came two key associates: Bastian Hello, to run North American's Cape Canaveral facility, and John Healy, to take charge of the Apollo spacecraft. "They Martinized the top levels of management," complained Crossfield, a loyal Storm Trooper. But the shake-up produced results, as the company's performance came up to the mark.

The Apollo fire represented a severe setback, and the Soviets soon experienced a similar tragedy with Soyuz. It had undergone three unmanned qualification flights, one of which blew up on the launch pad. Only one mission managed to return safely, and it had to be fished out of the Sea of Azov after it sank there. Nevertheless, in April 1967 the Soviets set out to launch two of them, in another dual mission. One would carry a single cosmonaut, Vladimir Komarov, who had commanded the three-man Voskhod mission of 1964. The second Soyuz was to carry its own three-man crew and would rendezvous and dock with the first. Two cosmonauts then would make their way into the first Soyuz; both of these craft then would return to earth.

Komarov flew to orbit, but one of his solar panels failed to deploy. The resulting power shortage left him without the energy the craft would need to perform the rendezvous, so the second Soyuz stayed on the ground. "Devil-machine, nothing I lay my hands on works!" he scowled, as he wrestled with balky onboard systems. He managed to stabilize his spacecraft, but he returned from orbit after only a day.

His parachute system had seen extensive tests in unmanned orbital missions. This time, though, the main chute failed to deploy while a reserve chute became tangled with the first one. His craft hit the ground at high speed and killed him. His death resembled that of the Apollo crew; whereas Apollo lacked the quick-release hatch that had been a feature of Mercury, Komarov's Soyuz had no personal ejection seat such as that in Vostok. Soyuz was a three-man ship, and as with the earlier Voskhod, ejection seats would have required too much weight and room.

Yet if a simple parachute landing could kill a cosmonaut, the planned circumlunar mission was vastly more dangerous. The Zond, a variant of Soyuz, would re-enter the atmosphere at particularly high speed. If it came in too steeply, it would experience excessive drag and deceleration, subjecting its crew to g-forces of sufficient magnitude to injure or even kill. Hence, the capsule had to spend enough time in the upper atmosphere to bleed off its velocity safely, prior to terminal descent.

Apollo faced the same problem, and solved it with a re-entry craft that could develop a modest amount of lift. Astronauts could use this lift to shape their path during atmosphere entry, maintaining optimal altitudes on the way in. Moreover, the arrangement followed the practice of Gemini, which also had used lifting re-entry to control the descent and to land at a specific spot. But Zond had much less lift during entry, and faced a particularly demanding task.

Returning from the moon, Zond would have to hit a corridor, an acceptable band of altitudes as little as six miles from top to bottom. In this corridor, Zond would lose some of its speed to atmospheric drag and would slow from its incoming seven miles per second to around five. It would exit briefly into space and then return for a conventional entry, as if from orbit. The margin for error would be narrow indeed. Too low and the cosmonauts would face excessive g-forces. Too high and the craft would skip into space like a flat stone over a pond, losing little velocity. Before it could return to the atmosphere for a second pass, the crew would run out of oxygen and die.

Hence, before Zond could qualify to carry cosmonauts, it first would have to demonstrate through actual flight test that it could find the corridor—under automatic control. But if it succeeded, Moscow might yet beat the Yankees to the moon. The circumlunar mission would merely send its crew on a quick loop around the farside, but this could nevertheless qualify as the first manned lunar voyage. Later achievements—first men to orbit the moon or first men to land on its surface—then might appear as no more than refinements.

During the fall of 1967, two Proton launches sought to carry out the unmanned test flights. On the first, one of the six first-stage engines failed; on the second, it was one of the four engines in the second stage. The resulting unbalanced thrust doomed both attempts. But early in March 1968, the booster worked successfully, flinging its spacecraft, Zond 4, to lunar distance. It did not approach the moon; on the contrary, it flew in the opposite direction, for a simple trajectory that could yield a clean exercise in atmosphere entry. The craft did not do this well; it entered much too steeply, because of an onboard malfunction that resulted in an attitude-control error. Falling far from its recovery area, Zond 4 blew up in flight as a ground controller triggered an onboard charge of explosive, a precaution against having it fall into the wrong hands. Still, the flight constituted an important step toward the moon.

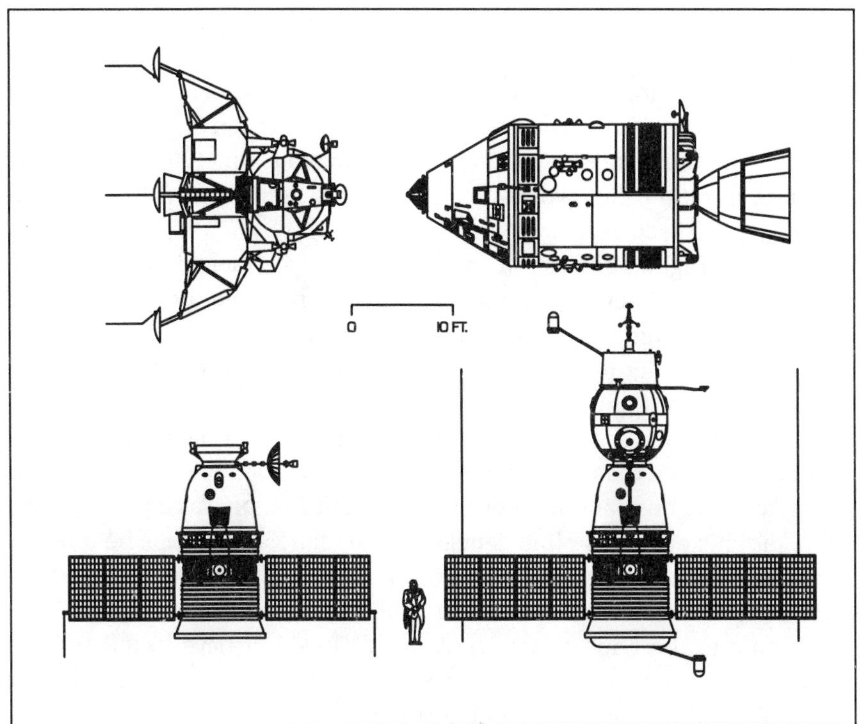

Manned spacecraft of 1970. ***Top,*** **U.S. lunar module and Apollo; Apollo spacecraft is in position to dock with the LM.** ***Bottom,*** **Soviet Zond and Soyuz, with figure for scale. (Dan Gauthier)**

At the same time NASA was recovering from the Apollo disaster of a year earlier. Prior to that fire, it had already qualified the S-IVB and the Apollo spacecraft, in successful flights of the Saturn I-B. Then in November 1967, the complete Saturn V scored its own success, brilliantly vindicating George Mueller's insistence on all-up testing. Its first and second stages were both flying for the first time, each with five engines. All worked properly, as this immense rocket, 363 feet tall, thundered into space atop a bright pillar of yellow-white flame. In a CBS News studio, the commentator Walter Cronkite shouted, "Oh, my God, our building is shaking! Part of the roof has come in here!" The roar was as loud as a major volcanic eruption, and people saw the ascent in Jacksonville, 150 miles away.

Five months later, in April 1968, the Saturn V flew again, in another unmanned test. The results threatened Apollo with a major setback.

The problems involved Rocketdyne's J-2 engines, in the second and third stage. The No. 2 engine in the S-II showed a falloff in combustion-chamber pressure, then lost all pressure and shut down after 263 seconds of operation. This caused the adjacent No. 3 engine to shut down also. An onboard computer compensated for the loss of the two engines, swiveling the three surviving rocket motors to maintain balance and directing a longer burn time. The single J-2 in the third stage then showed a similar falloff in performance, but continued to operate for 170 seconds, the planned duration. But when this engine was commanded to restart, it could not be made to do so.

On this flight, although the spacecraft indeed reached orbit, three of six J-2s had malfunctioned, either shutting down prematurely or failing to obey restart commands. This meant the engine was unreliable. Despite successes in earlier test flights, it could not be counted on to carry astronauts on a lunar flight.

Paul Castenholz was the J-2 project manager, and he knew he faced trouble. He had a casual and athletic air that accorded well with his reputation as an intense and hard-driving man. Further, he was a perfectionist who kept his eye on the fine details. "At the end of the flight," he remembers, "we went into a big meeting room at Cape Canaveral." Ranged before him were the directors and managers of the entire Apollo effort, who had come to the Cape to see the launch. "I said, 'I have to tell you we don't know what happened. But we will immediately go on a 24-hour schedule, and we will keep you informed.'[5]

"It was probably my lowest point as a rocket engineer," he adds. "I was thinking, 'What could be worse than this?' I remember driving down the street, coming home, thinking 'This is terrible! This is the wrong place to be!'" Back at Rocketdyne, Castenholz organized a round-the-clock work schedule. Sam Hoffman, the president, called up every person he could spare from other programs. Castenholz brought in cots, so people could stay in their offices overnight.

Why was the problem so serious? Not only did no one know why the engines had failed; there was no clear path toward finding the reason. There was no failed hardware to examine; the S-II had burned up in the atmosphere, while the S-IVB was in orbit and out of reach. And data from telemetry was meager. The J-2 had performed well on the earlier flight of the Saturn V, as well as during four Saturn I-B launches over the previous two years. Thus, flight by flight, data channels had been taken away from the J-2s and reassigned to other systems. Nor had a J-2 ever failed in this fashion in ground test.

However, the available telemetry sufficed to point to a culprit: an auxiliary fuel line, which might have leaked or ruptured. Rocketdyne's managers laid on J-2 test firings that subjected this line to unusual strain, but it held. This suggested that the little information at hand might be leading them down a blind alley.

The break came within only a few days. Castenholz recalls talking with a colleague, Marshall McClure, who asked, "Would it be different in space than on the ground?" They were spending a lot of time watching movies of J-2 tests on the stands at Santa Susana, and when Castenholz looked at those films again, he saw what he had not previously noted or viewed as significant. Ice was building up on many of the fuel and oxygen lines, as they carried their supercold liquid gases. But in space, there would be no way ice could form.

Why was the ice significant? Many J-2 propellant lines were corrugated to make them flexible, like a vacuum-cleaner hose. Being flexible, they could vibrate. If they vibrated too strongly, a line might break. But at Santa Susana, ice building up within the corrugations—known as bellows—could damp the vibrations and keep them from increasing in strength. That might be why a line would work properly in ground test, but would weaken and break after a few minutes of rocket operation in space. The suspect auxiliary fuel line "was covered with a stainless-steel mesh," says Castenholz. "So it didn't jump out at you that it was a bellows. But when you watched the film, you could see the ice form."

Rocketdyne had a specialized test chamber, in which Castenholz could study these components under vacuum. His engineers put eight of the fuel lines into this facility, operated them under conditions duplicating those of a J-2 firing in space—and saw all eight of them rupture. Hence this line was indeed the culprit, by breaking in space and causing the engines to shut down. The problem had evidently been a hit-and-miss affair, occurring in some J-2 motors but not in others. Nevertheless, the fix was simple: a different type of stainless-steel pipe. This minor but crucial adjustment amounted to the king's horseshoe nail, for want of which the kingdom was lost. But now, with this part in hand, Apollo could proceed at full throttle.

The next Saturn V flight, the third, would carry astronauts, and NASA expected to launch both the Apollo spacecraft and the lunar lander. This rocket then would carry all the equipment for a moon landing, but it would not head for the moon. Instead these craft would remain in earth orbit, conducting rendezvous exercises in preparation for the main event. However, during the summer it became clear that the lander would not be ready to fly for several months. Nevertheless, the third Saturn V was being made ready at Cape Canaveral, together with its Apollo spacecraft.

In August George Low, the director of Apollo at Houston, proposed to boldly fly with that combination—and go for the moon. He won quick support from his boss Robert Gilruth, the director of the Manned Spacecraft Center, as well as from other key leaders, including von Braun. But it would be another matter altogether to win formal approval at NASA headquarters in Washington, for Low's plan was premature. The program still had not carried out the intended mission of the three astronauts who had died in the fire: to operate the spacecraft in earth orbit by flying it aboard a Saturn I-B, sans lunar lander. Such a mission was on the calendar as Apollo 7, but it would have to succeed before anyone could truly commit to go to the moon.

Meanwhile, in September Moscow gave NASA a push with Zond 5. It showed a considerable advance over Zond 4, earlier in 1968, for it flew around the moon, approaching within 1,200 miles. It then began its return, maneuvering to hit the corridor in the atmosphere and to achieve the proper altitude. On the way back, it obtained a fine photo of the earth. An error by a ground controller led again to an excessively steep re-entry trajectory, but the craft came down in the Indian Ocean and was recovered. Sir Bernard Lovell tracked its flight at Jodrell Bank, and when the craft reached Moscow, it held two turtles who had come back safely. An impressed James Webb described this flight as "the most important demonstration of total space capacity up to now by any nation."

Two months later, Zond 6 did even better by successfully executing its planned return. It hit the corridor accurately, dipping within twenty-eight miles of the earth's surface while decelerating to 4.7 miles per second, which was just below orbital velocity. Briefly returning to space, it then made its second atmosphere entry, coming down inside the standard recovery area in central Asia. It had encountered problems: a faulty rubber gasket that brought about loss of cabin pressure, along with parachute problems that caused the vehicle to crash rather than land gently. But it had flown around the moon, reaching as close as 1,500 miles, and had excellent photos to show for it. And with the successful re-entry, it had carried off its most difficult task.

The Zond effort now had demonstrated all the features needed for a safe manned circumlunar flight, though not on the same mission. But the way lay open for Moscow to put the pieces together and to commit to a manned mission, perhaps as soon as the next flight. To underscore their readiness, the Soviets also completed a successful Soyuz mission in October, with the cosmonaut Georgi Beregovoi spending three days in orbit. And Korolev's great N-1 moon rocket was not just waiting in the wings; it was already at Tyuratam. The first flight-ready version had rolled out the previous May, and reached the launch pad in November.

But NASA was also ready. Apollo 7 saw a success of its own that same October, as Wally Schirra, Walt Cunningham, and Donn Eisele flew the Apollo spacecraft to orbit and stayed aloft for eleven days. This achieved final exorcism of the ghosts from the Apollo fire and led within weeks to the commitment George Low had sought. The next flight, Apollo 8, would indeed go to the moon. In humanity's first voyage to that focus for so much hope, the spacecraft would not merely swing around in the fashion of Zond. It would fire its engine to enter lunar orbit, circling the moon ten times before lighting its rocket anew to break from the moon and return home.

Where were you on the morning of December 21, 1968? If your name was Frank Borman, James Lovell, or William Anders, you were reclining in your astronauts' couches in your spacecraft, high atop a Saturn V, ready to begin the expedition that would win the race and put America in the lead to stay. And if your name was Walter Cronkite, you were anchoring the live television coverage as you remained calm, cool, unflappable.

Then, at the moment of liftoff, even Cronkite was caught up in the passion of the event:

". . . twelve, eleven, ten, nine. We have ignition sequence start. The engines are on. Four, three, two," intoned Jack King, from NASA's mission-control center. And Cronkite added, "She's building up and you can see the fire."[6]

"Coming off, we have commit. We have liftoff," said King.

"She's clearing the tower, on a bright tower of flame," said Cronkite, his voice now resounding with emotion. "We're getting the shaking; the tremendous blast is hitting us now," as he struggled to make himself heard over the roar. "She's going straight; she's in the roll program; she's looking good—a bright sapphire flame! Seven and a half million pounds—they're on their way!"

9

Lunar Aftermath

Space Stations and the Shuttle

THE FLIGHT OF APOLLO 8, late in 1968, took place at the height of the Vietnam War. An important reason for Apollo, back in 1961, had been the wish to impress Third World nations with America's technical strength, thereby dissuading them from going Communist. The Vietnam commitment had grown from the same root, with both Vietnam and Apollo stemming from the leadership of President Kennedy. In his inaugural address, also in 1961, Kennedy had declared that the nation would "pay any price, bear any burden" to win the "long twilight struggle" against Communism. But the world had changed since then. The Cold War was easing, making it possible to envision withdrawal from Vietnam on terms short of victory. And because NASA had evolved as an instrument of the Cold War, it too would change.

The waning of the struggle had begun in the wake of the Cuban missile crisis. Within months, American and Soviet diplomats joined to write a treaty to outlaw the testing of nuclear weapons aboveground. This represented an important step toward arms control: It demonstrated that the superpowers could turn from confrontation to negotiation. In this atmosphere, with nuclear war ceasing to loom as an imminent threat, the nation would redirect its attention toward domestic concerns. The central issue, which emerged during 1963, was race.

It is no coincidence that the civil-rights movement came to the fore amid this easing of tensions between the superpowers. Nations do not pursue sweeping programs of reform when under threat of war; such reforms must await a time of peace. The modern civil-rights era dates to 1954, when the Supreme Court handed down its landmark ruling that segregation of schools was unconstitutional; yet it took another decade to build public concern to a critical level that would enact the appropriate legislation. Then these civil-rights laws became part of a broader movement of change in peacetime: the Great Society.

The war in Vietnam, another Cold War relic, escalated through the 1960s despite the easing of tensions between the superpowers. The American public supported the war effort, amid reassurances from President Johnson and other leaders that the conflict would soon end. Then in February 1968 the North Vietnamese and the Viet Cong, their allies in South Vietnam, undertook a major offensive. It failed militarily, as American counterattacks routed the foe. But it scored a brilliant success by convincing the United States that the Communists were in the war to stay and would not surrender. In the wake of this offensive, Washington responded not with further escalation, but with talk of disengagement. During the presidential campaign, Richard Nixon promised to end the war. When he won the election, it was clear that the United States would indeed pull out of Vietnam, even if the details were uncertain.

And just as the waning of the Cold War would bring withdrawal from Vietnam, it would bring a similar retreat from the moon. At the very moment when Apollo achieved full success and when NASA had the moon within its grasp, changing national priorities would prevent this agency from pursuing a follow-up program of extensive lunar exploration leading toward manned flight to Mars. Rather than going forward as a focus for the nation's hope, the space program would have to find its role in a prosaic world where the glow of Kennedy's challenge had faded, with many people viewing Apollo as a waste of money.

Apollo had briefly faced controversy amid the diplomatic thaw of 1963, as influential leaders decried its expense and questioned its value. Congress had then cut half a billion dollars from NASA's budget. But in the wake of President Kennedy's assassination, the controversy faded. The goal of a moon landing became politically sacrosanct, for it would be a monument to this martyred leader. Nor was it surprising when NASA was swept along in the expansiveness of the Great Society, for LBJ had taken the lead as far back as 1957 in pushing for a vigorous space program. He had advised JFK to pursue Apollo, and he accepted the goal as his own once he reached the White House.

Apollo also soared on the fuel of a booming economy, as the gross national product jumped from $500 billion in 1960 to $660 billion only five years later, amid very low inflation. This made it easy to cover the cost of the moon program, as NASA's budget leaped from $1.2 billion in fiscal 1962 to a peak of $5.9 billion for fiscal 1966. At that peak, the agency was spending close to one percent of the GNP. Employment totaled 411,000 among the contractors and 36,169 in NASA itself. In the nation as a whole, over a million people lived in households that drew their livelihood from the space program. This population approached that of Nebraska, topping Utah, New Mexico, and New Hampshire.

Like a growing state, the space program had its eye on the future and was pursuing projects that offered opportunities beyond those of Apollo. High among them was the X-15, and while that project dated back to the 1950s, its long shadow would extend well into subsequent decades.

The X-15 set speed and altitude records even when fitted with its initial engines, a pair of XLR-11 rockets from Reaction Motors that together provided 16,000 pounds of thrust. Then, in its last major project, Reaction Motors built the follow-on XLR-99 engine for the X-15, whose 57,000 pounds matched the thrust of the V-2. In addition, to stretch its performance, this rocket plane did not take off from a runway, but rode beneath the wing of a B-52 carrier aircraft. Its most dazzling flights took it to 67 miles in altitude and to 4,520 mph, or Mach 6.7. No piloted aircraft would exceed these marks until the space shuttle.

The X-15's flight program spanned nearly a decade, from 1959 to 1968, and made a particular contribution by showing how a high-performing rocket plane could be kept from tumbling out of control. Flight tests with earlier craft, during the 1950s, had shown that they could easily fall prey to what aerodynamicists called an inertia-coupled spin. In *The Right Stuff,* Tom Wolfe wrote of "pilots going into the final dive, the one that killed them, and the man would be tumbling, going end over end in a fifteen-ton length of pipe, with all aerodynamics long gone, and not one prayer left, and he knew it, and he would be screaming into the microphone: 'I've tried A! I've tried B! I've tried C! Tell me what else I can try!' "[1]

It could happen to the best of the best. It had happened to Chuck Yeager in 1953, when he took the X-1A to Mach 2.44, as the plane fell out of the sky, from 75,000 feet. Its violent motions threw him about the cockpit and nearly knocked him unconscious, as this aircraft dropped below 30,000 feet before he recovered his wits and halted his spin. Two years later, it was Milburn Apt in the X-2. He reached Mach 3.2 before going out of control, but he had the chance to save himself by separating the nose of the X-2 and deploying a parachute. But he failed to jump and use his own chute and died when the nose hit the ground.

For the X-15, the solution introduced four large fins, mounted like those of an arrow, along with a reaction-control system that oriented the craft using small jets, like those of Mercury and Gemini. These took over from the ailerons and rudder when the air became too thin. Reaction controls depended on the use of cockpit instruments, which weren't foolproof. When they malfunctioned during a flight by Mike Adams in 1968, he came down from a fifty-mile peak and entered the atmosphere with his plane facing sideways. He went into a spin, fought his way out of it, then encountered violent pitching that subjected him to fifteen g's. He couldn't move his arm to reach the controls, only inches away, and died when the aircraft broke up in flight.

Still, this was the only fatal accident in over a hundred missions above one hundred thousand feet. The X-15 showed that a rocket-powered plane could fly to space and return safely. It opened the door to the development of winged boosters that would reach orbit and then come back by landing on a runway.

Another promising NASA initiative, pursued jointly with the Atomic Energy Commission, was building nuclear-powered rocket engines. They worked by pumping hydrogen through the core of a reactor, at temperatures up to four thousand degrees. The hydrogen would become very hot and expand vigorously, flowing through a nozzle to produce thrust.

Nuclear rockets were very promising because their performance far surpassed the best chemically-fueled versions. The measure of a rocket's performance is its exhaust velocity. The F-1, burning kerosene and oxygen, gave 8,500 feet per second. The J-2, fueled with hydrogen and oxygen, yielded 13,700 fps, greatly enhancing its ability to send Apollo missions to the moon. The nuclear-rocket effort aimed at 26,500 fps, nearly twice the performance of the J-2. Such an engine could power a manned mission to Mars.

The project originated in 1955 and represented one more that NASA had inherited from the Air Force. Initial work, as early as 1959, built a series of experimental rocket reactors called Kiwi, after the flightless bird. During the mid-1960s the project conducted ground tests of a true nuclear rocket called Nerva (Nuclear Engine for Rocket Vehicle Application). Aerojet built it, with a reactor from Westinghouse; it ran for up to a full hour at rated power, 1,100 megawatts. By 1968 the program was ready for the next step: developing a flight-rated engine with 75,000 pounds of thrust. It would not lift off from earth; it would need the Saturn V for that. But following boost to earth orbit, such an engine would fly to Mars with ease.

But while nuclear rockets might represent the future, Apollo stood for the here and now. The flight of Apollo 8, late in 1968, proved to be all that anyone could want. After riding aloft to orbit, the astronauts occupied themselves with onboard checkouts for two and a half hours. Then came the command from Houston: "All right, you are go for TLI." This laconic directive meant that Frank Borman, the commander, would fire his S-IVB stage for five minutes and eighteen seconds to execute Trans Lunar Injection—and fly to the moon.

The moonship accelerated to 24,226 mph, a speed at which it broke nearly free of earth's gravity. Seen through a hatch window, the world became a blue and white sphere. The gap between clouds and surface disappeared; instead, the cloud formations hugged that surface like paint. Following TLI, the size of the planet shrank perceptibly from minute to minute. Over the next three days it would start as a basketball, then become a baseball, and finally diminish to the size of a large bright marble that glowed in the emptiness of deep space.

The astronauts entered the moon's shadow as they prepared for the rocket burn that would drop the spacecraft into lunar orbit. Anders saw a sky full of bright stars—and a sharply defined arc within which he could see no stars at all, but only blackness. With a sense of eeriness, he realized that this gap in the starry fields was the moon.

They spent much of the day in lunar orbit, then proceeded with a live television broadcast:

Borman: The moon is a different thing to each of us. My own impression is that it's a vast, lonely, forbidding-type existence—a great expanse of nothing that looks rather like clouds and clouds of pumice stone.

Lovell: My thoughts are very similar. The vast loneliness up here is awe-inspiring, and it makes you realize just what you have back there on earth. The earth from here is a grand oasis in the big vastness of space.

Anders: The horizon is very, very stark. The sky is pitch-black and the moon is quite light. The contrast between the sky and the moon is a vivid dark line.

On this Christmas Eve, much of the world listened as the radio circuit carried the flat voices of these astronauts:

Anders: In the beginning God created the heaven and the earth. And the earth was without form and void, and darkness was upon the face of the deep. And the spirit of God moved upon the face of the waters, and God said, "Let there be light." And God saw the light, and that it was good. And God divided the light from the darkness.

Lovell: And God called the light Day, and the darkness he called Night. And the evening and the morning were the first day. And God said, "Let there be a firmament in the midst of the waters, and let it divide the waters from the waters." And God made the firmament, and divided the waters which were above the firmament. And it was so. And God called the firmament Heaven. And evening and the morning were the second day.

Borman: And God said, "Let the waters under the heavens be gathered together in one place, and let the dry land appear." And it was so. And God called the dry land Earth. And the gathering together of the waters called he Seas. And God saw that it was good.

Borman then concluded: "And from the crew of Apollo 8, we close with good night, good luck, a Merry Christmas, and God bless all of you—all of you on the good earth." The poet Archibald MacLeish soon responded: "To see the earth as it truly is, small and blue and beautiful in that eternal silence where it floats, is to see ourselves as riders on the earth together."[2]

Yet, despite the success of this achievement, the astronauts had not truly reached the moon, for they had not set foot on its surface. Nor would Apollo 9 or Apollo 10 do this; the missions had the task of exercising the lunar lander together with the manned spacecraft, first in earth orbit and then in orbit around the moon. But Apollo 11 would carry out the landing, and the names of its craft suited this historic national event. The astronauts Neil Armstrong, Buzz Aldrin, and Mike Collins called them *Columbia* and *Eagle*.

The first men to land on the moon lifted off from Cape Canaveral on July 16, 1969, seven months after the flight of Apollo 8. Large portions of the world watched and listened as their Saturn V soared aloft amid a shattering roar, riding a pillar of fire that was brighter than white-hot molten steel. Four days later they were in lunar orbit, and Armstrong and Aldrin entered the lander, which NASA called the LM, lunar module. Upon separating it from *Columbia*, the mother ship, Armstrong radioed, "The *Eagle* has wings." It also had twelve minutes' worth of propellants for use during the descent to the lunar surface. The airless moon provided no way of coming down by parachute; the astronauts would have to keep their engine burning every foot of the way.

Now, at 34,000 feet in altitude, they heard the high-pitched sound of the main alarm. The computer flashed a warning and signaled that it was overloaded. Armstrong's heart raced; he knew that if this persisted, they would have to abort the landing. He cleared the problem; it quickly recurred. After this happened several times, Houston transferred some tasks to a ground-based computer, a quarter-million miles away. The difficulty eased; the threat of abort diminished. *Eagle* continued toward a landing in the Sea of Tranquillity.

The onboard computer steered the craft as it continued to descend. At a thousand feet, Armstrong saw that they were dropping toward a crater several hundred feet across, surrounded by an extensive field of large boulders. He took over control from the computer, flying the lander as he looked for clear and level ground. He saw it, but by then he was getting very low on fuel.

Now, as he came close, his rocket exhaust blew at the lunar dust and obscured his view. "This blowing dust became increasingly thicker," he later said. "It was like landing in a fast-moving ground fog." He kept his eyes on rocks that remained visible through the blur. Sixty feet to go—and sixty seconds' worth of fuel. Still no touchdown, and now only thirty seconds left.

"Drifting to the right a little," said Aldrin. "Contact light. Okay, engine stop." Probes on the landing pads touched the surface. The two explorers shut down the engine amid a quick flurry of activity; then they turned to each other and clasped hands. "Houston, Tranquillity Base here," said Armstrong. "The *Eagle* has landed." Houston replied, "Roger, Tranquillity, we copy you on the ground. You got a bunch of guys about to turn blue. We're breathing again. Thanks a lot."

Armstrong then described the landing area: "Out of the window is a relatively level plain with a fairly large number of craters of the five- to fifty-foot variety and some ridges, small, twenty or thirty feet high, I would guess. We see some angular blocks out several hundred feet in front of us that are probably two feet in size. There is a hill in view." That was fitting; they had climbed a high hill indeed.

It took several hours of preparation before they were ready to open the hatch and climb down a nine-step ladder to walk on the surface. Armstrong went first: "I'm at the foot of the ladder. Although the surface appears to be very, very fine-grained, as you get close to it, it's almost like a powder. I'm going to step off the LM now."

He had planned in advance the deathless words he now spoke: "That's one small step for man, one giant leap for mankind." He had meant to say "one small step for *a* man," but in the excitement of the event, he muffed his line.

"The surface is fine and powdery," he continued. "I can pick it up loosely with my toe." He looked around at the moonscape: "It has a stark beauty all its own. It's like much of the high desert of the United States. It's different, but it's very pretty out here."[3] Then Aldrin joined him, with a description of his own: "Magnificent desolation." They set up a flag with a wire support to unfurl its colors. They collected stones and soil samples, moon rocks that geologists would treasure, and Armstrong unveiled a plaque on a landing leg:

HERE MEN FROM THE PLANET EARTH
FIRST SET FOOT UPON THE MOON
JULY 1969, A.D.
WE CAME IN PEACE FOR ALL MANKIND

Eagle was a two-stage craft; the first stage, which had powered the descent, would remain on the moon as a permanent and enigmatic monument. Nearly a day after the landing, the second stage lifted off to rendezvous with *Columbia*, where Mike Collins had been waiting in lunar orbit. With the three astronauts reunited, they cast adrift the spent *Eagle* and returned to earth to receive that planet's accolades.

In NASA, the triumph of Apollo spurred hope that follow-on programs would also go forward rapidly. High on the list was a space station, and among its chief proponents was von Braun. He had run his Marshall Space Flight Center as a rocket-development facility that had built the

Saturn-class boosters, but he knew that MSFC would need new projects or it would fade in importance and in the size of its staff. Moreover, he had cherished thoughts of space stations even amid the fast-paced missile projects of the 1950s. Rocketdyne's Sam Hoffman remembers that when von Braun came to visit him in his home near Los Angeles, "he had a big blueprint of a space station, which he spread out on my living-room floor. That's what interested him; the missiles and other things were just steps along the way."

Von Braun's hopes gained important support after 1965, as NASA's George Mueller initiated a new Apollo Applications Program. An initial emphasis centered on a search for activities that could provide work for the Saturn I-B, which could not reach the moon but nevertheless held out good prospects for operations in earth orbit. Von Braun's advanced-projects group proposed a focus for these operations: an initial space station that would develop out of a spent S-IVB stage. After riding to orbit aboard a Saturn I-B, astronauts would fly up to it atop their own -I-B, carrying experiments and furnishings for a crew. They would convert this stage into a home and laboratory, living and working inside it.

As this post-Apollo effort took shape, other officials came forward with further proposals. These included an Apollo Telescope Mount that would carry an orbiting solar observatory, with instruments to study the sun at ultraviolet and X-ray wavelengths, which the atmosphere screens out. The ATM then would demand its own pair of Saturn I-Bs: one to carry it aloft for attachment to the orbiting station and a second to bring the crew.

This mission plan revolved around the "wet workshop," the S-IVB whose main tank would carry liquid hydrogen but later serve as the astronauts' quarters. Its 9,550-cubic-foot interior offered plenty of room, while solar panels, mounted to the exterior of this stage and deploying in orbit, would provide five kilowatts of power. An alternative approach, the "dry workshop," called for a complete space station that would be assembled on the ground, with the ATM attached from the outset. This version would avoid the problems of building a station whose interior had to hold rocket propellant. It would carry more sophisticated equipment, along with extra oxygen and crew provisions that would allow the crews to stay in space longer.

In final form, the space station's weight came to 165,000 pounds; it demanded a Saturn V for its launch. For a time this requirement appeared out of reach; NASA's James Webb had had to exercise his full powers of persuasion to convince Congress and the Budget Bureau that Apollo needed at least fifteen of these launch vehicles, and he would tolerate no suggestion that any of them might go over to Apollo Applications. But during the first half of 1969, the growing imminence of full success made it increasingly likely that a Saturn V could indeed be made available, and by the end of June, Mueller succeeded in shifting the program emphasis from wet to dry. On July 22, while Apollo 11 was on its way home, the decision became final. The space station, called Skylab, would indeed use a Saturn V, while three Saturn I-Bs, launched in sequence, would carry Apollo spacecraft with three-man crews that would live inside it.

Meanwhile, as plans for Skylab took shape, it was none too soon for NASA's planners to envision the *next* space station. The concept that emerged projected a permanently occupied facility with a crew of twelve. However, NASA's studies showed that for such a project, the Saturn I-B wouldn't do. Each flight cost $120 million, and while this price was acceptable for Skylab, which needed only a few such launches, the next space station would require so many flights that much of the budget would go for nothing more than the logistics of resupplying this orbiting lab.

In response, NASA sought new approaches to low-cost spaceflight. The Air Force was already active in this area, and its point of departure revolved around efforts of the 1960s that had developed large solid rockets for use as boosters. In contrast to liquid rockets that were sensitive and delicate, the big solids featured rocket casings that a shipyard—specifically, the Sun Shipbuilding and Dry Dock Company, near Philadelphia—would successfully manufacture.

In September 1964, Aerojet fired a 120-inch-diameter motor near Miami, producing 600,000 pounds of thrust. Lockheed meanwhile had one measuring 156 inches, which produced 1.2 million pounds for two and a half minutes. A similar rocket from Thiokol topped three million pounds early in 1965. However, the project that people had their eyes on was NASA's. It featured a motor with a diameter of 260 inches, built by Aerojet.

It took some doing just to ignite such a behemoth. The answer called for a solid rocket that itself developed a quarter-million pounds of thrust, producing an eighty-foot flame that would ignite the surface of the big one's solid propellant all at once. This igniter rocket needed its own igniter, a solid motor that weighed a hundred pounds and generated 4,500 pounds of thrust. The 260-inch motor was kept in a test pit with its nozzle pointing upward. A night firing in February 1966, again near Miami, shot flame and smoke that reached 7,500 feet into the air; people saw it nearly a hundred miles away. Another firing, in June 1967, set a new record with 5.7 million pounds of thrust.

At the same time, the Air Force was building a new heavy-lift launch vehicle by mounting two 120-inch solids to the sides of the Titan II. This two-stage rocket already had the advantage of using storable liquid propellants, which reduced its complexity. The two solids gave 2.4 million pounds of thrust at liftoff; the complete package placed a third stage atop the Titan II, with the ability to restart in space and to transfer its payload from one orbit to another. This assemblage, three stages plus two solids, was the Titan III. It first flew in June 1965, carrying 21,000 pounds to orbit; and while that payload had nothing more than lead bricks, real spacecraft would soon follow.

The Martin Co. was the rocket's builder. A few years later, as the Titan III became operational and put together a record of experience, Martin advertised that this booster could launch payloads at a cost of $435 per pound. This beat the $650 per pound of the Saturn V, which the latter achieved by spreading its launch cost over as much as a quarter-million pounds of payload. Furthermore, the Titan III had plenty of room for more growth. By replacing its 120-inch strap-ons with 156-inch units, it might boost up to 100,000 pounds to orbit, with little increase in launch cost. That would drop the cost per pound toward $100.

NASA also wanted $100 per pound, and turned to its contractors for ideas by commissioning a series of studies early in 1969. One useful approach came from Max Hunter of Lockheed, who had been the chief design engineer on the Thor more than a decade earlier. He now proposed an Integrated Launch and Re-entry Vehicle (ILRV) that would cut costs by achieving reusability. A Saturn I-B or Titan III could fly only once before splashing into the ocean, but the ILRV would fly repeatedly, like an airplane. It would place the expensive elements of a booster within a core vehicle: the rocket engines, electronics, onboard computers, crew life-support systems, and the like. But propellants for the main engines would go into a big set of expendable tanks that would attach to the core's exterior. The tanks would drop off as the ILRV approached orbit, and would burn up in the atmosphere. But they would not be expensive, and the core of the ILRV, modest in size, could come down through the atmosphere to land on a runway.

At the Manned Spacecraft Center in Houston, Maxime Faget, who had defined the shape of the Mercury and Gemini spacecraft a decade earlier, proposed to achieve even lower costs through full reusability. His concept would throw away nothing, not even inexpensive tankage. It called for a two-stage rocket, with both stages forming winged boosters that would burn hydrogen and oxygen. The lower stage would carry the upper one on its back, lifting off from Cape Canaveral and accelerating to several thousand miles per hour. The upper stage, featuring a payload bay, would then fly onward to orbit. Both stages would carry all propellant internally, and both would land on runways for reuse.

NASA expected that such craft would shuttle to and from orbit on a frequent schedule. They thus became known as space shuttles. Along with other elements of the future—Nerva, Skylab, follow-on space stations—these concepts fed into a major post-Apollo planning exercise that got under way in February 1969, shortly after the inauguration of President Nixon.

Eight years earlier, JFK had directed Vice President Johnson to define new goals in space; LBJ had responded by endorsing the manned moon landing. Nixon now told his own vice president, Spiro Agnew, to chair a blue-ribbon commission that would set further goals. Its members included Robert Seamans, who now was Air Force Secretary, and Lee DuBridge, the longtime president of Caltech who was Nixon's science adviser. The new NASA Administrator, Thomas Paine, was also a member. On the morning of the Apollo 11 launch, Agnew showed up at Cape Canaveral and announced that NASA would land astronauts on Mars.

In September his commission, the Space Task Group, issued a report calling for a space program that would exceed Apollo both in scope and expense. Its cornerstone would indeed be a manned flight to Mars, at a cost of some $100 billion. In contrast, Apollo had spent a total of $21.35 billion through the first landing on the moon. At a time when the gross national product was just approaching the level of $1 trillion, the cost of this Mars effort would equal the economic activity of the entire state of California, for a full year.

The report proposed other projects as well: a shuttle and a twelve-man space station, along with a 100-man space base. The plan also called for a station in lunar orbit and a vigorous program of further flights to the moon that would culminate in a station on its surface, too. Nerva would also go forward and would make flight to Mars possible as early as 1983. The plan presented three options, the most ambitious one calling for NASA's budget to climb to $8 billion by 1976.

Looking for a down payment, NASA approached the Office of Management and Budget with a request for $4.5 billion for fiscal 1971. The agency had become accustomed to virtual carte-blanche treatment during the Apollo years, but its officials now received a cold bath in the Sea of Reality. The OMB rejected the request out of hand, chopping nearly a billion dollars from this proposed budget. Further cuts came during subsequent months, and NASA wound up with $3.4 billion for that year. With allowance for inflation, this represented a drop of 54 percent from the peak funding of 1966.

As the fall and winter deepened and the decade of the 1960s came to an end, NASA began to lower its sights. In the Nixon administration's judgment, such projects as the space base, the lunar stations, and the Mars expedition were simply too expensive. Nerva, with no mission in prospect, would also die. That left the space station and the shuttle. Still, with these projects as a focus for future plans, NASA had a basis for hope.

Agency officials described the shuttle and station as a single interdependent program. They proposed to develop the shuttle simultaneously with the station, at a cost of about $5 billion for each project. NASA did suggest other uses for the shuttle: launching and servicing unmanned satellites, environmental studies conducted from orbit, military reconnaissance, and rescues in space. But these were clearly secondary to its role in support of the space station and as a springboard to new manned adventures.

The "shuttle/station" pitch was forthright enough, but it nearly killed both projects. Congressman Joseph Karth (D-Minn.) of the House committee on science and astronautics, usually an ardent supporter of the space program, claimed that NASA was seeking to extract from Congress a piecemeal commitment to what he called "its ultimate objective" of sending astronauts to Mars. In April 1970, Karth introduced an amendment to block appropriations for the shuttle/station. The amendment failed by the narrowest possible margin, a 53–53 tie. In July, Senator Walter Mondale proposed a similar amendment, which went down to defeat by only 32 to 28.

Nevertheless, Apollo now was hitting its stride. NASA had laid out a program projecting ten moon landings in addition to Skylab, for it was well aware that this lunar program still could arouse enormous public interest. This became dramatically clear that April, as Apollo 13's astronauts faced near disaster. Their spacecraft, carrying the lunar lander, suffered a damaging explosion while en route to the moon, which knocked out their main power supply. Lacking electricity, they couldn't fire their main engine. If this had happened a day later, after they entered lunar orbit and dropped off the lander, all of them would have been stranded with no way to get home.

Fortunately, they still had the lander, which would serve as a lifeboat. It had a good engine, able to execute the burns that would put them on a safe course around the moon and back to earth. But it had battery power for only a two-day stay on the moon, and the crew had to stretch this to four days, the time it would take to return. This meant shutting down every possible item of equipment, including the computer. It meant that the cabin would become a damp and chilly cave, with temperatures in the forties and moisture condensing to form globs and sheets of water on windows and instrument panels.

The crew had plenty of oxygen, but the lander lacked enough canisters of lithium hydroxide to remove the carbon dioxide that the crew was exhaling. Additional canisters were in the spacecraft, but the two vehicles had been built by different contractors, and a canister from the spacecraft wouldn't fit the housing within the lander. It had the wrong shape, amounting literally to a square peg in a round hole. But ground controllers in Houston were aware of the items of equipment they had on board, and told them how to rig a new housing from odds and ends:

cardboard from the covers of a flight-plan book, plastic storage bags, and gray duct tape. In their high-tech world, this makeshift secured the margin of survival, as they succeeded in lowering the concentration of carbon dioxide to a safe level. The three astronauts of Apollo 13—Jim Lovell, the commander, along with the rookies John Swigert and Fred Haise—made it through the crisis and came back.

Around the world, hundreds of millions of people felt the suspense as they followed the news. Many prayed, and the pope offered a prayer of his own. At television store windows across the nation, passersby stood ten and twenty deep. Nixon had planned to make a major address on troop reductions in Vietnam; he put it off until later in the month. The successful splashdown, carried live, reached what may have been the largest worldwide television audience in history. And when it was clear that the astronauts were safe, women broke into tears while men cheered.

NASA was also upgrading Apollo's mission equipment, providing extra supplies to allow men to stay on the moon for up to three days. A geologist, Harrison Schmitt, was on the roster as a crew member for an upcoming flight. In addition, future astronauts would have a lunar rover, a battery-powered automobile with bucket seats and power steering. It could roam as far as twenty miles from the lander, though NASA would limit this radius to six miles so the crew could walk back if the car broke down.

However, each Apollo mission was costing $400 million. Such a sum, endowing a university and invested at 6 percent, could pay the salaries of three hundred professors while leaving well over $10 million per year for a construction fund, all without touching the principal. By contrast, Apollo's lunar science mostly fell within the narrow realm of field geology. Its moon rocks could support the work of only a small array of specialized research centers, notably the Lunar Receiving Laboratory in Houston and the Lunatic Asylum run by Gerald Wasserburg at Caltech. And these centers' scientific results weren't all that impressive, amounting largely to physical descriptions of the rocks themselves. They shed little light on such interesting and significant questions as the origin of the earth, or of the solar system.

The budget cuts of 1969 had already forced NASA administrator Tom Paine to scrub one Apollo mission and stretch out the flight schedule for the rest. The summer of 1970 brought further reductions in funding. In September, with these new cuts hitting home, he scrubbed two more missions, which meant that Apollo 17 would be the last. Paine tried to make the best of the matter by noting that this decision would free up two Saturn Vs, thus offering new opportunities that would give NASA flexibility in its future planning. But he was whistling past the graveyard: Neither of the boosters would receive another assignment. Both would wind up on public display, at the NASA centers in Houston and at Cape Canaveral.

There had been interest within NASA in flying a second Skylab. One of those Saturn Vs could have served as its launch vehicle, and there still were a few Saturn I-Bs to carry astronauts to it. But it would cost $1.5 billion while offering little more than a repetition of the approved Skylab, and the money just wasn't in the budget. The program would build a second and fully functional Skylab space station for use as a backup, but it never flew. In 1976 NASA would donate it to the National Air and Space Museum, where you may see it today.

Just then, late in 1970, NASA truly faced a winter of discontent. The end of Apollo was approaching, while Skylab would fly with no follow-on. The shuttle/station was hanging by a thread, vulnerable to even a slight increase in the adverse sentiments that were sweeping through Congress. Senator Mondale had spoken for much of the public when he had attacked the shuttle/station: "I believe it would be unconscionable to embark on a project of such staggering cost when many of our citizens are malnourished, when our rivers and lakes are polluted, and when our cities and rural areas are dying. What are our values? What do we think is more important?"[4]

But Paine and his colleagues had one more card to play. Their twelve-man space station could not operate without the shuttle, but the shuttle might stand on its own, as an all-purpose launch vehicle. By offering low launch costs, lower by far than any expendable booster, the shuttle might make these expendables obsolete and go on to carry every future payload. It would do this by introducing airplane-like reusability as its route to low-cost launches, thus becoming all things to all people. This prospect might yet allow NASA to proceed with the shuttle as its next major manned program, and to maintain itself as the agency it had become and wanted badly to remain.

The argument was both ingenious and ingenuous. It was brilliant, for it presupposed that the shuttle, which existed only on paper, could draw on the Apollo experience to achieve capabilities no rocket had ever demonstrated. Each shuttle vehicle was to undergo its turnaround and preparation for a new flight in only two weeks. A fleet of these craft would achieve economies of scale through frequent use, launching all space missions of the nation.

In addition, the approach lent itself to the formal cost-benefit analysis of economics. By making appropriate assumptions—estimates of development cost, cost per flight, number of flights per year, effective interest rate on funds spent up front for development—economists could assert that a shuttle not only would save money when operational, but would actually save enough to pay back those entire up-front costs. In addition, NASA and the Air Force both had lengthy wish lists of payloads that people hoped someday to build and launch. By assuming that these wishes would all come true, NASA could win political support from those payloads' sponsors, while setting forth an impressive-looking list of specific missions and spacecraft that the shuttle would carry.

But the argument was also ingenuous, for it really was beside the point. It took the view that the high cost of launchers stemmed from their once-only use, so that a reusable shuttle could mean large savings. But in the words of Adelbert Tischler, an influential propulsion specialist and colleague of Abe Silverstein, "Cost is people." It was taking 20,000 employees to check out and launch the Saturn V with its Apollo moonship, and because the shuttle would draw directly on the background of Apollo, it was an exercise in hand-waving to say that this new craft would require substantially fewer people.

An unmanned shuttle, guided to a runway under ground control, might indeed have called for less than this cast of thousands. But such thoughts were beyond the pale. NASA, after all, had ridden to glory on its manned missions; people in the agency were fond of saying, "No bucks without Buck Rogers." And as a practical matter, any major new project had to fit the institutional needs of the Manned Spacecraft Center in Houston, where the astronauts were kings.

Payloads raised further questions. Those wish lists would require money if they were to come true, and with NASA's budget on a steep downward course, it was not clear where the funds would come from. To get around this embarrassing issue, the shuttle's advocates took to arguing that its low cost would inspire a flood of new uses. These hopes drew on experience; the airlines, after all, had stimulated a surge of new business by cutting their fares. But NASA's eventual plan envisioned a fleet of shuttles that would fly sixty times per year, carrying as much as four million pounds to orbit. In 1969, the best year to date, the total of all civil and military payloads had come to only 442,360 pounds, most of which stemmed from four Apollo flights.

Still, with NASA's future riding on the success of its arguments, an initial task was to bring the Air Force on board, as a major customer with a strong space effort of its own. Right at the start, this left Air Force officials feeling that NASA was Santa Claus. The agency's unmaidenly eagerness contrasted sharply with the Air Force's experience in the previous decade, when it had struggled in vain to establish a manned space program of its own.

These efforts had begun following the launch of Sputnik and had centered on a six-ton reusable spaceplane, the X-20. Boeing won the contract to build it, in 1959, and it progressed sufficiently in development for the Air Force to pick six test pilots who were to fly it. It was to ride a Titan III to orbit, making a quick pass over a target with reconnaissance equipment. Then, after firing a retro-rocket, it would use highly swept delta wings to execute a hypersonic glide in the upper atmosphere, a maneuver known as dynamic soaring.

People therefore called it Dyna-Soar, and the name proved appropriate for it became extinct. In 1963 Defense Secretary McNamara decided that its military missions would be better accomplished with a small space station, the Manned Orbiting Laboratory. He canceled the X-20, but authorized the Air Force to proceed with MOL instead.

Two years later, Douglas Aircraft won the contract to build it. The design called for a cylinder, forty-one feet long and ten feet across, to fly aboard the Titan III. A Gemini spacecraft would come up to rendezvous and dock, carrying two astronauts who would use this station for reconnaissance. Then in 1969, with unmanned CIA satellites continuing to improve in capability, the Defense Department canceled MOL, too. For the Air Force, this added insult to injury. Not only would it lose its own space lab, but the demise alleviated concern in Congress that MOL would wastefully duplicate some features of Skylab, and helped ensure that this rival NASA project could go forward strongly.

Now, with NASA's space shuttle on the hot seat, Air Force Secretary Seamans found himself in an unusual and quite fortunate situation: NASA needed his service's business more than he needed a new launch vehicle. He would testify in a 1971 congressional hearing: "I cannot sit here today and say that the space transportation system is an essential military requirement." This put him in a position to drive a very hard bargain.

According to the ultimate agreement, the Air Force would contribute its political support and would agree to use the shuttle to launch its payloads. However, it would not put up money for the shuttle's development. NASA would not only pay the cost of development in full, but would also pay to build all the operational craft, even though the Air Force would in effect have full use of two of them. In addition, NASA would design the shuttle to meet the Air Force's specific requirements.

NASA's design studies had emphasized an orbiter, or upper stage, with a straight wing, simple and light in weight. But the wing lacked "crossrange," the ability to fly long distances to right or left during re-entry. The Air Force wanted enough crossrange to fly the shuttle for a single orbit, for reconnaissance, and then have it return to its launch site, to permit operation from a single secure base. This meant a delta wing for the orbiter, which would add weight and call for far more thermal protection.

Another set of demands involved payload weight and the size of the cargo bay. NASA had considered payload capacities as low as 25,000 pounds, with this bay measuring perhaps 14 feet wide by 45 feet long. The Air Force insisted on 65,000 pounds and 15 by 60 feet, which would sharply increase the size, weight, and cost of the entire shuttle. Nor would that service budge. It might not be able to conduct reconnaissance from

MOL, but it was responsible for launching the CIA's satellites, and its planners anticipated that future versions would need that much room and weight.

At a meeting held in Williamsburg, Virginia, in January 1971, NASA officials agreed to give the Air Force most of the crossrange it wanted, and all of the payload capacity. NASA also "decoupled" the shuttle and space station; the latter faded in the agency, while the former received its own project office. Manned flight to Mars? Perish the thought; at a budget briefing early in 1971, George Low, filling in for Tom Paine, said, "We have in our program today no plans for a manned Mars landing." Emphasizing the role of unmanned craft in exploring that planet, he then added, "I repeat, we have no plans at this time for a manned Mars landing mission."

The space shuttle effort was now on its own, justified in economic terms by the savings it would provide in launching satellites for an *unmanned* space program. This represented an about-face from NASA's position of only a few months earlier—and Congress bought it. Opposition in the House evaporated as Congressman Karth returned to the fold, declaring that he now was "enthusiastic" about the shuttle.

In the Senate, Walter Mondale remained unmoved. He argued: "NASA has repeatedly changed its design, purpose and justification, not to meet technological or scientific demands, but to make it politically saleable. NASA desperately wants this multibillion-dollar project and will seek any rationale to justify its development. What we see, then, is a classic case of a program and agency in search of a mission."[5] But NASA now had a solid phalanx of support from key Air Force and Pentagon officials. Mondale's motions to delete shuttle funding went down to defeat, 50 to 26 and then 61 to 20.

Having covered its political flanks, NASA and its contractors could get on with their designs. Max Faget's two-stage fully reusable approach was in the ascendancy, and for a time the most important design concepts came from North American Aviation—now merged with a manufacturing company, Rockwell Standard, to form the new firm of Rockwell International—and from McDonnell Douglas, the product of a 1967 merger between the two aerospace leaders. The parent corporations had held most of the main contracts for Mercury, Gemini, Apollo, and Saturn V, and their executives had long experience in giving NASA what it wanted.

NASA wanted, and apparently would get, a behemoth. Both designs called for craft that approached the size, weight, and complexity of a fully fueled Apollo mission. Both featured a winged first stage over 250 feet long, making it somewhat larger than a Boeing 747. It would boost the shuttle to a speed of 7,400 mph, then fall away to fly back to the launch site. The orbiter, some 200 feet long, then would continue onward.

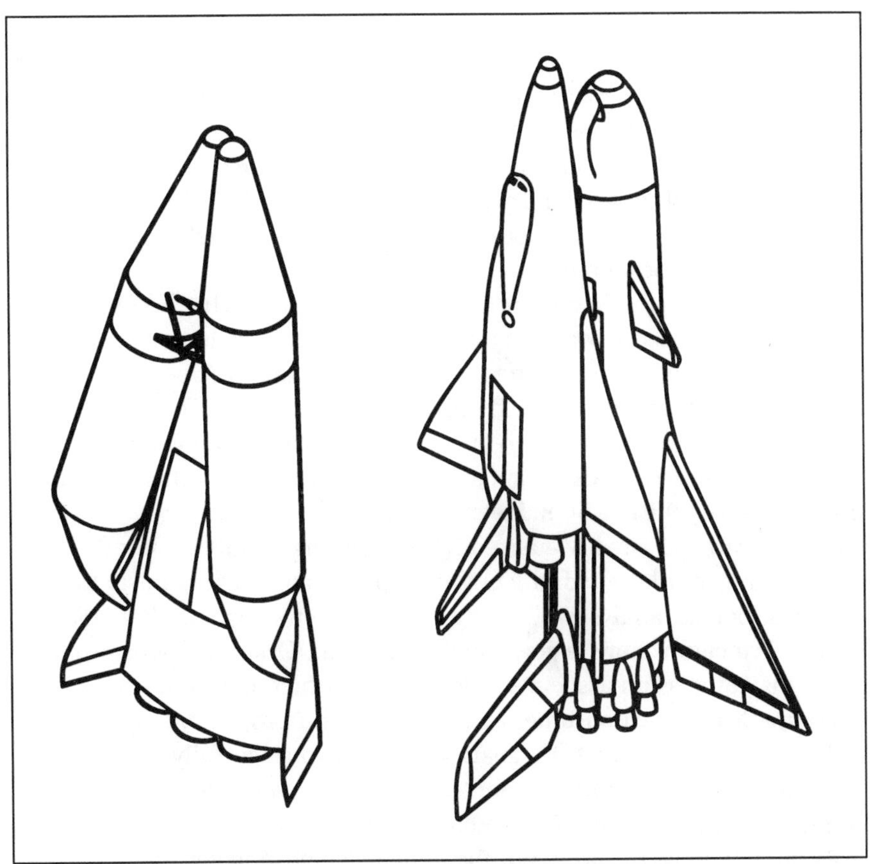

Space shuttle concepts of 1970: Lockheed's Integrated Launch and Re-entry Vehicle, with propellant tanks; North American Rockwell's two-stage fully reusable shuttle. (Dan Gauthier)

Loaded with propellants, the complete vehicle would weigh some 2,500 tons at liftoff, or five million pounds.

The concept was enough to take your breath away, particularly if you were an aerospace engineer. Within that profession, it was common knowledge that merely to double the weight of an existing aircraft represented a considerable jump. Boeing had done this in advancing from the 707 to the 747, had run into very serious problems in the process, and had set a new record for corporate indebtedness by turning to its banks for $1.2 billion in loans. Boeing also had worked on the supersonic transport, a Mach 3 airliner weighing 675,000 pounds. It drew on experience with military projects, including the XB-70 bomber. But the project was very risky, and in taking it on, Boeing insisted that the federal government put up ninety percent of the development cost.

The two stages of the shuttle were riskier still. Although the first stage would be up to seventy-five times heavier than the X-15, it would reach Mach 11, whereas this experimental plane had achieved only Mach 6.7. The orbiter would have somewhat the same mission as the X-20, which had never flown, but would again be seventy-five times heavier. But if all went well, the shuttle would fly for $4.6 million per flight, achieving a low, low launch cost of $70 per pound. The sticker price? A mere piffle: $9.92 billion for development.

At this point, it was appropriate to remember that what NASA was trying to do, supposedly, was to build a new launch vehicle that would improve on the existing unmanned versions that had developed from Thor, Atlas, and Titan and that were flying nearly every week. The people who would remember this were in the Office of Management and Budget, which scrutinized requests such as NASA's before sending them on to the president for inclusion in budgets submitted to Congress.

The OMB had already advised NASA to strengthen its economic rationale for the shuttle by contracting with professional economists to conduct cost-benefit analyses. The contract went to the firm of Mathematica, Inc., in Princeton, New Jersey, whose chairman, Oskar Morgenstern, had made his name as an associate of John von Neumann. And the news from Mathematica was worrisome. At face value, the projections agreed that NASA indeed could meet its goals with the two-stage fully reusable design. In fact, their reports amounted to a strong warning against it. The shuttle's technical challenges would almost certainly result in a large cost overrun, which would wipe out the anticipated advantages.

This accorded with OMB's view that NASA simply couldn't afford it. The agency's budget was leveling out at $3.3 billion per year. OMB expected it would stay there, and that this would have to cover the whole of NASA's activities. The shuttle of NASA's dreams would require up to $2 billion per year in peak funding, and OMB wanted to chop this to $1 billion, by cutting the total cost of development in half. Nor was OMB impressed with the view that the shuttle could achieve cost savings by flying a very large number of payloads. William Niskanen, head of OMB's Evaluation Division, described NASA officials' wish list as "unrealistic. They start at a number that strains credibility and go up from there."

Backed by analysis from Mathematica, Niskanen believed that the answer would lie in a simpler shuttle resembling Max Hunter's ILRV, with a first stage that might merely feature a pair of big strap-on solid boosters, as with the Titan III. Such a shuttle would cost more to operate and would not achieve $70 per pound, but that wasn't important, not with OMB placing little confidence on that projected cost in the first place. What made this approach attractive to OMB was that it would cut the all-important development price, the year-to-year budget line item that would require congressional appropriations. NASA's approach would incur huge up-front costs in return for the vague hope of large savings after everyone had retired. OMB wanted smaller near-term costs in exchange for the prospect of a somewhat greater cost of launch fifteen years down the road. And OMB held most of the cards.

NASA didn't like it, and at first it continued to press strongly for the two-stage fully reusable shuttle. Klaus Heiss, who directed the Mathematica studies, said: "In the first place, there was the irresistible urge to go for the most advanced design and technology possible. And then there was a deep-seated bureaucratic bias for two manned vehicles." When Heiss recommended the simpler design to NASA, "for a long time some people over there kept seriously telling us, 'We can't go that route, because we've got to have something for the Marshall Space Center as well as something for the Houston space center.' "[6] Marshall was to manage the development of the flyback first stage, the winged craft larger than a Boeing 747 that was to outperform the X-15. If it was to give way to anything as simple as solid-fueled strap-ons, this center would have much less to do.

Meanwhile, Tom Paine, the head of NASA, was resigning in protest of the budget cuts. In March 1971 NASA received a new man at the top: James Fletcher. His background included work with Ramo and Wooldridge, and although he had spent the last six years as president of the University of Utah, this gave him an independence of mind that would allow him to take a fresh look at the shuttle. He quickly realized that if he were to continue to insist on the two-stage fully reusable design, he might wind up with no shuttle at all; OMB was that strong. He therefore set out to make peace with OMB by meeting its demands, and if that meant stepping on toes in his own agency, he would do it. He repeatedly challenged the views of other NASA officials, saying, "I just don't buy your assumptions on this."

Within months, it was clear that the shuttle was in for a major redesign that would focus on cutting the cost of development. As one aerospace executive put it, "Some people saw it coming and some didn't, but whatever the case we all knew by July that the whole damned system had suddenly gone up for grabs."

Knowing that many officials wanted somehow to hold on to a flyback booster, NASA's contractors tried mightily to propose designs that could do this and still meet the OMB's ceiling of $1 billion per year for the development cost. Throwaway tankage quickly emerged as a promising approach, because the shuttle's orbiter would shrink in size and cost if it didn't have to carry its propellants internally. Boeing made a valiant try at turning the first stage of the Saturn V into a flyback booster, by proposing a design that would fit it with wings, landing gear, a pilot compartment, and ten turbojet engines. But while the shuttle might fly, this proposal didn't.

Pressed by a continued insistence on lower development costs, design contractors showed increasing ingenuity in coming up with novel layouts. This encouraged NASA, but it fed the skepticism of OMB. An OMB staff report, late in November, stated: "If NASA's resourcefulness to date in changing the shuttle's design is any guide, we have not yet begun to see what they could achieve if they really tried to optimize a system for $3–4 billion." Fletcher was trying to reach $5 billion; to get below $4 billion, he would have to chop the payload bay in half, to ten by thirty feet and 30,000 pounds in capacity.

This left him ready to offer further concessions. At year's end he responded to OMB by proposing a compromise, at 14 by 45 feet and 45,000 pounds. This would not suit the Air Force, but such a shuttle still could build a space station someday. And while the hope of such a station was dim at the moment, it was far from dead.

But Fletcher was about to win support from the man who counted most: Nixon. The president's view of space paralleled that of Kennedy and Johnson, for he saw it as a test of America's ability to lead. He greatly admired astronauts; he had personally greeted the returning crew of Apollo 11, not in Washington but on the deck of the aircraft carrier that had just recovered them in the Pacific. As he learned about the space shuttle, he became fascinated. When Fletcher met with him and brought a model, Nixon held on to it as if he would not give it back.

Early in January 1972, Fletcher met with George Schultz, the head of OMB. Fletcher was ready to argue for the 14-by-45-foot bay, describing it as the "minimum acceptable option." To his pleasant surprise, Schultz told him that Nixon had already approved development of the shuttle with the full-size bay, 15 by 60 feet. When the news reached Dale Myers, head of the shuttle program, he felt amazed.

In the end, early in 1972, NASA wound up with the shuttle design we know today, with propellants in a big external tank and two solid-fuel strap-ons 146 inches in diameter. It also was to have a new rocket motor, the Space Shuttle Main Engine. This would burn hydrogen and oxygen, while providing work both to Rocketdyne, the contractor, and the Marshall Space Flight Center, which would manage its development.

NASA projected a development cost of $5.15 billion, half the $10 billion sticker price of the two-stage fully reusable designs of 1971. The projected cost per flight went from $4.6 million to $10.4 million. With payload capacity holding firm at 65,000 pounds, NASA claimed that the new design would achieve $160 per pound. These numbers still were wildly optimistic, and Fletcher knew it, but he had his eye on Congress. "I wish we hadn't done it," another official later told *Science*, "but we were getting *so* many votes. . . ."

Early that January, Fletcher and his deputy, George Low, met with Nixon to receive a presidential blessing for the project. Nixon praised the space shuttle fulsomely: "It will revolutionize transportation into near space, by routinizing it. It will take the astronomical costs out of astronautics." Then, in a statement reminiscent of JFK, he quoted the poet Oliver Wendell Holmes: " 'We must sail sometimes with the wind and sometimes against it, but we must sail, and not drift, nor lie at anchor.' So with man's epic voyage into space—a voyage the United States of America has led and still shall lead."[7]

With this endorsement, NASA could proceed with the serious business of awarding the main contract for shuttle development. Rockwell International won, and an important reason was that this firm's space division had learned from the hard knocks of Apollo. Rockwell's proposal was especially strong in the areas of management and electric-power systems, reflecting lessons learned from the problems that had caused the 1967 fire and had laid Stormy Storms low. At $2.6 billion, it was the largest NASA contract in a decade.

NASA would not return to the glory days of Apollo, not on $3.3 billion a year. And it would have to get along without von Braun, who decided that this agency offered no new worlds to conquer and left it in mid-1972 to work in private industry. But the old gang from Apollo days was together again: Rocketdyne and Rockwell, Huntsville and Houston. And with modest modification, the immense Saturn V facilities at Cape Canaveral would launch the space shuttle as well.

At this time the Soviet Union was proceeding with its own new generation of launch vehicles and spacecraft. Following the two nearly successful Zond circumlunar flights of 1968, a launch in January 1969 failed when the second and third stages malfunctioned. This was the third attempt in the last six to experience such upper-stage problems, and the program stood down for six months while people worked to improve the reliability of its four-stage Proton. Those were the very months in which Apollo charged ahead, with a set of missions that culminated in the moon landing in July. By the time the Soviets were ready to try again, it was too late.

The flight that ensued, Zond 7, proved bittersweet. It was completely successful; in fact, it turned out to be the only such success in the entire Zond program. Like Zond 5 the previous September, it flew around the moon, passing within 765 miles of its farside, while remaining intact through the whole of its six-day mission. Like Zond 6, it executed its demanding re-entry properly; indeed, it did this so accurately that it landed within thirty miles of its target. But Zond 7 flew in August 1969, a month after Neil Armstrong had made his one small step. The flight therefore bore no propaganda value; it was too little and too late.

It was a measure of America's lead, a lead that rested on the powerful Saturn V, that Moscow tried repeatedly but failed to achieve what Apollo gained merely by accident. The goal of the Zond program, after all, was to loft two cosmonauts into a loop around the moon and have them return safely. In nine launches, they never succeeded in qualifying their spacecraft for this task. Indeed, they would not even repeat the achievement of Zond 7; the last shot in the series, in October 1970, merely repeated the near-success of Zond 5, including its excessively steep re-entry over the Indian Ocean. By contrast, when Apollo 13 carried its three astronauts on exactly the same type of circumlunar flight, it had a crippled spacecraft that executed the flyaround purely as an emergency measure, to achieve a safe return from lunar space.

Nevertheless, just as NASA had progressed from the Saturn I-B to the Saturn V, so Russia was moving ahead from the Proton booster to Korolev's final rocket, the vast N-1. This moon rocket's first stage mounted its thirty engines in pairs. An onboard Engine Operation Control System (KORD, in Russian) was supposed to sense any malfunction. It then would shut down the faulty engine along with its mate, located diametrically opposite, to maintain a thrust balance.

The first launch took place in February 1969. During the initial minute of flight, the first-stage engines throttled back to reduce loads on the vehicle. Then at T + 66 seconds, these engines returned to full power more quickly than planned. The resulting stress ruptured a liquid-oxygen line and started a fire. KORD failed to stop the engine in which the problem was centered; instead it overreacted and shut down all thirty rocket motors. The N-1 fell some thirty miles downrange.

Vasily Mishin, Korolev's successor, was reassuring: "This is normal for a first launch." He placed his hope in the second N-1, which made its attempt in July, two weeks before Apollo 11. Mishin wanted to use this rocket to set records, and he succeeded, but not in a way that would cause anyone to boast. This N-1 became the largest launch vehicle ever to blow up on its pad.

The rocket lifted off and rose several hundred feet. Then a stray piece of metal entered the liquid-oxygen turbopump of engine No. 8, causing it to explode. This broke onboard cables in the electrical circuitry, damaged adjacent engines, and started a major fire. Again KORD shut down all propulsion. This moon rocket, fully fueled and weighing as much as a naval destroyer, fell back onto the launch complex and exploded in an enormous fireball.

Mishin had expected to encounter failures and to learn from them, and he now had enough lessons to earn him a Ph.D. KORD was extensively redesigned, while the N-1's first stage received a Freon firefighting system. Nikolai Kuznetsov, builder of the engines, laid on a major program to improve their reliability—including installation of the simple filters that could have screened out the bit of metal that had destroyed the rocket and launch pad. The pad was marked for rebuilding, while a second launch complex, already under construction, went forward to completion. Even so, the N-1 would not fly for another two years.

Nevertheless, Mishin was not at a loss. He still had his R-7 booster with its Soyuz spacecraft, while the Proton had shown a semblance of reliability late in the Zond program, with four successes in five launches. Dmitri Ustinov, still a power in the Kremlin, now ordered Mishin to use the R-7 and Proton to conduct his own counterpart of the Skylab program. It would focus on a small space station, Salyut—a "salute" to Yuri Gagarin. Unfortunately, Gagarin wasn't around to receive the compliment. He had died in 1968, when he flew his jet fighter into the powerful wake turbulence that trailed for a considerable distance behind another fighter plane. Gagarin's own plane went out of control, and he died in the crash.

A handsome tribute, Salyut weighed over 40,000 pounds. With Soyuz attached, the combined craft totaled 3,500 cubic feet of interior space, over one-third as much as Skylab. Like its American counterpart, Salyut would emphasize long-duration flights. As a warmup, Moscow took back the record for time in orbit, as Soyuz 9 kept three cosmonauts aloft for eighteen days in 1970. Then in April 1971, ten years after Gagarin's flight, Salyut 1 rocketed into space. Soyuz 10 then came up and docked with it, though a faulty hatch kept the crew inside their own craft. Six weeks later came Soyuz 11, carrying Georgi Dobrovolsky, Vladislav Volkov, and Viktor Patsayev. They entered this space station and stayed there for over three weeks.

As their mission was ending, they fired a set of explosive bolts to separate a re-entry capsule from the rest of the Soyuz. They still were over a hundred miles up as the explosions jarred open a valve that was intended to equalize the cabin pressure with the pressure outside the spacecraft. The men were wearing no pressure suits; they did not even have any aboard. It was possible to shut the valve by hand, and Patsayev may have lost his balance while struggling in weightlessness to do this, for he received a bad bruise on his face. But the air in the cabin streamed away before anyone could get the valve closed, and the crew perished. Although spacefarers had died before, these were the first to lose their lives in the vacuum beyond the atmosphere.

"That valve had been checked hundreds of times in test units and it had been used on all our previous craft," Mishin later said. "It had always worked fine. It never occurred to anyone that such a simple device could fail." Like the N-1, the Soyuz spacecraft went back for redesign. Space suits for future cosmonauts were an obvious necessity, and to make room for them and their oxygen supply, the Soyuz was changed from a three- to a two-man craft. This reversed the modifications that Feoktistov had made to Voskhod in 1964, which had created the world's first three-man spacecraft. A decade would pass before the Soviets would fly again with a crew of three.

During the same week in which the cosmonauts died, the N-1 made another try. As it cleared the tower, unanticipated aerodynamic forces made it roll rapidly on its axis. This caused a breakup of the support structure between the second and third stages. The third stage fell off, a substantial rocket in its own right that carried a mockup of the lunar moonship. It crashed and blew up near the launch pad. The rest of the N-1 continued onward, not knowing for the moment that it had lost its head. The rolling continued; the guidance system, overtaxed, failed while trying to stop it. The first-stage engines had been performing satisfactorily, but KORD now shut them all down. The amputated moon rocket arced through the sky like an enormous artillery shell. When it hit the ground, twelve miles from the launch site, it, too, blew up and left a crater a hundred feet across.

Another year and a half went by. Then in November 1972, fifteen years after the first two sputniks, the N-1 was ready for another try. This time it nearly worked, flying normally for the first ninety seconds. At that moment six engines in the center of the first stage shut down as planned, to reduce the vehicle's acceleration while the other twenty-four rocket motors continued to thrust. Unfortunately, the shutdown was too abrupt, and a water-hammer effect ruptured some propellant lines. A fire broke out; the Freon fire-extinguishing system did not activate; engines began to explode at T + 105 seconds, and two seconds later, KORD shut down those remaining. It was too late, for now the entire rocket exploded. Its first stage had come within ten seconds of the moment of normal cutoff, when the second stage was to have ignited.

This record of total failure contrasted painfully with von Braun's record of complete success. In over thirty launches of Saturn-class rockets between 1961 and 1975, including thirteen Saturn Vs, all reached orbit or accomplished their suborbital objectives. With the important exception of the second Saturn V, in April 1968, all upper stages operated properly. The reason could be summarized in one word: testing.

Apollo, to those who worked in the program, was preeminently an exercise in testing. It began literally at the level of nuts and bolts. Charlie Feltz, a leading Storm Trooper, once told the House Space Committee that the iron ore for those fasteners came from a particular section in a specific open-pit mine in the Mesabi Range, near Duluth, Minnesota. The bolts then took eleven steps to manufacture, and the product had to be certified at every step, through meticulous tests. This certification applied to the ingot smelted from the ore, the billet forged from the ingot, and the steel rod extracted from the billet, as well as the bolts themselves that were milled from the rod. The fasteners that resulted were some fifty times more costly than the ones that Feltz might buy in a hardware store, but this was what it took to send astronauts to the moon.

The Russians didn't do it this way. In particular, although Kuznetsov ran his rocket engines on test stands, the Soviet lunar program had no big test stand that could accommodate the N-1's first stage, with its thirty engines. By contrast, at the Mississippi Test Facility, even the largest complete Saturn V stages could undergo static operation. In Mishin's words, "We tested in pieces and did not even dare to think of firing all thirty motors in the first stage as a full assembly. Then the pieces were assembled, without guarantees, of course, that they were properly run in.

"The Americans invested $25 billion in the program, and they reached the moon. We had almost ten times less."[8] In the words of another specialist, "The Americans spent $15 billion on the creation of an experimental base; we spent only about $1 billion." The Russians failed to reach the moon because they tried to get there on the cheap. And this, in turn, reflected precisely the merits of the two competing societies, which their rivalry in space subjected to its own test.

President Kennedy had gotten it right in 1961: As long as this competition would remain at the level of the rockets and other hardware that existed at the time, Moscow could win. The Titan II could match Korolev's R-7, and Gemini would outperform Voskhod, but not by much. Similarly, a competition between Proton and the Saturn I-B, with a lunar flyaround as the prize, might have become a real horse race. But at the level of Saturn V, Apollo, and the N-1, the contest truly did test the abilities of the two superpowers to come up with the enormous sums of money needed to carry out a serious effort. *Newsweek*, early in 1961, had compared the space race to the potlatch ceremony of the Kwakiutl tribe in the Pacific Northwest, who vie to throw the most valuable objects into a fire. But this lunar potlatch was what the nation wanted, and what it got.

In addition, while the Soviet Union was now emphasizing the flight of its Salyut space stations, the United States was ready to lead in this area as well. Skylab, four times heavier than Salyut, was truly a home away from home, with the spaciousness of a three-bedroom house. Its length of 118 feet, with an Apollo spacecraft attached, made it longer than a Titan II, making it an early example of an orbiting craft that was larger than a good-sized launch vehicle. Its crews would enjoy sleeping bags in private compartments, a wardroom with a dining table, a shower for use once a week, and a zero-g toilet that used air suction. Astronauts could shave, and their cuisine included lobster, filet mignon, and butterscotch pudding. NASA considered adding wine to the menu, but decided not to, out of concern for protests from the Women's Christian Temperance Union.

The space station went into orbit in May 1973, amid serious trouble. On the way up, a thermal shield ripped loose, taking a main solar panel and jamming another one so it could not unfold. The missing thermal shield put Skylab at risk of overheating, thus spoiling its stored food and medical supplies, ruining photographic film, and causing polyurethane insulation to decompose and release deadly gases. In addition, although the Apollo Telescope Mount carried a separate set of solar panels, the loss of Skylab's panels would cut the power supply in half.

It was Apollo 13 all over again, as engineers on the ground struggled to come up with quick fixes that could save this $2.5-billion program. A sunshade was the most immediate necessity. It took the form of a large deployable rectangle of plastic film, resembling an umbrella. When furled, an astronaut could stick it out through a hatch in the side of Skylab, open it, and pull it flush against the outer wall. Then, as onboard storage batteries began to fail, a power-supply problem became acute. The astronauts Pete Conrad and Joseph Kerwin made a four-hour spacewalk, the longest to date, and freed the jammed solar wing.

The crew then settled down to a 28-day stay, during which they performed medical tests. Kerwin was a physician, while the third member of the crew, Paul Weitz, could perform minor surgery or extract a tooth. Other activities centered on solar astronomy, earth observations, and experiments that studied the behavior of liquid metals in zero gravity. Later in 1973, a second set of visitors stayed for fifty-nine days, and this mission went so well that although a third crew had originally been slated for a stay of similar length, NASA managers extended it to eighty-four days. The last two flights carried solar astronomers—Owen Garriot, then Ed Gibson—and both made detailed observations of sudden releases of energy known as solar flares, with Gibson following one from the moment of its birth.

Life in weightlessness had its quirks. Food tasted bland, and crew members took to dousing the servings in onion seasoning and salt. Air bubbles in the food and water caused another problem, described by Bill Pogue, an astronaut on the third mission: "We have to pass so much gas. I think passing gas five hundred times a day is not a good way to go. The only redeeming feature is that everybody is passing the same amount." But everyone came through in fine shape, which suggested that there was no limit on how long people could live in weightlessness. By exercising vigorously, all the astronauts stayed healthy.

Then there were the earth observations, which involved photos of earth made at a number of wavelengths, including heat-sensitive infrared. Infrared photos showed possible sources of geothermal energy in the western United States, along with upwellings of cold water in the Caribbean that could influence the development of storms. A scientist from a drought-stricken region in Africa used Skylab images to search for water, while geologists spotted potential reserves of oil and copper.

Still, though this work had undoubted value, Skylab would have no follow-on. With no Mars mission in prospect, the worth of long-duration medical studies was minimal. Other orbiting observatories would fly in the future, viewing earth and its oceans as well as the sun, but these would be unmanned, far less costly, and in operation for years. Skylab helped to bridge NASA's gap between Apollo and the space shuttle, making good use of specialists who might have been laid off, and of equipment that might have vanished into museums. But its cost could have provided a four-year budget for the National Science Foundation, which was funding much of the nation's basic research. Though Skylab produced a decent scientific return, its high expense contrasted sharply with the modest cost of such ongoing enterprises as the study of Antarctica and the exploration of the deep sea. It was too costly to stand on its own merits, or to justify further flights to this space laboratory.

Apollo had one last hurrah in mid-1975, as the final launch of a Saturn I-B carried its spacecraft to a linkup with a most unusual target: a Soviet counterpart, Soyuz 19. This joint mission stemmed from a 1972 agreement between Nixon and Premier Alexei Kosygin, and it represented a beguiling space fantasy: The very institutions and rocket equipment that a few years earlier had raced to the moon, and had sought to demonstrate national superiority, now joined in space to show how well the superpowers could cooperate.

It was true that Moscow and Washington were no longer at each other's throats, but theirs was a strange friendship—if it even deserved that name. In 1972 the Russians had pulled off the "great grain robbery," using their knowledge of the commodities market to make massive purchases that sent the price of wheat soaring, thus contributing noticeably to inflation. A few years later they would invade Afghanistan to prop up a Communist regime, an act that renewed the Cold War's chill. Still, they weren't putting missiles in Cuba or threatening war over Berlin, which definitely counted as a step forward. But although the two spacecraft docked in orbit and remained together for nearly two days, the two nations' programs made no such rendezvous. Like the Vostok craft of the early 1960s, Russia and America momentarily approached each other in space before separating to pursue their courses.

In Moscow, the Soviets were charting a new direction amid a shakeup in the nation's leadership. The man responsible for that direction was Dmitri Ustinov, who had spotted Korolev's talent thirty years earlier. He remained powerful in Moscow: He was Secretary of the Central Committee and was on his way to a new term as minister of defense. He had lost interest in the moon in the wake of Apollo's successes. But he currently was looking favorably upon Valentin Glushko, the longtime engine-builder who had steered clear of Korolev's moon-landing effort.

At the same time, Ustinov had lost patience with Vasily Mishin, whose long-running attempt at building Korolev's N-1 moon rocket had produced craters on the wrong planet. The program had continued to go forward even after the four launch failures from 1969 to 1972, but in May 1974, Ustinov decided that Mishin, its director, had to go. He obtained the assent of Mishin's boss and of the head of state, Leonid Brezhnev, and then replaced Mishin with his longtime rival, Glushko.

At that moment, two complete N-1 launch vehicles were at Tyuratam, representing as impressive a display of lunar capability as the Soviets would ever possess. Both would be ready to fly within the next few months. With support from the Kremlin, Glushko canceled the planned launches and suspended the program. Then, in 1976, he terminated it outright. Nor would he permit the two N-1s to go on exhibit; he ordered them destroyed.

These decisions came as a terrible blow to the project staff. One senior manager later wrote that "no one gave any thought at all to the honest labor of thousands of people who devoted the best years of their lives to the N-1. They not only took no consideration of the people, they did not even offer any explanations. Together with the N-1 they relegated to the scrap heap its builders as well, many of whom certainly experienced such a psychological blow that they could never create anything of equal value again. And these were the best personnel of Korolev's design office."[9] But Moscow would build no monuments to failure; it preferred to toss the N-1 into a memory hole.

Some N-1 components would find new uses. Stages, cut in half lengthwise, became garages or storage buildings. Propellant tanks saw new life as a recreation center, as a summerhouse for relaxation, and as part of the water-supply system for nearby apartments; a dish-shaped dome was turned into a solarium for suntanning. Then, with the N-1 disposed of, Glushko could indulge to the full his preference for advanced propellants. He had led Chelomei into the realm of storables, over Korolev's heated objections. Now he turned to liquid hydrogen, planning a new class of heavy-lift launch vehicle that would use this propellant and would compete with America's space shuttle. After a decade of work, the result would be the Energiya booster and the Buran spaceplane.

The Salyut orbital stations were meanwhile emerging as centerpieces of the space program. There were military versions that conducted reconnaissance and were counterparts of the Manned Orbiting Laboratory, featuring a similar internal volume. Those that Moscow would talk about, of course, were slated for civilian uses. These bore a general resemblance to Skylab and hosted similar onboard activities: medical studies of long-duration flight, earth observations, and solar studies.*

However, the Soviet manned spacecraft were far simpler than their American counterparts. Soyuz, for instance, lacked onboard computers, advanced inertial guidance, and backup systems for heating and cooling. Cosmonauts made few onboard decisions; almost all activities were controlled from the ground, including even turning off the lights at bedtime. In addition, several early Salyuts went into low orbits that decayed rather quickly, and new ones had to fly at a fairly frequent rate.

The Salyut 1 mission, in 1971, had resulted in the tragic loss of the crew of Soyuz 11. But their deaths had stemmed from a problem in Soyuz, not Salyut, and the latter appeared ready for further activity. The next one, Salyut 2, reached orbit in April 1973. Though it shared the Salyut name with Mishin's craft, it was an entirely different space station called Almaz (Diamond), built by Mishin's rival, Chelomei. It was a military spacecraft, with a large telescope that took up room all the way from floor to ceiling. It viewed its targets through an optical sight that moved to compensate for the motion of the earth's surface, yielding sharp images.

The crew was to take photos with this instrument and develop them onboard, transmitting some of these images by television. These cosmonauts could also send back the film itself by using recoverable capsules, each with a retro-rocket, heat shield, parachute, and radio beacon. The crew quarters included viewports as well as a sleeping area, a dining table, and a comfortable chair. Fearing attack by American space interceptors, the designers had also planned to equip it with a rapid-firing antiaircraft gun.

No one ever got to enjoy these amenities. The craft had a maneuvering rocket that exploded before the crew arrived, showering nearby space with pieces of debris and puncturing the station's hull. A follow-up launch, a month later, put a civilian Salyut into orbit, but it quickly failed. The Soviets doggedly persisted, and in June 1974 Salyut 3, another military type, reached space and operated properly. The two cosmonauts who stayed for two weeks in July, Pavel Popovich and Yuri Artyukhin, were the first to live aboard a Salyut and return to tell about it. It helped that Popovich was among his country's most experienced cosmonauts; he had flown in Vostok 4 during the dual mission of 1962.

---

*In contrast to more general astronomical studies, solar astronomy offered the advantage of very short photographic exposures. This was important, for the movements of crew members, aboard their orbiting labs, would have prevented telescopes from pointing accurately at distant stars and galaxies for long-duration photography.

Next came Salyut 4, a civilian version, which flew to orbit just after Christmas. This mission initiated a program of long-duration flights that sought to pick up where Skylab had left off. Soviet doctors had access to the medical results of Skylab, as part of the exchanges that preceded the Apollo-Soyuz linkup six months later, but they wanted to run their own tests. The first crew to reach this station, Alexei Gubarev and Georgi Grechko, repeated the initial Skylab mission by staying for twenty-nine days, while conducting a heavy schedule of medical and solar studies. They also enjoyed a feature of this Salyut that cosmonauts would treasure: a small onboard garden, called the Oasis, where they tried to grow green peas. It was only the size of a suitcase, and the peas died within weeks, but on a long flight, it looked like a promising way to keep up their spirits.

This mission confirmed the results of Skylab by showing that a thirty-day flight was not the way to go. It left people only partially acclimated to zero gravity, while subjecting them to all the discomforts of the return to the earth's gravity. Longer flights were obviously in order, and in April 1975, the cosmonauts Vasili Lazarev and Oleg Markarov rode a Soyuz to make the attempt. The third stage malfunctioned; ground controllers thought the problem was with their instrument, and Lazarev got angry enough to curse them before they realized the mission was in trouble. It now faced an emergency landing, as these cosmonauts became the first spacefarers to abort while en route to orbit. They would land a thousand miles downrange.

Their first concern was that they could land in China, which might put them in prison. As they followed their trajectory, one of them asked plaintively, "We have a treaty with China, don't we?" Then came re-entry: a sharp plunge into the atmosphere that bore no similarity to the shallow and gentle descent from orbit that had been on the flight plan. The forces topped 18 g's as an onboard meter went off scale, but though battered, the crew survived. Parachutes opened, and the spacecraft landed in the Altai Mountains, on a snowy slope. The round capsule then rolled down the mountainside, heading for a cliff. Just in time, the parachute lines snagged on some scrubby trees and saved them. Rescuers arrived, and the cosmonauts realized that now they truly were safe, for those people were Russian.

Six weeks later the backup crew, Pyotr Klimuk and Vitaly Sevastyanov, tried again. Both were veterans; Sevastyanov in particular had flown on the eighteen-day mission of Soyuz 9 in 1970. They reached a Salyut 4 whose onboard systems were beginning to deteriorate. Life in a space station demanded control of the humidity, for moisture in a man's breath totals several pounds per day, which had to be removed from the air. This required nothing more than cold plates to condense the moisture, but on Salyut 4, they didn't work properly.

The analyst Jim Oberg notes that after a month in space, the cosmonauts had plenty to complain about. "The windows are still fogged over," one of them remarked. "The green mold is halfway up the wall now. Can't we come home?" The answer was *nyet;* like soldiers in the Great Patriotic War, they were to stay at their post and do their duty for the motherland. The green mold continued to spread and the spacefarers continued to ask if they could come down. Finally, late in July, they received the order they wanted. Their flight overlapped that of Apollo-Soyuz, thus requiring ground control to deal simultaneously with two separate missions. The sixty-three days they spent in space confirmed Moscow's commitment to long-duration flights in the civilian Salyut.

After 1975, then, both nations were set firmly on their new courses. Both had abandoned the moon, turning away from the immense Saturn V and N-1 rockets that their lunar programs had brought forth. But both had powerful launch vehicles—Titan III and Proton—that could launch heavy spacecraft, advancing well beyond the modified ICBMs that had included the R-7 and Titan II. NASA was fully committed to the space shuttle as its principal new endeavor, though its officials still had plans for a space station, and were biding their time. Their Soviet counterparts had their own initiative, featuring Salyut space stations, though Valentin Glushko was preparing to build his own shuttle. And while manned flight continued to draw the most attention, unmanned spacecraft were carrying out the real work of space—and producing a good share of the excitement.

10

Electrons in the Void

The Unmanned Space Programs

THE VILLAGE OF ANDOVER, MAINE, lies in a hilly region near the New Hampshire border, in wooded country the poet Robert Frost would have appreciated. In the early 1960s its local telephone service featured party lines and hand-crank phones. Still, this low-tech remoteness suited Bell Labs, for it meant that there would be little radio interference in the area. Bell's management picked it as the site for its main satellite-communications center. Here they built a fully steerable horn-shaped antenna 177 feet long inside a protecting radome.

Telstar, the first satellite to advance beyond Echo, reached orbit in July 1962. It was a commercial venture in its entirety, for AT&T, which owned Bell Labs, even paid NASA $3 million for the Thor-Delta launch vehicle. Fifteen hours after launch, as Telstar flew along a northeast arc from the Caribbean to Spain, its controllers tried to transmit the world's first transatlantic television broadcast. Television signals travel in straight lines and do not follow the earth's curving surface; even when broadcast from atop the Empire State Building, they reach a range of only about fifty miles. But with this satellite ready to pick up the signals and to amplify and retransmit them, they might span the Atlantic with ease.

The show was to feature transmissions between Andover and two European ground stations, at Goonhilly Down in Cornwall and Pleumeur-Bodou in Brittany. The American segment featured a videotape of the Andover radome with the U.S. flag flying in the foreground. It included recordings of "America the Beautiful" and "The Star-Spangled Banner," along with a view of AT&T's chairman, Frederick Kappel, as he gave a statement. The French picked it up nicely and responded in kind, with a tape of the actor Yves Montand, a female vocalist, and a guitarist. The words FIRST TV TRANSMISSION FROM FRANCE flashed on American screens, as the unknown and somewhat puzzled guitar player strummed his instrument.

However, technical problems at Goonhilly brought a delay on the British side. Richard Dimbleby, the BBC's star anchorman, was on hand, but as the night wore on he had nothing to report. "I said this six times tonight," he noted as midnight approached, "but I'll say it again. We'll get the picture next time even if it kills us." Finally, after 1:00 A.M., Kappel's face appeared momentarily—without audio.

But a few days later, all was in order. The CBS network was running a late movie when Charles Collingwood appeared on the screen and said, "We interrupt this program. The British are ready to bounce a program off Telstar." They proceeded to present a view of the control room at Goonhilly, with its technical managers. In this impromptu fashion, the world took the first major step toward global TV.

The broadcasts held particular interest for Washington, where Congress was about to pass the Communications Satellite Act. A year earlier, President Kennedy himself had taken the lead by issuing a policy statement. "I invite all nations to participate in a communications satellite system," he declared, setting the stage for federal involvement at a moment when the promise was clear but the institutional base was wide open for definition. The 1962 act established the Communications Satellite Corp.; the historian Walter McDougall would describe it as "a chartered company of the sort founded by European princes in the age of the mercantilist state." *Fortune* magazine wrote that Comsat at first had "no management, no physical facilities, no agreement on what it proposes to make, no underwriters, and no history or immediate prospect of profits." Still, the intent of Congress was clear. Comsat was to control U.S. satellite communications and to pursue their development.

Moreover, the law was aimed right at AT&T, a major corporate power that operated much of the telephone system as a regulated monopoly. In 1953 it had arranged to lay an undersea cable that would carry thirty-six simultaneous phone conversations, and had dealt directly with the British government rather than relying on an intermediary such as the State Department. It was quite prepared to set up and run its own global communication-satellite system, and it had a specific proposal that would feature fifty Telstars, to ensure that at least one would always be in view. But the Kennedy administration had no intention of letting this corporate behemoth freeze out its potential competitors. Senator Estes Kefauver, who was closely involved in the matter, said that to let AT&T have its way was "no more appropriate than defining free enterprise like the elephant dancing among the chickens, who shouted, 'It's every man for himself.' "

Among these chickens, one of the more noteworthy was Hughes Aircraft, where the engineering manager Harold Rosen was nurturing hopes of a system that would leap beyond Telstar. The Telstar concept called for satellites that would orbit at altitudes of a few thousand miles, circling the earth every several hours. This followed the cautious, step-by-step approach of the Bell System, for such orbits were well within the capability of the Thor-class launchers its managers expected to use. However, because their satellites would move across the sky, the system would require fully steerable radio telescopes in the ground stations, which would be complex and costly. The Andover antenna, for instance, weighed 370 tons and had to pivot to follow a spacecraft as it crossed the sky in any direction.

Rosen wanted to take a big leap upward by placing his spacecraft in geosynchronous orbit, 22,300 miles above the equator. At that altitude they would take twenty-four hours to circle the globe, and would appear to hover motionless over particular locations. Ground-station antennas then might point constantly in a satellite's fixed direction, making these installations much simpler and less costly. Nor would Rosen need AT&T's dozens of orbiting craft. Each geosynchronous orbiter would have nearly half the earth's surface in view and could connect ground stations separated by over 11,000 miles. Three such satellites could suffice to provide a complete global network.

The science writer Arthur C. Clarke had proposed this arrangement as early as 1945. In a brilliant leap into the future, he determined that a power of 1200 watts at the spaceborne transmitter would permit "small parabolas about a foot in diameter" to receive the signals on the earth. Solar energy would provide the power, and while the solar cell (ironically, a product of Bell Labs) lay a decade away, he saw hope in "photo-electric developments." He envisioned his communications satellites as manned space stations, each with an onboard crew to replace vacuum tubes as they wore out.

Though Clarke drew his inspiration from the German V-2 and from his wartime background as a radar officer, his paper amounted to speculative futurism in the spirit of Hermann Oberth and Konstantin Tsiolkovsky. But in contrast to the hit-and-miss character of their prophecies, Clarke scored a bull's-eye in predicting the most important commercial use of space, while offering introductory technical detail. He might even have patented it. In particular, while other visionaries were writing of journeys to distant planets, Clarke proposed that the future of astronautics would involve craft that would hover over one position and thus in effect would go absolutely nowhere.

However, Clarke's article appeared in *Wireless World*, a little-read British publication. John Pierce, working at Bell Labs, did not learn of it until he was well along in his own studies. Clarke reiterated his ideas in his 1952 bestseller, *The Exploration of Space*, but Pierce missed that one, too; his taste in reading ran more to *Proceedings of the Institute of Radio*

Engineers. Nevertheless, it took only fourteen years for advances in electronics and rocketry to put Clarke's vision within range of being realized.

On an overcast morning in August 1959, a Thor-Able thundered aloft from Cape Canaveral, placing a 142-pound spacecraft, Explorer 6, into an elongated orbit that ranged from 157 to 26,400 miles in altitude. It was not a communications satellite; its mission involved detailed studies of space near earth, including its radiation environment. But it featured solar cells and reliable solid-state electronic circuitry, both of which would be essential for communications. Its 26,400-mile apogee was higher than geosynchronous orbit; a small solid rocket, firing at the proper moment, could have circularized the orbit and turned Explorer 6 into a true 24-hour satellite.

Its success strongly encouraged Rosen, who was already nurturing his concept. He offered it to the Pentagon; they turned it down. Then he approached NASA, which christened it Syncom and awarded a contract in August 1961. Like Echo, it was to demonstrate a principle rather than serve as a practical communications link; it weighed only seventy-one pounds and featured a single two-way telephone channel. But again as with Echo, the project strengthened NASA's involvement in the field of satellite communications, giving Hughes Aircraft the means to compete with AT&T on the latter's chosen ground.

Syncom 1 flew atop a Thor-Delta in February 1963. It soared to its planned altitude; an onboard timer ignited its apogee motor—and twenty seconds later a nitrogen tank exploded, destroying it. Syncom 2 flew in July—and it worked; it reached its planned orbit and its electronics operated successfully. Then a year later, NASA placed Syncom 3 high above the International Date Line, ready to offer initial service across the Pacific. And within this burgeoning commercial world, the builders of Syncom quickly found themselves at odds with the NBC television network.

It was the first geosynchronous satellite with the ability to carry a TV transmission, and it reached orbit in time to cover the 1964 Olympic Games, live from Tokyo. However, NBC had bought exclusive rights to this coverage and was planning to show videotapes flown in by jet, with sponsors such as Schlitz beer paying the costs. Federal officials put pressure on NBC; following a call from Undersecretary of State Averell Harriman, Chairman David Sarnoff agreed to air fifteen minutes a day of satellite coverage—at 7:00 A.M.

But routine transoceanic TV was only a few months away. The institutional infrastructure was achieving definitive form, as the Comsat Corp. took a controlling interest in a global consortium: the International Telecommunications Satellite Organization. When Intelsat prepared to launch its first commercial satellite, as a counterpart to AT&T's Telstar, Hughes Aircraft had what was needed: a modified Syncom with 240 telephone channels. Right at the start, this single spacecraft provided a valuable supplement to the five transatlantic cables that were then in service, which together offered no more than 412 channels. This satellite, called Early Bird, reached its station high over the Atlantic in April 1965.

Its TV coverage opened with considerable drama on May 3, as the surgeon Michael DeBakey operated on a heart while fascinated colleagues in Geneva looked over his shoulder. NBC's Chet Huntley teamed with the BBC's Dimbleby. A panel discussion, "Town Meeting of the World," linked statesmen in New York, London, and Paris for a discussion of Vietnam. Later in May, ABC announced that it would carry several sports events live from Europe, including the Grand Prix at Le Mans.

As Early Bird proved itself in day-to-day service, it opened the way for Hughes to establish itself as the world's premier builder of communications satellites. Working with NASA, the firm built a series of Applications Technology Satellites that tested and demonstrated new design approaches. Antennas received particular attention. Early Bird and its initial successors were spin-stabilized, and although their antennas had some directionality, they wasted much of the limited onboard power by transmitting it uselessly in all directions. The answer turned out to be a "despun" antenna, which would point fixedly at a particular continent while the body of the spacecraft continued to rotate for stability. NASA demonstrated such antennas aboard ATS-1 and ATS-3, launched in 1966 and 1967. Despun antennas became operational in 1968, when the firm of TRW built the Intelsat III satellites as successors to the Early Bird.

Intelsat reached its goal of global coverage during 1969. By then it had three spacecraft hovering over the Atlantic and three over the Pacific. A seventh, high above the Indian Ocean, linked London and Tokyo at midyear and completed the initial system. Its spacecraft were upgraded versions of Syncom and Early Bird, but as demand burgeoned and prices fell, everyone knew that bigger satellites were necessary, and Hughes responded with the Intelsat IV series. Each of them provided four thousand telephone circuits, with a later version raising the number to six thousand. Thor-class boosters weren't powerful enough for them; they demanded the Atlas-Centaur, one of NASA's largest. Then a decade later even the Intelsat IV proved inadequate and gave way to Intelsat V, with twelve thousand circuits, fifty times more than in Early Bird.

The Intelsat consortium also went from strength to strength. At the outset, in 1965, it had forty-five nations as members; the number doubled during a decade. The roster included bitter enemies: Israel and Egypt, India and Pakistan. It included the wretched of the earth: Bangladesh, Haiti, and much of sub-Saharan Africa. Intelsat also served South Africa, even though that country was well on its way to becoming an international pariah. Within this consortium, a hundred nations were cooperating in space, year after year, and making money at it. This performance contrasted sharply with the realm of manned flight, which had boasted the highly publicized Apollo-Soyuz mission of 1975. This too had brought about cooperation in space between the superpowers, but only briefly and at considerable expense.

The Soviets, for their part, were doing things their way. Their resources included a good part of the Arctic Circle, and with much of this nation lying at high northerly latitude, a geosynchronous satellite was at a strong disadvantage. When parked over the equator, a Russian antenna would have to point toward it at a low and ground-grazing angle, and would pick up radio interference. Moscow's solution introduced an elongated orbit like that of Explorer 6, which angled sharply to the north. Even in arctic regions, an antenna then could look at the spacecraft by pointing nearly straight up.

The first such communications satellite, Molniya (lightning), went into orbit two weeks after Early Bird, with a period of twelve hours. It spent most of its time near apogee, and although it did not hang stationary in the sky, it moved slowly for hours at a time. Twice each day, it could offer a lengthy period of uninterrupted communication. Molniya spacecraft became standard for use in the Soviet Union, supplementing those of Intelsat. They also provided the world's first domestic communications-satellite system.

Indeed, while Intelsat provided international links, these satellites offered considerable promise in a number of countries that combined large areas with thin telecommunications. In those countries, a spacecraft could leap past the costly and lengthy process of building microwave links and coaxial-cable networks, as in the United States and Europe. Third World nations found the prospect irresistible, and Indonesia took the lead. It was one of the world's most populous nations, but it had large islands, mountainous and heavily forested, and included the jungles of Borneo. NASA launched this country's Palapa satellite in 1976. Indonesia later broadened the coverage to include the Philippines, Malaysia, Thailand, and Singapore.

Other countries followed with similar arrangements: Brasilsat, India's Insat, Mexico's Morelos. Canada already had its own domestic system, Anik, and Australia followed with Aussat. In addition, twenty-two nations in northern Africa and the Middle East formed a regional group, the Arab Space Communication Organization. Its members included the Palestine Liberation Organization, and its management insisted on boycotting companies such as Hughes that did business with Israel. But France's firm Aerospatiale proved acceptable, and built the Arabsat spacecraft that duly entered service.

In the United States, domestic operation got started through cable TV. Cable television went back to 1949, serving rural areas where TV signals were weak, or where tall hills blocked reception down in the valleys. A large antenna atop a mountain picked up clear signals from distant cities, while cable carried them to subscribers. It represented no more than a niche in the overall market of TV viewers, run by local operations that did not originate their own programs.

Then in 1972 the Federal Communications Commission announced a new policy that opened the door to domestic satellite service. RCA responded with Satcom 1, which reached orbit in 1975, and a group of entrepreneurs immediately began using it to transmit Hollywood movies to the nation's cable services, as a new type of pay TV. This channel, Home Box Office, proved very popular and set the stage for the development of cable TV as a significant new industry. Other channels soon followed HBO: Cinemax, Showtime, the Movie Channel. Children could watch the Disney Channel, Discovery, and Nickelodeon. Ted Turner launched CNN, and for true news junkies, C-SPAN provided full-time coverage of Congress. ESPN took the lead among sports services. Viewers could also check out the Weather Channel and Financial News Network, or watch for Madonna on MTV.

For a while, this plethora of new programs was freely available to anyone with an appropriate receiver. Comsat for a time had required hundred-foot radio telescopes in its ground stations, but increased wattage in new satellites meant that home viewers could pick up their signals with dish antennas that were only ten or twelve feet across. These were not the "small parabolas about a foot in diameter" of Clarke's 1945 vision, but they proved to be within the financial reach of up to two million homeowners. As satellite dishes sprouted in people's yards, particularly in rural areas, the cable companies took to encrypting their signals, hoping that viewers would spend some $400 for a decoder and then pay a monthly fee. This created another new industry, featuring pirate computer chips that could unscramble the codes illegally. People then could continue to point their antennas and receive the programs at no cost.

The Pentagon also emerged as a major user of satellite communications. It entered the field with a bang in mid-1966, as a Titan III put up seven initial spacecraft in a single launch, with eight more following a few months later. They introduced secure channels for message traffic that proved far more effective than the previous arrangements, which had used unreliable shortwave radio. By allowing commanders to learn what they needed to know, these spacecraft complemented reconnaissance satellites such as Discoverer.

The Discoverer spacecraft were "bucket droppers"; they loaded their exposed film into bucket-like recoverable capsules and dropped them through the atmosphere for retrieval by aircraft. The first thirty-eight missions were publicly announced, but early in 1962 the program faded to black and vanished from public view. However, it did not disappear; Lockheed kept building Agena upper stages with their payloads, ground crews continued to launch them aboard Thors fired from Vandenberg Air Force Base, and the CIA continued to run the program under the name Corona.

The photography, which was highly classified, carried the designation Keyhole. The first thirty-seven Discoverers had gone through three successive types of camera, which now were given the names KH-1 through KH-3. All photographed the earth below in swaths, creating a sequence of panoramic views by using a lens holder that scanned from side to side rather than looking straight down. In addition, the cameras all achieved large-motion compensation, to prevent the movement of the earth, sliding past at five miles per second, from blurring the exposures.

The KH-1 and KH-2 were quite similar, for they required the spacecraft to fly at one predetermined altitude. If it didn't, the photos would begin to blur. This accuracy in altitude proved hard to achieve, and while the camera had been designed for resolution of twenty to twenty-five feet, at its best it achieved only thirty-five- to forty-foot resolution. Clearly, the system needed to be able to provide image-motion compensation over a range of orbital altitudes. This refinement came with the KH-3, which first flew aboard Discoverer 29 in August 1961. It also included a lens of larger aperture, which permitted the use of fine-grained film. The KH-3 thus improved the resolution to as sharp as ten feet.

Discoverer 38 introduced the KH-4 camera, called Mural. It amounted to two KH-3s, one pointing forward and the other pointing toward the rear. As the spacecraft flew past a target, the cameras would shoot it from two different angles, and when viewed through a stereoscope, the images would appear three-dimensional. Then during 1963, the KH-4A system introduced dual buckets. Each spacecraft could then operate in orbit for up to two weeks, returning the film in two batches. The KH-4A first flew aboard mission No. 69, in late August, and quickly became standard. Beginning in mid-1964, the CIA used it on a regular schedule of at least one flight per month.

In addition, Corona broadened its scope through a series of flights that performed geodetic mapping, accurately locating targets with respect to known benchmarks. This sought to address such issues as learning just where Moscow was, in a way that would permit this information to direct the flight of an ICBM. The program, called Argon, returned mediocre results, though the overall Corona effort produced geodetic data of very high value. Still, Argon had its uses. The Army built its mapping camera, the KH-5, and NASA later used this wide-angle instrument to map the moon.

The Corona program also sponsored Lanyard, another subprogram. Its satellites carried a high-resolution camera, the KH-6, that sought to take close looks at possible antiballistic missile sites near Tallinn in Estonia. The missiles in question proved to be of the antiaircraft variety, though Lanyard didn't disclose this. Indeed, it sent back little of value. But it offered a particularly sharp focus, and CIA officials hoped that Lanyard might emerge as a close-image system to complement mainstream Corona, with its panoramic views.

But the Air Force had its own program in space reconnaissance and came in with its own camera and spacecraft, the KH-7 and Gambit. Gambit, a two-bucket satellite, first flew in July 1963 and soon took close looks indeed, swooping well below a hundred miles. Two of its flights set records for low altitude, at seventy-five and seventy-six miles, and resolution was as sharp as eighteen inches. For the Air Force, the gambit of Gambit proved very successful. It outperformed Lanyard, and after only three flights of that CIA rival, all during 1963, the CIA canceled that program.

Gambit also launched a trend toward larger spacecraft and more powerful boosters. Corona, Argon, and Lanyard all used variants of the Thor-Agena, but Gambit at first used the Atlas-Agena. Then in mid-1966 the Titan II went into service, and it began tossing up a new version of Gambit that carried an improved camera, the KH-8. Meanwhile, the standard Corona spacecraft was receiving its own finishing touches with yet another camera system, the KH-4B. It replaced its predecessors' KH-3s with new models that sharpened the resolution from ten feet to five. The KH-4B first flew on mission No. 120 in September 1967. Working alongside the new Gambits, these Coronas continued to fly until 1972.

What did they show? They photographed all Soviet ballistic-missile launch complexes, following existing as well as new missiles through development and deployment. In particular, they found and repeatedly observed a major center at Plesetsk, near the northern city of Arkhangelsk. Plesetsk specialized in reconnaissance satellites and other military spacecraft. At its height, it accounted for more than half of all space launches in the entire world, with Tyuratam a distant second, and Cape Canaveral and Vandenberg far behind.

Corona also was first to see Severodvinsk, the main construction site for ballistic-missile submarines. This made it possible to monitor the launching of new types of subs, and to follow them through to operational deployment. The CIA could also observe the rapid growth of the surface navy. Coverage of aircraft plants and air bases kept analysts up-to-date on bombers and fighters, while other coverage allowed Army experts to learn the nature of the tank forces that NATO would face if the Soviets were to invade Europe.

Corona photography uncovered the construction of true antiballistic-missile sites near Moscow and Leningrad, along with the radar installations that supported them. Other photos located antiaircraft batteries and made it possible for the Strategic Air Command to find routes for its bombers that could avoid these missiles. In the field of geodetic mapping, Corona supplanted Argon and became the main source of data for the Defense Mapping Agency's military charts.

As recently as the mid-1950s, the Soviets had been able to fool the Yankees concerning their strength in the air, and to touch off a major Washington flap over a supposed "bomber gap," merely by flying the same aircraft around twice at an air show. By contrast, a 1968 intelligence report contained the unequivocal statement: "No new ICBM complexes have been established in the USSR during the past year." As early as June 1964, Corona had photographed all twenty-five of the complexes then in existence. If there had been any new ones, Corona would have seen them.

"I wouldn't want to be quoted on this," LBJ told a gathering in 1967. "We've spent $35 or $40 billion on the space program. And if nothing else had come out of it except the knowledge that we gained from space photography, it would be worth ten times what the whole program has cost. Because tonight we know how many missiles the enemy has and, it turned out, our guesses were way off. We were doing things we didn't need to do. We were building things we didn't need to build. We were harboring fears we didn't need to harbor." The analyst Jeffrey Richelson describes space reconnaissance as "one of the most significant military technological developments of this century and perhaps in all history. Indeed, its impact on postwar international affairs is probably second only to that of the atomic bomb. The photo-reconnaissance satellite, by dampening fears of what weapons the other superpower had available and whether military action was imminent, has played an enormous role in stabilizing the superpower relationship."[1]

In addition, the CIA built an advanced bucket dropper, known variously as Hexagon, Big Bird, and KH-9. The first one went up in June 1971 and quickly sent Corona to pasture, replacing its spacecraft entirely in 1972. Still, bucket droppers had their limits. In addition to requiring physical recovery of their film, it took up to a month for images to make their way from the satellite to a photo interpreter's desk. The Six-Day War of 1967, in which Israel overran its enemies, showed that a major conflict could begin and end while film from a spacecraft was en route to Washington.

Real-time photo intelligence was the answer, but the obvious approach, television, lacked the high resolution that the CIA needed. Then in 1970 two Bell Labs physicists, William Boyle and George Smith, invented the charge-coupled device. It could substitute for photographic film while capturing light up to twenty times more effectively. As an electronic device, it returned its images as streams of bits, well suited both to real-time transmission and to computer processing that could bring out fine detail. What was more, a satellite with a CCD could stay up almost indefinitely, for it would use no film and hence would never run out.

The reconnaissance satellite that resulted was named Kennan; its optical system, the KH-11, featured a telescope ninety inches in diameter. The first one went up in December 1976 and remained active for two years, with later ones operating for as long as a decade. Resolution was as sharp as six inches, and a set of leaked photos of a Black Sea shipyard showed what this meant. The spacecraft was five hundred miles away; yet the photos came close to showing the cables on cranes at dockside. In addition, the combination of large, highly precise optics and CCDs was very useful to astronomers. They provided a technical basis for the Hubble Space Telescope.

Weather satellites represented a different type of reconnaissance. They didn't need advanced photographic cameras or film recovery; a simple TV camera could show a lot. The first of them, Tiros 1 in 1960, stirred great excitement among meteorologists even though it didn't stick around for the hurricane season. In the words of one manager, its photos "showed that the earth's cloud cover was highly organized on a global scale. Coherent cloud systems were found to extend over thousands of miles and were related to other systems of similar dimensions." Moreover, these cloud patterns corresponded to major weather systems, including fronts and low-pressure regions that brought storms and rain. It was as if the atmosphere was using clouds to draw its own weather maps.

Tiros 3 went up in mid-1961 and tracked five hurricanes that autumn. It provided valuable support to aircraft observations in following the destructive Hurricane Carla, which killed forty-six people along the Gulf Coast. It also became the first spacecraft to discover a hurricane, spotting Esther two days before conventional reportage might have revealed its existence. In addition, the satellite photos made it possible to determine the strength of winds by noting the degree of organization of a hurricane's clouds, in spiral patterns.

These spacecraft also helped mariners by keeping track of ice in navigable waterways. During the early spring of 1962, Tiros 4 repeatedly saw the ice pack in the Gulf of Saint Lawrence, which surrounded Prince Edward Island but gradually retreated. This fulfilled the hope of Hermann Oberth, who had written in 1923 that eyes in orbit "would notice every iceberg and warn ships."

However, the early Tiros craft were of only limited use for an operational system. They were drum-shaped, rotating at around ten revolutions per minute for stability. Unfortunately, they were a little too stable; they tended to point the camera in a single direction, which was fixed with respect to the stars. Much of the time they faced away from earth, looking at the blackness of space. As a result, each Tiros could view only about one-fifth of the earth's surface each day. But NASA was not at a loss. It had a second-generation program, Nimbus, whose spacecraft were to point continually downward. Nimbus was slated to enter service as an operational system, bought and paid for by the Weather Bureau.

The Weather Bureau was weak (it received little support from its parent, the Commerce Department), and NASA was approaching the height of power. But Weather Bureau officials became unhappy as Nimbus fell behind schedule. Its launch was originally planned for 1961, but its contractor, General Electric, ran into serious problems with its attitude-control system. This raised the prospect that NASA would stop launching Tiros satellites while Nimbus remained on the ground and then would saddle Weather Bureau's meager budget with its costs.

By itself, the bureau lacked the clout to do anything about the matter. But the Defense Department shared the concern that NASA might abandon the useful Tiros to pursue the technically sweeter Nimbus, and it offered to step in with a competing weather-satellite system. Armed with this support, the Weather Bureau stood up to mighty NASA, declaring in September 1963 that it would no longer go along with Nimbus. It wanted more Tiros-class spacecraft, and also a voice equal to NASA's.

The bureau's objections gained further force in 1964; although Nimbus 1 finally reached orbit in August, it shut down after less than a month. But RCA, builder of Tiros, was now ready with its own improvement. It introduced a fast-acting TV camera that could make good images while pointing off to the side of the spacecraft as it rotated. A Tiros satellite then could roll along its orbital path like a wheel, taking its shots at the moments when the camera was looking straight downward. In addition, the spacecraft could fly in a polar orbit, and thus would combine earth-pointing imagery with complete global coverage. The first to do this, Tiros 9, took 480 photos of the sunlit earth during a 24-hour period in February 1965. Pieced together to form a mosaic, the photos covered the whole world except for Antarctica.

An operational system, employing two similar satellites, entered service a year later on behalf of the Weather Bureau. One of these craft employed Automatic Picture Transmission, which allowed small earth stations to receive photos on demand, using a simple antenna and a commercial facsimile recorder. APT had shown its value in experimental flights aboard Tiros 8 and Nimbus 1; now it brought satellite weather observations within reach of most of the world. Within a few years, more than four hundred such stations were in use, in over forty countries.

Subsequent spacecraft incorporated infrared imaging, which allowed them to photograph the earth at night. Whereas standard images relied on reflected sunlight, infrared systems observed the long wavelengths emitted by land, sea, and clouds because of their heat. The photos showed cloud patterns clearly, thus providing true round-the-clock coverage. Two such satellites, in polar orbits, now could produce a montage of the entire world, every six hours.

This accomplishment gave new life to the promise of long-term predictions. Prior to 1940, the meteorologist Carl-Gustav Rossby, at MIT, had written equations to predict the motion of atmospheric waves, hemispheric in scale, that influence the world's weather. His colleague Jerome Namias, drawing on this work, went on to issue the first five-day forecasts. The paucity of worldwide weather data made such forecasts chancy at best, but operational satellite services improved their accuracy.

Having conquered the weather from a few hundred miles up, an obvious next step would provide observations from geosynchronous orbit. However, the needs of meteorology differed from those of communication. A geosynch spacecraft, hovering over the equator, could not see the polar regions. This caused few problems in TV and telephone service, because virtually all demand was concentrated in the temperate latitudes, far from the poles, where the world's people lived. But the Arctic and Antarctic spawned much important weather, and it was important to keep close watch on those areas. In addition, with APT in use, weather forecasters around the world already had the advantage of simple and inexpensive ground stations. Geosynch orbits would not improve them further.

Hence, in contrast to the rush to geosynch among communications specialists, this move took much longer for the meteorologists. To prime the pump, NASA installed a black-and-white camera in the geosynchronous ATS-1 of 1966, and a color camera in ATS-3 a year later. At first meteorologists wanted to learn just what they might see that was new, and they saw a lot. ATS-1 viewed a complete hemisphere; one set of photos showed six simultaneous typhoons and similar storms, five in the Pacific and one in the Caribbean. Verner Suomi of the University of Wisconsin, who developed this camera, remarked that "here the weather moves—not the satellite." Frequent photos from these craft, as they remained stationary high overhead, made it easy to follow the growth and dissipation of major storms. Using time-lapse images, scientists such as Suomi prepared movies, some of them in color, that showed the complete lives of hurricanes and other forms of weather.

An operational system took several more years to develop, for it encountered delays amid NASA's budget cuts of the early 1970s. Even so, NASA contracted with Ford Aerospace Corp. to build two prototypes, the Synchronous Meteorological Satellites. These reached orbit in 1974 and 1975.

Meanwhile, in 1970, the Weather Bureau had been absorbed into a large organization, the National Oceanic and Atmospheric Administration, within the Commerce Department. NOAA had given its name to a new type of polar-orbiting spacecraft that replaced earlier Tiros-class satellites, circling the earth at altitudes of up to a thousand miles. The agency now also took responsibility for follow-on geosynchronous orbiters, dividing the contracts between Ford Aerospace and Hughes. This ongoing series, Geostationary Operational Environmental Satellite (GOES), viewed entire hemispheres of the earth in both visible light and infrared, and took pictures every thirty minutes. The combination of GOES and NOAA craft brought weather observation to its definitive form. Now weather photos from space began to appear in the daily newspapers, with coastlines and state boundaries drawn in by computer.

These photos showed more than cloud patterns. They displayed snowpack in mountains, ocean currents such as the Gulf Stream, and the behavior of ice in polar regions, thus providing data of great importance to hydrology, oceanography, and glaciology. Other advances proceeded from aerial photography in the infrared. By using scanners and color film that was sensitive to the right wavelengths, investigators found they could map areas of forest, vegetation, and agricultural land, and even discern whether the trees or crops were healthy. Infrared images also provided new detail for geological maps, which showed fault lines and major formations.

William Pecora, director of the U.S. Geological Survey, had been an early advocate of using satellites to make such observations. He won strong support in the Interior Department, the parent of USGS, which pushed NASA into action. The Nimbus program provided a point of departure, for its satellites were operating as test beds for new instruments. In 1969 NASA shifted the emphasis of this program from weather to the new field of earth observations.

An appropriately equipped version of Nimbus, later called Landsat 1, flew to polar orbit in mid-1972. The next dozen years saw four more such spacecraft. Then, during the mid-1980s, the Landsat program approached operational status and the Reagan administration handed it over to Eosat, a commercial company that represented a joint venture between Hughes and RCA. In addition, France pitched in with its own commercial system called Satellite pour l'Observation de la Terre (SPOT), launching its first satellite in 1986.

Skylab astronauts took similar photos in the course of a few months, but the Landsats and SPOTs returned pictures year after year. They caught clues that hinted of global warming: glaciers surging in Alaska; Antarctic ice shelves cracking to form massive bergs. They also provided data on other environmental issues: the southward spread of the Sahara; the clearing of Amazon rain forest. Landsat images even found use in the well-mapped United States, by supporting land-use surveys and inventories of wetlands.

The changes they followed were often slow; a single good set of photos at times could retain its value for several years. In addition, aircraft could perform updated surveys. As a result, Landsat spacecraft could be few and far between; there was no demand for rapid-fire launches. But there was plenty of demand for the photos. At first the USGS made them available through its office in Sioux Falls, South Dakota. Eosat then added greater convenience to the service by offering them at a center near Washington, D.C., complete with consultants.

With these uses, the unmanned satellites of NASA and other agencies formed a sharp contrast with Apollo. Apollo had grown up amid the shadow of a feared Soviet superiority. But with its leadership in communications and weather and earth observations, the United States was demonstrating the substance of American superiority. Moreover, Apollo differed from these applications somewhat as a royal procession differs from a newly built highway. The moon landings were events to admire from a distance, but everyone could take advantage of the new services: TV, telephony, and weather forecasts. And while Apollo impressed the nations of the Third World, the new satellites helped give their people the advantages of the First World. Everybody wanted TV and a phone, while the weather services entered societies that had trusted to folklore.

These opportunities also led Europe to launch a major push into space. Its nations, sheltering under America's nuclear umbrella, had little need of strategic reconnaissance. Nor were they tempted by the vainglory of manned missions. But space science, weather, communications, and earth observation were other matters entirely. Initial ventures featured European payloads atop American boosters; the first of them, Britain's Ariel 1, flew in April 1962 aboard a Thor-Delta emblazoned with both the Stars and Stripes and the Union Jack. But it wasn't long before Europeans began to build their own launch vehicles.

An initial point of departure grew out of Britain's postwar hope of retaining its centuries-old role as a great power. This nation developed its own nuclear weapons and jet bombers, and went on to build Polaris submarines. In addition, in 1954 the Tories, led by Sir Winston Churchill, started work on an intermediate-range ballistic missile called Blue Streak. But by 1960 it was too costly and too marginal. Amid howls from the opposition Labour Party, Prime Minister Harold Macmillan opted to cancel it.

But Blue Streak came back to life in January 1961, following a meeting between Macmillan and the French president, Charles de Gaulle. De Gaulle had a fondness for cross-Channel partnerships; he soon would

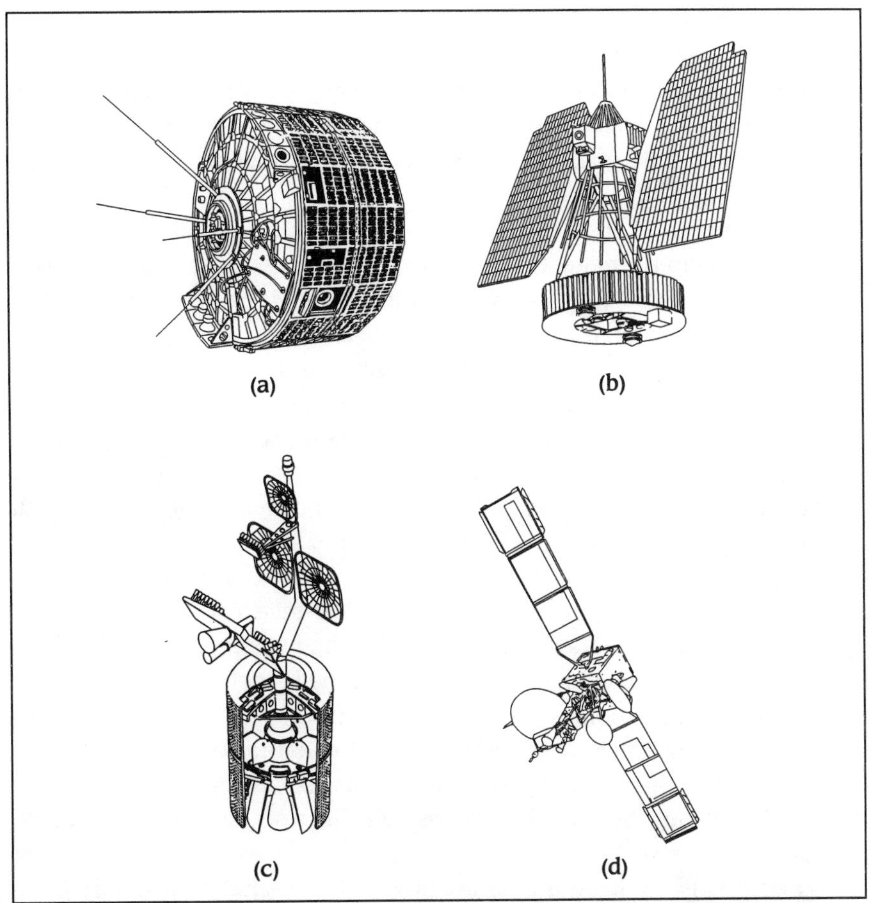

Unmanned spacecraft: (a) Tiros weather satellite; (b) Landsat for environmental studies; (c) Intelsat IV-B communication satellite in cutaway view; (d) Intelsat V. (Don Dixon)

endorse the British-French collaboration that would build the Concorde supersonic airliner. Following that meeting, he realized that Europe could become a third space power. Sir Peter Thorneycroft, Britain's minister of aviation, then took the initiative in setting up the European Launcher Development Organization (ELDO). The major nations of western Europe all joined a year later, with the goal of building the three-stage Europa. Britain's Blue Streak would serve as the first stage, with France and West Germany contributing the second and third stages respectively. Italy would build the early satellites.

Meanwhile, two physicists, Italy's Edoardo Amaldi and France's Pierre Auger, were leading a similar initiative in the field of space science. Both had been active in the formation in 1954 of CERN—the European Center for Nuclear Research; they now sought a similar cooperative effort in space. They won enthusiastic support from a British counterpart, Sir Harrie Massey. Then in February 1960, representatives from eight nations met informally in Auger's apartment in Paris and agreed to go ahead. The result, in June 1962, was the formation of the European Space Research Organization (ESRO).

ESRO built a spacecraft-development center near Amsterdam and a control center in Darmstadt, along with a sounding-rocket range in northern Sweden. But this consortium bit off more than it could chew in its first two major projects: an astronomical satellite and a mission to a comet. The projects languished during the 1960s amid much grumbling because France received a disproportionate share of the contracts. The orbiting telescope wound up needing considerable help from NASA before it finally flew in 1978 as the International Ultraviolet Explorer. The comet probe was even less successful at first. It did not really take shape until 1980, when it emerged as the Giotto mission to Halley's Comet.

ELDO faltered as well. Its Europa program aimed at building a rocket that could put a one-ton payload into a 300-mile orbit, which would give it the power of a Thor-Agena. The Thor had taken only a year to advance from contract signing to hardware on the launch pad, and the Blue Streak project gave Europa a head start. But Europa had no prime contractor analogous to Douglas Aircraft, with the authority to provide overall management. Each contributing nation was responsible for its own part. Even so, the Blue Streak flew successfully in tests, which gave people hope.

Accordingly, in 1966 a coordinating body, the European Space Conference, decided to upgrade Europa into a vehicle that could put two hundred kilograms into geosynchronous orbit. This capability still was Thor-class, but the only thing that went into orbit was the budget. It ran 350 percent over the initial estimate of £70 million. Beginning in 1968, ELDO tried seven times to achieve a successful launch, but failed with both the early and the later versions of Europa. The third stage brought particular difficulty; by turns it shut down prematurely, failed to ignite, and suffered an electrical failure. As with Moscow's N-1, these mishaps played into the hands of political leaders who wanted to start anew. In May 1973 the directors of ELDO decided to cancel the project.

The fecklessness of ELDO contrasted painfully with the success of France, which pursued an independent national effort. This involved more than de Gaulle's pursuit of *la gloire;* it offered real advantages, particularly in telecommunications. The historian Walter McDougall tells of a friend who watched the Apollo 11 moon landing in 1969 along with a family of farmers in the south of France. The TV announcer reported that President Nixon was about to talk with the astronauts, and Madame excitedly called the family to watch: "Look! He is going to telephone the moon—and we cannot even get a line to Paris!"

De Gaulle, like his British counterparts, had a strong interest in an independent *force de frappe*—a strategic striking force. He too built nuclear weapons, advanced jet aircraft, and nuclear submarines. Also in 1965 his three-stage Diamant rocket (the "diamond") put a small satellite into orbit and made France the third nation to enter the space age on the thrust of its own booster. At 62,000 pounds this thrust was modest indeed, putting Diamant in a class with the Jupiter-C. Even the V-2 had had nearly as much power. But it showed that France could do what ELDO only hoped to achieve.

This rocket flew from a site in Algeria's Sahara, which Paris was under orders to vacate at the end of 1966. The government turned to French Guiana on the northern coast of South America, which was not a colony but was legally a *département* of France itself. It was best known as the site of Devil's Island. Here Alfred Dreyfus, condemned through perjury, had served four years before supporters won his release. Other *bagnards* (prisoners) had hobbled naked in leg irons or lived in ancient cells two feet wide. They also faced yellow fever, malaria, and sadistic guards. The worst of the lot were guillotined and tossed to the sharks.

France shut down the prisons in 1946. Still the country remained primitive, with heavy jungle and brutal humidity along with jaguars, alligators, and piranha. Descendants of slaves lived in the bush, speaking a language of only 340 words. Yet for space shots its location was ideal. It not only had extensive stretches of open water to the north and east but its equatorial latitude offered rockets the full advantage of the earth's rotation, which provided an eastward velocity of over 1,000 mph while a booster still sat on its launch pad. Located at Kourou, north of the capital, Cayenne, this *centre spatial guyanais* attracted a colorful motley of workers from Brazil, Martinique and other Caribbean islands, and over a dozen other nations. It opened for business with sounding rockets in 1969 and began launching the Diamant to orbit during 1970.

A French-German communications satellite project, Symphonie, soon showed that Europe would have to do much more. The project dated back to 1967, as an initiative independent of ESRO; its builders hoped to launch it using Europa. When this proved impossible, they turned to NASA. NASA was willing—but only on condition that Symphonie not compete with Intelsat's spacecraft. It was not to operate commercially.

Europe had no choice but to accept, though to the proud French, this was galling indeed. "Symphonie made it clear to everyone that Europe needed its own launcher," recalled Frédéric d'Allest, a leading space official in France. "We were not about to remain dependent on anyone for launch services. The issue was clearly sovereignty."

With France at the forefront, European leaders held a succession of high-level conferences that brought about a complete revamping of the space program by 1975. ELDO and ESRO ceased to exist, giving way to a new European Space Agency with more clout. ESRO had made a modest but significant start in space science, building seven satellites that flew to orbit aboard NASA boosters between 1968 and 1972. But ESA now took a decided turn toward applications.

France already had its own weather-satellite project, Meteosat; ESA took it over and gave it a European flavor. The new agency also inherited an important new communication project, the Orbital Test Satellite, with two follow-on efforts clearly in view: the operational European Communications Satellite as well as a variant, Marecs, for maritime service. In 1978 OTS reached orbit, again courtesy of NASA, and carried Sky Channel, Europe's first pay-TV link.

In addition, ESA gave strong support to a new booster-development effort that would eliminate its dependence on American launchers and give the Yankees a good run for their money in the bargain. However, this time people would steer clear of the politically correct multiculturalism of ELDO. The new rocket instead took form as a French enterprise, with a name, Ariane, that virtually waved the *tricouleur*. With first and second stages burning storable propellants and a third stage fueled with hydrogen and oxygen, Ariane sought to leap past the Thor-class boosters to match the lifting power of an Atlas-Centaur, placing a full ton into geosynchronous orbit.

Europe's decisions also showed that ESA would cooperate with NASA's space-shuttle program—and wait for it to falter. Following up a 1973 commitment by ESRO, ESA endorsed a German initiative, Spacelab, that would build an orbiting research center for installation in the shuttle's cargo bay. Spacelab was no Skylab; it was not a free-flying space station, for it would depend on the shuttle for electric power and crew facilities. It reflected a certain amount of horse trading: By endorsing West Germany's Spacelab along with the largely British Marecs, ESA won those nations' assent to France's Ariane. In addition, Spacelab represented a ploy whereby Europe might ride on the shuttle and take advantage of its opportunities, if any, while steering clear of sharing in its costly development.

But ESA was also ready to challenge the shuttle. NASA officials were freely predicting that this reusable launch vehicle would serve all customers and would drive down the cost of spaceflight, making obsolete expendable boosters such as Thor, Atlas, and Titan. Nor would this agency leave the matter to chance, by maintaining a free market that would allow its users to choose from a variety of rockets. Instead, once the shuttle was operational and its flight schedule was up to speed, NASA would halt production of the expendables and go over completely to the shuttle, putting

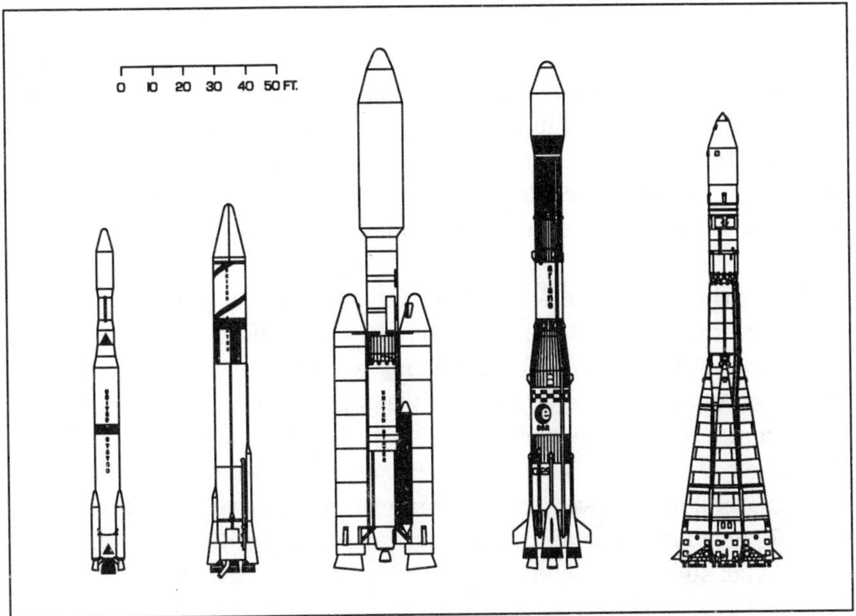

Expendable boosters of 1980: U.S. Delta, Atlas-Centaur, Titan III; Europe's Ariane; Soviet Molniya. (Dan Gauthier)

all its eggs in this single basket. The commitment to Ariane amounted to anticipating that this strategy would fail, that the world would continue to need expendables, and that Europe intended to offer them even if NASA would not. When Ariane made a first flight with brilliant success, just before Christmas of 1979, it announced that Europe was now seriously in space, ready to compete with the Americans on their chosen ground.

Yet in another key area, exploration of the planets, the United States remained unchallenged. The Jet Propulsion Laboratory emerged as the world's principal center, with its director, William Pickering, providing the leadership. He had been born in a small fishing village in New Zealand, where he became a local celebrity by building a crystal radio that picked up music from Australia. He went to high school in the capital, Wellington. Then in 1929 a wealthy uncle in Los Angeles brought him to L.A., where he enrolled at Caltech. He earned a Ph.D. in physics in 1936, in a graduating class that included John Pierce and Dean Wooldridge, and stayed on to become a professor of electrical engineering. Soon he was spending increasing amounts of time at the nascent JPL, where he became a leader in the new field of telemetry. He taught his last class on campus in 1950, taking over as director of JPL four years later.

The lab he now headed belonged formally to Caltech, but resembled Army arsenals like Redstone or Picatinny. Its raison d'être revolved around battlefield missiles: the Corporal and later the solid-fueled Sergeant. This was part of a pattern whereby major universities were proud to do their part in the Cold War, much as MIT, Berkeley, and the University of Chicago had shown their leadership during World War II. But when Pickering took the helm in 1954, he persuaded Lee DuBridge, the president of Caltech, to agree that Sergeant would be JPL's last major military missile. Instead, Pickering would turn his eyes to guidance systems, and then toward space.

His early opportunities unfolded through Wernher von Braun, as JPL developed radio-inertial guidance for his Jupiter IRBM. Von Braun was also building the Jupiter-C for nose-cone tests; JPL pitched in with small solid-fuel rockets for the upper stages, along with a tracking and telemetry system. Then Sputnik went up and von Braun won permission to use the Jupiter-C to launch satellites. Jet Propulsion Laboratory built the first of them, Explorer 1, along with its early successors, while expanding the tracking network. After that, a major role in space for JPL was assured.

The lab still was little known; when the Los Angeles *Examiner* ran the headline CALTECH SATELLITE CIRCLING EARTH, *Newsweek* remarked that "California found a local angle." But with von Braun's support, JPL soon would do more. The Pentagon's Advanced Research Projects Agency staked it to an 85-foot radio telescope near Goldstone Dry Lake in the Mojave Desert, as the nucleus of the Deep Space Network that would track the nation's lunar and planetary missions. When von Braun got permission to fire two small probes into interplanetary space, JPL again pitched in with upper stages, tracking, and the spacecraft themselves. The second of them, Pioneer 4, flew in March 1959; it became the first American craft to escape the earth's gravity and enter solar orbit.

By then, JPL had left the Army's embrace to become part of NASA. Pickering took the lead in pressing for an ambitious program of unmanned flights to the moon, Venus, and Mars. This suited senior NASA managers such as Keith Glennan and Abe Silverstein, who had begun to plan seriously for manned lunar missions during that year. Late in 1959 the lab won approval to build Ranger, a series of lunar spacecraft that would fly on the Atlas-Agena. They would feature solar panels and a high-gain antenna that would have to point continually at the earth, even though the lunar probes could rely on batteries and a low-gain antenna. There was a reason: Those features would be essential for planetary work. Characteristically, Pickering was looking ahead.

But the culture of the lab still remained that of a center for battlefield missiles, where failure in test flight was acceptable because other rounds would be available for future launches. As a result, Ranger became the exercise that taught JPL how to build successful spacecraft. The first two missions flew during 1961 but never got out of low earth orbit, as the

Agena second stages failed to ignite for their main burns. The responsibility lay with Lockheed and the Air Force, the builders of Agena, but subsequent problems would rest entirely at Pickering's doorstep.

Ranger 3 flew in January 1962. A failure in the main guidance of Atlas forced a switch to a backup system with less accuracy, which meant the spacecraft would miss the moon by 20,000 miles. Still the opportunity remained to exercise this craft in deep space, and ground controllers scored a useful success as they executed a midcourse correction. This was the first time that anyone had adjusted a lunar trajectory by firing an onboard rocket. But when these controllers tried to order their probe to point at the moon and to transmit TV images, its onboard computer failed, rendering it useless.

Ranger 4, in April, was even more of a disappointment. This time the Atlas-Agena flew with such precision that this spacecraft would hit the moon even without a midcourse correction. But its computer's master clock failed, leaving it unable to execute a timed sequence of operations that were to include deployment of its solar panels and high-gain antenna. Nor could it accept and act on commands from the ground. A NASA official lamented, "All we've got is an idiot with a radio signal." Three days later it struck the lunar farside, dead as a doornail. It was the first American craft to physically reach the moon, but by 1962 this alone was far from enough.

Ranger 5, in October, resulted in much of the same. An hour into the flight, a short circuit robbed it of power from the solar panels, leaving only a small onboard battery. Project directors nevertheless tried to perform a midcourse correction, as an engineering test. But before the spacecraft could execute the necessary maneuvers, another short circuit finished it off. It would have taken some fine images of the lunar surface, for it swept past the moon at a distance of only 450 miles. But by then it too was dead.

These failures left the lab severely chastened. Pickering's people were a gung-ho group; in the words of James Van Allen, "They have tremendous *esprit*. It's almost offensive. It's like the Marines." In particular, they had brought about a far-reaching change in JPL, turning it from a rocket center into a lab whose mission lay in electronics. But the Ranger failures had all involved onboard circuitry, which was very worrisome. A review panel issued a scathing report, describing the lab's approach as "shoot and hope."

Pickering replaced the program's top managers, ordering a stand-down that lasted over a year. He gave additional authority to the new project manager, Harris "Bud" Schurmeier, and made a sweeping commitment to testing, quality control, and attention to detail. The craft received a major redesign, stressing simplicity and reliability. Finally, in January 1964, everyone was ready to try again with Ranger 6.

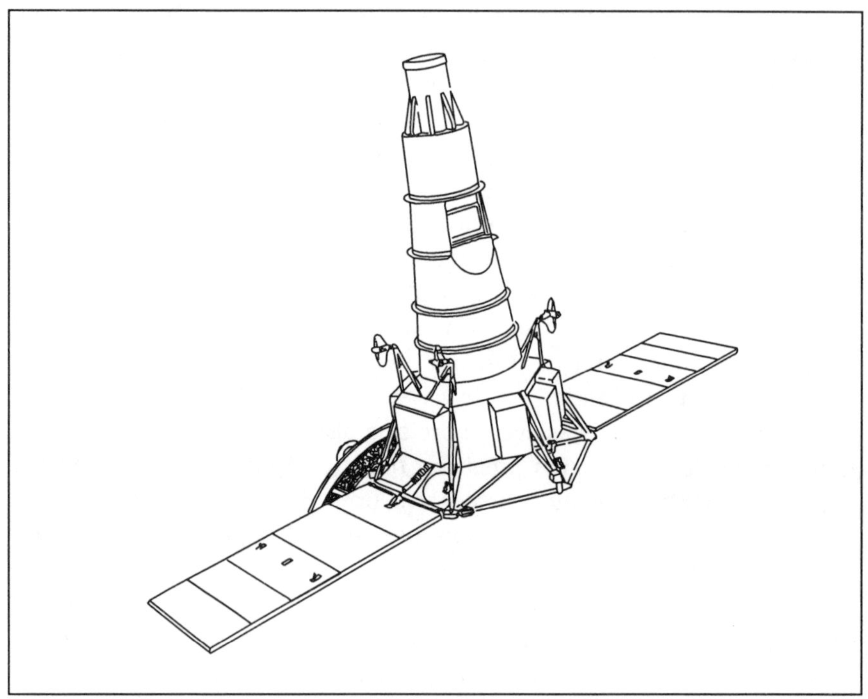

Ranger lunar spacecraft. (Don Dixon)

Its Atlas-Agena put it onto a good trajectory; it then executed a fine midcourse maneuver and sailed toward its target. Minutes before impact, its TV cameras warmed up. Then came the chilling words of an announcer: "There is no indication of full power video." Ground controllers sent up commands, but to no avail. Massive electrical arcing during launch had destroyed the high-voltage power supply. The probe crashed near the Sea of Tranquillity—the sixth failure in as many tries.

Now the lab was really up against it. NASA's Robert Seamans ordered another high-level review, which again raked JPL and Ranger over the coals. On Capitol Hill, Congressman Joseph Karth held a hearing. Heads rolled anew in Pasadena, as Pickering responded to the pressure by taking on a new deputy director and giving him responsibility for day-to-day technical and administrative activities, with authority to act on his own. In New Jersey, where RCA was building the TV cameras, managers tightened procedures for dealing with problems, while assigning a host of additional people to the work. There was a clear understanding that a failure of the next one might mean cancellation of the entire program, with devastating consequences for JPL.

But Ranger 7 performed flawlessly, transmitting pictures all the way as it made a swan dive into Mare Nubium. "Excellent . . . excellent . . .

signals to the end," said the announcer. "IMPACT!" The sudden release of tension brought a roar from both news reporters and employees; some broke into tears. "I was emotionally choked up," said one senior manager, "and I don't get that way very often. For those of us who had lived Ranger for so long, it was kind of a spiritual happening."

This was a turning point, both for JPL and for the lunar and planetary program. The historian Clayton Koppes notes that "the laboratory's problems had been less in design than in carrying out the precision engineering and quality assurance necessary for spacecraft which had to operate nearly perfectly every time. God was in the details—in spotless cleanliness, in thorough testing, and in ruthless follow-up to make sure each failure report was corrected rather than accepted on faith. After Ranger JPL was a different organization."[2]

This success came in July 1964. Two more Rangers flew early in 1965, with the last of them sending its photos directly to network television. The nation's viewers saw its images of the crater Alphonsus on early-morning TV, above the words LIVE FROM THE MOON. And already JPL was deeply involved in the more difficult task of carrying out a soft landing on the moon, entirely under automatic control. It had contracted with Hughes Aircraft to build Surveyor, the craft that were to do this, but Hughes had run into the developmental problems that the lab knew only too well. Pickering gave his contractor plenty of support, and showed that JPL's lessons could be transferred to industry.

Surveyor did not make the first such landing; the Soviets got there in February 1966 with Luna 9, which sent back televised views of its surroundings. But Surveyor 1 followed in June, and over the next year and a half, this program scored four more successes in six additional tries. Its panoramic moonscapes gave Apollo astronauts a clear preview of what they could expect to see.

In addition, JPL built a series of spacecraft called Mariner, as variants of Ranger, that flew to the nearby planets. Venus, a target for Russia as early as 1961, drew similar attention from Pickering. Despite centuries of astronomical observation, scientists knew almost nothing about it, for a thick and never-breaking blanket of clouds perpetually hid its surface and lower atmosphere. Because this planet lies well to sunward, it was reasonable to view it as a world of torrid temperatures. Yet even this was uncertain; some specialists argued that the bright clouds reflected so much sunlight that its surface actually was cool. Other speculations went further. The astronomer Fred Hoyle proposed that it was covered with "oceans of oil. Venus is probably endowed beyond the dreams of the richest Texas oil-king."

During the summer of 1962, NASA launched two probes at that planet, in the reasonable hope that at least one would make it. The one that did,

Mariner 2, flew past Venus in mid-December, as the first successful mission to another planet. It carried no camera, but infrared and microwave radiometers gave it the means to measure Venus's temperatures. These proved to be high indeed, with the surface at 800 degrees Fahrenheit, hot enough to melt lead. In addition, temperatures were almost as high on the planet's night side, showing that it had a very thick and dense atmosphere that could hold and transport a great deal of heat.

By piercing the clouds of Venus, Mariner 2 started to dispel the clouds of speculation, as planetary scientists began to see this planet as a world that had taken a bad turn at a very early date. It was nearly the size of the earth; but our planet had an ocean of water, while Venus smothered beneath an atmosphere of carbon dioxide with nearly a hundred times the surface pressure of ours. It was tempting to believe that the two worlds had been formed initially as virtual twins, which meant that the earth must have had a Venus-like complement of carbon dioxide, with Venus owning a major supply of water. Yet it was not hard to understand how the evolution of the two planets had taken such different courses.

An important clue came directly from terrestrial geology. Geologists knew that the earth holds vast quantities of carbon dioxide, not as a gas in the atmosphere, but in the form of limestone and carbonate rock. Chemical reactions, occurring within the oceans, had extracted the gas from the atmosphere and produced the rock. But one could understand Venus by assuming that it was never cool enough for its water to condense. Lacking oceans, its carbon dioxide would have entered the atmosphere freely. There it gave rise to the greenhouse effect, allowing sunlight to penetrate but trapping heat from Venus's surface to produce those fearsome temperatures.

But what happened to the water? Hot steam is very active chemically; it would have reacted with carbon monoxide in the atmosphere and with hot surface rock, oxidizing them both. In addition, solar ultraviolet would have split apart the water molecules, releasing oxygen for further reactions of this kind. The hydrogen from the water formed lightweight molecules that escaped into space, leaving a desiccated world of aridity and suffocating heat.

The earth might have lost its oceans somewhat similarly, but its atmosphere had protective features that Venus lacked. Earth had a cold stratosphere that acted as a barrier, keeping the water vapor at lower altitudes. In addition, life arose early in the earth's history, producing an oxygen-rich atmosphere. This oxygen formed an ozone layer, which screened out most ultraviolet by absorbing it. On such subtleties had hung the fate of worlds.

Mars was another such world. To view it at its closest approaches was like looking at a one-foot globe two miles away. Even so, the best telescopes might have mapped it in detail, showing features as small as twenty miles across. But the motions of the earth's atmosphere, which cause stars to twinkle, blurred all attempts at photography. Still there were times when astronomers, squinting through eyepieces, caught rare moments of atmospheric quiescence that allowed them to glimpse some of this detail. They then would sketch rapidly, knowing that seeing conditions would soon deteriorate. In this fashion, scientists succeeded in drawing Martian maps.

The maps showed light- and dark-colored areas with names as strange and fanciful as in a dream: Trivium Charontis, Phaethontis, Eridania. What were they? The maps also had linear markings showing the famous canals, with names such as Deuteronilus (the Double Nile). Did they correspond to anything geographical or were they mere artifacts of eye and mind working at the limit of visibility? What was Solis Lacus (the Lake of the Sun)? Why had Syrtis Major become noticeably darker during 1954? Was there sand in Arabia or water in Euphrates?

In an initial attempt at finding answers, JPL's Mariner 4 made a flyby in July 1965. It demanded all the care of Ranger, and for this reason it carried a dummy ultraviolet photometer. The real one had been prone to high-voltage arcing, and it had to be left behind. However, the builders of Mariner could not simply remove this instrument. Instead they installed a replica of the same weight, polished to give the same reflectivity, and engineered to use the same electric current.

To those who hoped there was life on Mars, this mission's photos came as an unwelcome surprise. They showed craters amid landscapes that were quite moonlike. The Martian atmosphere proved to be thin, with less than one percent of the surface pressure of the earth's. This view of Mars was reinforced in 1969, as two powerful Atlas-Centaurs launched Mariner 6 and Mariner 7, a pair of follow-up missions. They photographed the planet from a distance, getting the sharp images that ground-based telescopes had never attained, then swooped past for more close-up shots. And these showed . . . more craters.

But these three spacecraft together had looked at only a small portion of the surface. Seeking a complete survey, JPL sent Mariner 9 to Mars in 1971 and fitted it with a retro-rocket so it could go into orbit. The mission planners hoped to follow a seasonal change, for when spring comes to southern latitudes, the south polar cap begins to shrink and a wave of darkening sweeps northward, moving at twenty miles per day. But Mars also experiences major seasonal dust storms, and when Mariner 9 arrived that November, it found the planet shrouded from pole to pole. This Mars-wide storm was the worst astronomers had seen in a century of viewing. At ground level it was fierce, and a Soviet probe took the brunt of it. Hoping to one-up the Yankees, Moscow landed a spacecraft on the Martian surface. But it transmitted for only twenty seconds before high winds and blowing dust shut it down.

As the dust cleared, however, a new and fascinating world came into view. Though Mars has only half the diameter of the earth or Venus, it displayed volcanoes and valleys far larger than any terrestrial counterparts. Here was the great chasm of Valles Marineris, four to five times deeper than the Grand Canyon, eight times as wide, and long enough to extend across the entire United States. Here too was the volcano Olympus Mons, so enormous that astronomers had seen it from earth. Ringed by tall cliffs, it was 375 miles across, enough to stretch from Los Angeles to San Francisco. Its peak reached 86,500 feet, nearly three times the height of Mauna Kea, the highest mountain on earth. A low volcano, Alba Patera, spanned a thousand miles, while in some places lavas had flowed for 400 miles, the distance from New York to Cleveland.

What titanic forces had produced such immensity? The mobility of Martian lavas raised few eyebrows among geologists; they knew of flows that had shown similar fluidity. But the canyons and volcanoes were another matter, and the photography provided a clue: Mars had not experienced plate tectonics. The crust of the earth was an assemblage of great plates, which rode atop moving masses of rock in the mantle, but the crust of Mars existed as a single unbroken whole. Tectonic motions may have tried to break it; perhaps Valles Marineris represented an unsuccessful attempt to crack the crust into two such plates. But with the crust holding firm, volcanoes such as Olympus could rise to truly towering heights.

For these volcanoes drew their magma from deep sources that did not shift position. The unmoving crust then had allowed the magma to accumulate, forming not only Olympus but also the equally high Arsia Mons. From the summits of either of them, one would clearly see the planet's curvature. The volcanoes of Hawaii were similar. But owing to plate tectonics, the Pacific seafloor had moved past the magma source. This source thus threw up a chain of seamounts during a hundred million years instead of a single great mountain.

By 1971 the planetary program had outgrown what JPL alone could handle. NASA's Langley Research Center now was actively involved, and its contractor, Martin Marietta, was building the Viking spacecraft, which were to land on Mars's surface. They featured orbiters as well, with cameras better than those of Mariner 9. When the two Viking missions reached Mars in mid-1976, these orbiters found evidence of water.

A Mariner 9 photo had depicted a dry riverbed 350 miles long, sinuous in shape and showing tributaries. Viking now demonstrated that water had once flowed on a vast scale, forming fast-moving floods. In places, these broad and shallow streams had encountered a large crater and had flowed around it, to form teardrop-shaped islands bigger than Manhattan.

Most of the water was long gone, leaving an arid world with frozen carbon dioxide in the polar caps. Still, Mars in its heyday had been far more earthlike than people anticipated.

No TV viewers were watching when Viking 1's lander set down on the surface, in the early morning of July 20. You had to be in a NASA center, perhaps at JPL itself. Nor did it transmit its pictures quickly and in their entirety; rather, you watched as a scanner built each one up as an array of vertical lines, from left to right. Within minutes after touchdown, Viking was ready to present its first view. As it took form, it disclosed a sandy and rock-strewn landscape—a desert plain with dunes. The horizon was only two miles away. Color images followed, showing a dark yellowish-brown Mars landscape. The sky was pink, from suspended dust. Later views showed the Martian sunset, with the rocks incarnadined in the fading glow.

Both landers also searched for life by employing automated chemical labs, exposing soil samples to moisture and adding nutrients. Two of the experiments yielded results that indicated active chemicals in the soil, but that were hard to reconcile with biology. A third, the pyrolytic-release test, sought to learn whether microbes would grow in response to the nutrients. It came up with a tantalizing sign of life, for its results suggested that up to several hundred such cells might have developed. The fourth experiment, a gas chromatograph-mass spectrometer (GCMS), looked for pre-existing organic compounds by the simple technique of heating the soil strongly and searching for their chemical signatures. The GCMS was sufficiently sensitive to detect organics at a level of parts per billion, a concentration up to a thousand times less than that of the earth's desert soils. Yet this instrument found nothing.

Was there life on Mars? Norman Horowitz, chairman of Caltech's biology department, noted that "the pyrolytic release instrument had been rigorously designed to eliminate nonbiological sources of organic compounds." Significantly, its results, which were consistent with life, lay below the levels that the GCMS could detect. However, a general view among the project scientists was that the GCMS was the one to trust. When Gerald Soffen, the chief scientist, learned of its results, he said, "That's the ball game. No organics on Mars, no life on Mars."*

Certainly there was nothing to support earlier hopes, which had invoked rapid plantlike growth to explain Martian darkenings, including

*A small number of rocks from Mars are known to have fallen to the earth as meteorites. Studies of one such rock, published in *Science* in August 1996, suggest the presence of microscopic fossils that could have resulted from microbes, when that planet was young. However, the reported findings remain ambiguous. *Science* describe them as "evidence—but not proof—of ancient life on Mars. The claim has excited skeptical fascination among scientists but has made no converts so far."

that of Syrtis in 1954. Instead, these seemed to result from nothing more than windblown dust. Indeed, because the winds would distribute organic compounds widely, and because such organics exist in meteorites and survive their impacts with the surface, it appeared that solar ultraviolet destroyed these compounds as they arrived. Mars not only lacked organics; it was actively hostile to their existence.

While NASA-Langley was proceeding with Viking, JPL was seeking new worlds to conquer, and had found a new way to reach them. NASA by then was using the nation's most powerful rocket, the Titan III-Centaur. It had launched Viking and had the thrust to reach Jupiter as well as the outer planets. But these missions would take a long time: sixteen years to Uranus, thirty to Neptune. In the words of Homer Stewart, the head of JPL's advanced-planning office, "The management problems in organizing and carrying out a direct 30-year mission to Neptune (sheer boredom on the part of the participants) look great enough to deter even the most determined explorer."

The new approach relied on the fact that when a spacecraft flew past a planet, the planet's gravity, combined with its motion around the sun, could deflect the probe in its trajectory and give it extra energy. When Mariner 4 had flown past Mars, for instance, it had picked up over three thousand feet per second in velocity. Jupiter could do even more, for it was the largest of the planets. It could add the boost of an extra rocket stage, entirely for free, and could reduce the flight time to Neptune to as little as eight years.

There was more. When the JPL analyst Gary Flandro undertook studies of such Jupiter-assisted missions, he found that the outer planets were about to enter a rare alignment. A single spacecraft could fly past all the major ones—Jupiter, Saturn, Uranus, and Neptune—in only nine years! Each planetary encounter could add energy, speeding the spacecraft to the next. Furthermore, the launch dates—only a dozen years in the future—were just right: far enough ahead to allow for careful planning and preparation and close enough to be well worth pursuing. Astrologers had long declared that planetary alignments control the affairs of individual people, and this one would certainly influence the future of JPL. The opportunity was too good to pass up.

Flandro's work had begun in 1965. Further studies followed, and with his customary expansiveness, Pickering proposed a venture that would build a flotilla of craft, two to follow the course of Jupiter-Saturn-Pluto and two others for Jupiter-Uranus-Neptune. However, while the celestial portents were propitious, the budgetary ones were not, for this $750-million plan came to the fore just as NASA was in the throes of the post-Apollo cuts in its appropriations. Pickering had to settle for two spacecraft rather than four, and to aim no farther than Saturn. The new program took the name Voyager.

Nonetheless, NASA came up with two supporting projects. One grew out of a low-key partnership between its Ames Research Center and the firm TRW, which had continued the Pioneer missions of the late 1950s. Pioneer 6 through Pioneer 9 had probed interplanetary space; now Pioneers 10 and 11 would reach for Jupiter. Moreover, Pioneer 11 gained the task of using the gravity-assist technique to fly onward to Saturn. The missions, which flew in 1972 and 1973, served as pathfinders. Jupiter in particular had a powerful magnetic field, trapping a zone of intense radiation that could damage a spacecraft's instruments. Saturn's rings carried enormous numbers of small solid bodies that could destroy a probe that came too close.

For both Pioneer 11 and the Voyagers, the successful use of gravity assist would require careful aim during the Jupiter encounter, for this planet had to act as a slingshot, shooting the spacecraft onward to Saturn with high accuracy. To gain experience, JPL carried out a practice run with a separate mission, Mariner 10. It used Venus to fling it toward Mercury; indeed, its post-Venus orbit allowed it to fly past Mercury three times, with cameras and other instruments alert.

Mercury, the closest planet to the sun, had shown even fewer details than Mars in astronomers' telescopes. Mariner 10 photographed much of it in sharp crispness, disclosing a heavily cratered world that closely resembled the moon. The images also gave evidence of the enormous violence of the impacts that had formed its features. One such collision had formed a great basin called Caloris. On the other side of Mercury, at Caloris's antipode, lay a curious patch of jumbled terrain. Donald Gault, a NASA scientist, proposed that the impact had produced intense seismic waves that converged onto this antipode, throwing up the jumble in an earthquake that shook the entire planet.

As Voyager approached launch, it earned a reprieve. The first spacecraft would indeed fly only to Jupiter and Saturn. But NASA officials decided that the second might fly past Saturn to Uranus and Neptune. So it was that these spacecraft rode their Titan III-Centaurs, in August and September 1977, as they set forth on expeditions as ambitious and potentially fruitful as any we will see in our lifetimes. Their high launch velocities took them past the orbit of the moon in less than twelve hours, as they covered a distance that for the Apollo astronauts required days. And while they traversed the far greater distance of half a billion miles that measured the way to Jupiter, three California scientists—Stan Peale, Pat Cassen, and Ray Reynolds—proceeded with their own preparations.

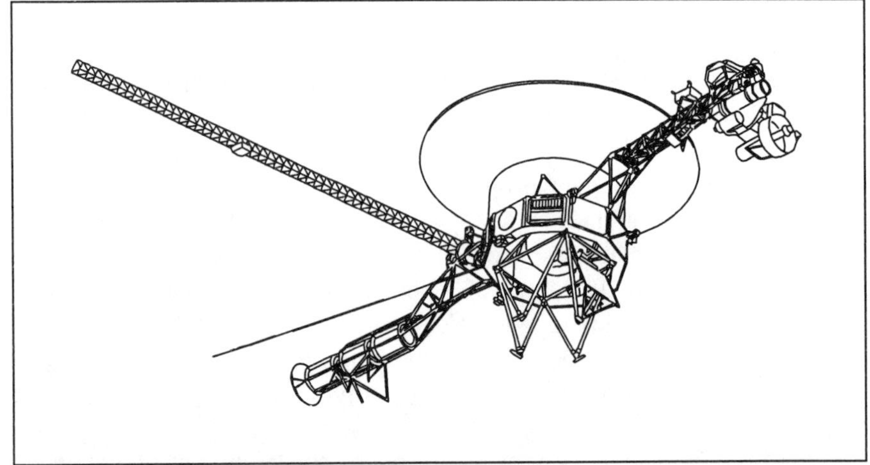

Voyager spacecraft. Long boom at left holds a magnetometer. Power source is at lower left; instruments are at upper right. (Don Dixon)

Their work involved mathematical studies of the system of Jupiter and its principal satellites, which formed a miniature solar system in its own right. Callisto, the farthest from Jupiter, was as large as Mercury; Ganymede was larger still. They strongly influenced each other's motions. Europa's orbital period was close to twice that of Io, the innermost satellite, while Ganymede's period was twice Europa's. In addition, Io acted as an electric generator as it orbited within Jupiter's magnetic field. It produced a flow of electricity that reached across space to this planet, with three million amps of current and two billion kilowatts of power. And these scientists expected more.

Io circles nearly as close to Jupiter as our moon does to the earth, and Jupiter has the mass of over three hundred earths. Hence, the effects of Jupiter's powerful gravity are mighty indeed. The moon's gravity raises tides in our oceans; that of Jupiter produces tides in the body of Io itself, distorting it and making it slightly football-shaped. Peale, Cassen, and Reynolds now proposed that the influence of Europa would combine with that of Jupiter's tides, causing Io to flex like a rubber ball that you squeeze in your hand.

Io was only slightly larger than the moon; left to itself, it would be geologically dead, with a heavily cratered surface. But this flexing would generate heat, and the investigators proposed that Io was molten throughout, except for a thin solid shell. "Io might be the most intensely heated terrestrial-type body in the solar system," they wrote, adding that "widespread and recurrent surface vulcanism would occur." Their paper, titled "Melting of Io by Tidal Dissipation," arrived at the journal *Science* in January 1979. Its editor, Philip Abelson, ran it in the March 2 issue, with Voyager 1 due to fly past Jupiter only three days later.

This spacecraft had been taking photos at long distance for weeks, and as it approached encounter, the activity became frenetic. Bradford Smith, leading a group of imaging scientists, remarked that "it may sound unprofessional, but a lot of people up in the Imaging Team area are just standing around with their mouths hanging open watching the pictures come in."

People were particularly interested in the Great Red Spot, a storm in Jupiter's atmosphere that had been raging for over three hundred years and could swallow the earth with ease. Close-up color photos included oval shapes of smaller storms—one a mottled white, the other light blue. Around them swirled a vast field of whirls and loops in red, white, crimson, orange, light brown, and gray. The views resembled abstract art, and someone said, "Are you sure van Gogh didn't paint that?"

Io, no more than a spot of light in the best telescopes, showed brilliant colors from close up. As its pictures grew more vivid, someone said, "I knew it would be wild from what we saw on approach but to anticipate anything like this would have required some sort of heavenly perspective. This is incredible." Io appeared garishly red and orange. To Brad Smith, it looked like a pizza. Actually, its true shade was a pale olive; the Day-Glo hues resulted from a poor color balance in the first processed images, as when a face appears green on your TV. Still, the pepperonis in this pizza were real enough. They were volcanoes.

Three days after the encounter, Linda Morabito, a JPL engineer, saw the first eruption. She was studying a shot of crescent Io, showing its night side, and noted an immense plume silhouetted against the black sky. Near the crescent's inner curve, another plume glowed brilliantly as it rose from the twilight and caught the bright sun. Io turned out to have at least eight active volcanoes, most of which were going strong four months later when Voyager 2 came by for another look. Io was indeed the most geologically active body in the solar system, more so even than the earth, and JPL had the pictures to prove this. At closest approach, Io had ceased to appear as a disk, for portions of its surface filled the entire field of view. The best photos had resolution of a few hundred yards. Some showed fresh flows of red sulfur. Other pictures showed the volcanoes themselves, including Pele, the largest. Shaped like a hoofprint upon the surface of Io, it was six hundred miles long, and it spewed its plume to a height of some 175 miles. No other planet had anything to compare with it.

During the same year of 1979 when both Voyager encounters took place at Jupiter, Pioneer 11 reached Saturn and navigated safely past it. Now the two Voyagers followed the trail it had blazed, reaching that planet in November 1980 and August 1981. Here were its rings, which the physicist Philip Morrison described as "of everything revealed by the telescope, the most meaningful to the most people." Voyager captured their delicate filigrees, as well as the vast shadow cast across them by the hull of Saturn, seventy thousand miles across.

All hope now rested with Voyager 2, which was to continue to the outermost planets. It had barely escaped being crippled, for its main radio receiver had failed in 1978. It had a backup, but a capacitor went out in this one, too. This probe could then receive commands only at a single specific frequency. It could not cope even with tiny Doppler shifts that the earth's rotation produced in the wavelength of ground-based transmitters. Jet Propulsion Laboratory specialists kept careful track of variations at the transmitter that would allow commands to get through, and the vital communication link stayed open.

Beyond Saturn, the two craft went on their separate paths. Voyager 1 flew past this planet on a course that made the best observations, but sacrificed the chance to go on to more distant worlds. But Voyager 2 approached Saturn along a different trajectory, and came away bound for Uranus. These were not Robert Frost's two roads that diverged in a yellow wood, but here too the choice of road made all the difference. Voyager 2 now followed its schedule faithfully, reaching Uranus in January 1986. It passed Neptune, the solar system's most distant outpost, in August 1989 and continued to work superbly.

From this most distant of perspectives, the sun itself lost significance, appearing in the sky as nothing more than the brightest of stars. Ahead lay the heliopause, where the physical influence of the sun gives way to that of interstellar space. Here the solar wind, a rapid outflow of particles from the sun, collides with a flow of rarefied gas that streams between the stars. Both Voyagers still were active, and during 1992 they received radio signals from this heliopause, far beyond the solar system. They flew toward it, as if the signals were beckoning. Like sailors venturing away from their home islands, these craft now approached the interstellar overwhelm, leaving the realm of the sun to enter the measureless reaches of the galaxy. They went forth into this immensity like two tiny rowboats, carrying cargoes of human hope, courage, ingenuity, and the will to explore.

11

Space in the Eighties

The Efforts Falter

NASA HAD NEVER LOST a manned mission. Across fifteen years, in its Mercury, Gemini, Apollo, and Skylab programs, it had a perfect record launching astronauts into space and returning them safely. This bred complacency, as the agency prepared to commit its future to the space-shuttle program. No one expected to see a shuttle explode in the sky over Cape Canaveral; no one anticipated that such a disaster would bring the space program to its knees. Yet already, during the 1960s, the Manned Orbiting Laboratory had provided a warning.

Evolving as an Air Force project, MOL consisted of a small space station with a Gemini spacecraft on top. The CIA liked its potential for reconnaissance and helped its design to develop as a counterpart of Moscow's Salyut military stations in subsequent years. Like the Salyuts, it was to fly with a camera and large telescope, the KH-10, six feet in diameter, with astronauts aboard to work with it.

President Johnson gave the project his approval in 1965, but by then MOL faced an alternative. It stemmed from the work of the CIA's Albert Wheelon, who had succeeded Richard Bissell as head of the agency's satellite program. In a world of the brilliant, Wheelon stood out; he had received a Ph.D. from MIT at age twenty-one, and had gone on to work with Ramo and Wooldridge. He now had strong backing from John McCone, the CIA's director, and he wanted to build a new unmanned reconnaissance satellite. The resulting project was known in the trade as Hexagon; other people called it Big Bird. After 1965, it was funded as a backup to MOL.

Both programs called for spacecraft of substantial size, ten feet across by about forty feet long. Their weights, at 30,000 pounds, were also similar. This was too much for the standard Titan III, so the Air Force ordered an uprated version, with bigger solid-fuel boosters.

The Manned Orbiting Laboratory camera, the KH-10, was built to yield resolution as sharp as nine inches. Hexagon's system, the KH-9, was not so acute, but offered its own advantages. It would broaden the width of the photographed swath to 360 miles, twice that of the Corona reconnaissance satellites of the 1960s. It would mount twin sixty-inch cameras, whose resolution would be two feet, close to the eighteen-inch sharpness of Gambit. Operating as a bucket dropper, Hexagon would carry four recoverable capsules for film return.

The Pentagon gave MOL strong support, but the Air Force had never carried through the development of a manned spacecraft, and did not know how to control its cost. By 1969 the projected cost was up from an initial $1.5 billion to $3 billion, and MOL found itself caught in the budget crunch of the Vietnam War. It was canceled in May.

The CIA and the Air Force had collaborated closely in MOL's development, but in the crunch, it failed to win support from Nixon's CIA director, Richard Helms. The reason is revealing. The nation's intelligence community could shrug off the loss of an unmanned reconnaissance satellite. But MOL would carry crew members, and Helms saw that this would greatly raise the stakes. In the words of the analyst Jeffrey Richelson, Helms's advisers "feared that an accident that cost the life of a single astronaut might ground the program for an extended period of time and cripple the reconnaissance program."

But three years later, with the space shuttle approved and under way, such thoughts were far from the minds of NASA's top people. These executives—James Fletcher, the administrator; George Low, his deputy; and Dale Myers, head of the shuttle program—now had the support they needed, both from Nixon and the Congress. An initial focus was Rocketdyne's Space Shuttle Main Engine (SSME). Paul Castenholz, who had rescued the J-2 a few years earlier, had prepared its design with considerable care.

"You start by making a parts list," recalls Sam Hoffman, president of Rocketdyne during the Apollo era. "That'll be your plan. Now what are the major components? Well, it's gotta have a thrust chamber assembly. It needs a turbopump assembly. It needs an injector. Also a throat and nozzle. I write down these words, leave a lot of blank space under each one, and knowing rockets I can list other subassemblies that make up each main component. Then I can further subdivide, down to the level of individual parts.

"Each part and subassembly has its own design, its own engineering drawing. Eventually I'll have five hundred or a thousand drawings listed in my plan. Then, knowing how many engineers I have, and how rapidly they can get the work out, I can estimate how long it'll take to prepare the design. If I need a special steel, I can order it as a long-lead item, so it won't be a bottleneck. If I need to get the design in faster, I can bring in more men. That's how my plan develops."[1]

The complete proposal for the SSME, which won the contract, filled a set of documents several feet in length. In addition, Castenholz had used $3 million in company funds to build a test version of the thrust chamber, regeneratively cooled and fitted with an injector similar to the one that would eventually fly. In February 1971 this engine had fired on a test stand for one-half second, enough to measure performance but not so much as to risk burning a hole in the side.

The SSME was to develop 470,000 pounds of thrust by burning hydrogen and oxygen. The J-2, burning the same propellants, had given only 230,000 pounds. In addition, the SSME was to operate at a pressure of 3,260 pounds per square inch, over two hundred atmospheres, versus 780 pounds for that predecessor. This extreme pressure would boost the engine's performance, while allowing it to produce its thrust within a compact package. But to make it work was far from easy. The first complete engine went on a test stand in May 1974, running again for one-half second. Then, over the next five years, the SSME program struggled through one major problem after another.

The problems started with the turbopumps. The fuel turbopump was modest in size, about four feet long by eighteen inches in diameter. It was the size of a large outboard motor, but to pump liquid hydrogen at the required pressure and flow rate, it had to develop 76,000 horsepower. This was more than had run such huge ocean liners as the *Mauretania* early in the century, in an era when their engine rooms and boilers extended most of the distance from stem to stern.

The first difficulty was that the turbine shaft was not held solidly enough in its bearings; it tended to jiggle with a circular motion. At 37,000 revolutions per minute, the jiggling quickly wore out the bearings. The solution was to stiffen them, but it took eight months to figure out how to do it properly.

Next came the matter of cooling and lubricating these bearings. No oil could be used; it was necessary to use liquid hydrogen. Since this was rather like lubricating a jet engine with water, it took some doing. One set of bearings turned out not to be getting enough hydrogen; it was overheating and then failing. The solution called for redesigning the channels that fed hydrogen to this bearing. That took another six months.

A third issue involved the turbine blades. One section of the turbine had sixty-three of them, each the size of a postage stamp. Each of them was generating six hundred horsepower—as much as an Indianapolis racing car. They were under considerable strain, and they tended to crack. The problem was traced to vibration, a well-known scourge of engineers, and finally corrected. This took another six months.

While work was progressing (slowly) on the fuel turbopump, the oxygen turbopump was also under development. It also met with serious problems, which were even more difficult to find and correct. When a turbopump encountered trouble, it would fail and shut down. A fuel turbopump failure simply caused the engine to lose power. Engineers then could disassemble it, seeking the cause of the failure. But trouble in the oxygen turbopump caused severe damage to the engine.

In earlier rocket-development projects, oxygen-pump failures had often resulted in the motor blowing up. The SSME was stronger; to contain its high pressure, it was built so stoutly that it rarely exploded. But it could, and did, catch fire. It was built of copper, nickel, and steel, which we do not regard as fire hazards. But at the temperatures and pressures of an SSME, in the presence of liquid oxygen, virtually anything would burn. These fires often destroyed so much of an engine that it was difficult to discover what part had failed, or in what sequence the fire had spread.

The oxygen turbopump suffered two major problems. The easy one took six months to find and fix. It involved a rapidly rotating seal that separated liquid oxygen from hot gases in the turbine. The seal was supposed to spin without friction, but it tended to rub against another engine part. This rubbing then produced heat by friction—heat sufficient to ignite the metal, as when Boy Scouts start a fire by rubbing sticks together. Eventually Rocketdyne licked this problem by choosing a different type of seal.

The more difficult problem was that the oxygen turbopump bearings repeatedly failed and burned up. In the end, a variety of expedients proved helpful. The turbine's rotating shaft was redesigned for better balance. Just as an unbalanced auto tire wears rapidly, the inadequately balanced turbine shaft, turning at 31,000 rpm, had worn so quickly as to cause the unit to fail. Designers not only rebalanced the shaft; they stiffened the bearing supports. Finally, the bearings and their races, or holders, were made bigger and rebuilt to carry heavier loads. After a year and a half, these turbopump bearings succeeded in passing their tests.

The last pump bearing failure came in March 1977, and by then the overall space-shuttle program was in serious trouble. From the outset, its funding had drawn on a strong dose of optimism, for it lacked financial reserves that could cope with major difficulty. Nor did it help when the Office of Management and Budget imposed over $300 million in cuts. In response, NASA instituted a strategy whose very title, "success-oriented management," suggested hope but concealed desperation. In contrast to Murphy's Law, that anything that can go wrong will go wrong, this approach assumed that everything would go right. In the words of an official, "It means you design everything to cost and then pray."

The approach had its roots in the all-up testing of Apollo, which had flown complete Saturn I-B and Saturn V multistage rockets on their first tries. However, all-up testing had grown out of the confidence born of extensive preliminary testing, within a lavishly funded program that had drowned its problems in money. By contrast, success-oriented management wound up resembling Russia's development of its N-1 moon rocket. For the SSME, it meant testing full-scale engines before their major components were proven and reliable, while stinting on spare parts. As *Science* put it, Rocketdyne "constructed everything to novel design, bolted it all together, and—with fingers crossed—turned on the power. At least five major fires resulted."

The original development plan of 1972 had called for the shuttle to make its first flight to orbit in March 1978. But when that month arrived, the engines were at the Mississippi Test Facility and were in no condition to do much of anything. Additional testing might have detected individual flaws that led to malfunctions, but with the entire engine under test, these costly rocket motors were in jeopardy every time a part failed. One such failure severely damaged the test stand itself, and because no one had had the money to prepare for this, the entire engine-test program was delayed for months until the stand was rebuilt.

Disaster struck again just after Christmas 1978, when an engine blew up. This time the turbopumps were not at fault; the new problem areas were the main oxygen valve and a heat exchanger. The valve problem was quickly solved through redesign. The heat exchanger was less simple. In the words of a senior manager, "The failure of the heat exchanger remains unexplained, and it gives you a very soggy feeling. These incidents occurring so late in the test program just do not inspire confidence."

Other failures occurred during tests in May and July 1979. By then, though, the engine was showing signs of reliability, and engineers proceeded to a new round of tests. The shuttle was to fly with a cluster of three SSMEs, and this work attempted to run such a cluster for eight minutes, the full duration of flight to orbit. In November, nine seconds into such a test, instruments detected a problem and triggered a shutdown. This broke a nozzle that was carrying hydrogen for cooling, immolating that engine's interior. A successful full-duration firing of all three engines did not occur until December.

Even so, not all was bleak. The program now had active support from President Carter, because of his strong interest in arms control. Diplomatic efforts in that area dated back to 1969, when Nixon had launched the Strategic Arms Limitations Talks with Moscow. The initial SALT treaty, in 1972, restrained the superpowers from building antiballistic-missile systems. Carter now was pursuing SALT II, which sought to place outright limits on these nations' total forces of missiles and bombers. Such an agreement would require reconnaissance satellites for verification, and the shuttle could launch them.

Back in 1970, NASA's Fletcher had turned to the Air Force for the political backing that had allowed him to initiate this program. Now the Air Force provided support that strengthened its prospects in its time of difficulty. The man who made the difference was Hans Mark, who had headed NASA's Ames Research Center and who had a strong commitment to the shuttle. Carter picked him as the Air Force undersecretary; Mark also was named to head the National Reconnaissance Office, a Pentagon center that managed the space-surveillance work of the entire U.S. intelligence community. Accepting his recommendations, Carter issued a presidential directive, PD/NSC-37 of May 1978, that firmly endorsed the shuttle. Carter followed this up by seeking more money for NASA in 1979, making it clear that he wanted the shuttle to receive enough funding to ensure its success.

Just then, with engine development still far from complete, the program was also in trouble over thermal protection. Earlier spacecraft, including Apollo, had used ablative heat shields to guard against the blazing temperatures of atmosphere entry. But these were heavy and allowed only onetime use. The shuttle was to be reusable; it needed thermal protection that could stand up to many re-entries, and it needed light weight. The solution called for tiles of matted silica fiber, bonded or glued to the shuttle's outer surface. These were to serve as insulation, radiating away the heat while preventing it from penetrating. This was vital; if the heat got through, it could burn up the vehicle and incinerate its astronauts.

Everyone worried that the tiles might peel off like roof shingles, and their installation required meticulous care. The process involved mounting two of them, which typically were six inches or less on a side, and then designing a third to fill the gap between them. Workers would carefully measure space for the third and then send the information to the tiles' contractor, Lockheed in Sunnyvale, California. There the third tile would be machined to order. It took one person three weeks to install four tiles, and it took 31,000 tiles to cover a single space-shuttle orbiter.

The first of them, *Columbia,* went from the assembly plant in California to Cape Canaveral in March 1979. With it went two thousand workers of Rockwell International, the shuttle's builder. They had to mount some ten thousand tiles that had not yet been installed; they also had to replace some 7,500 others that were damaged in transit. Then success-oriented management again reared its head, as belated wind-tunnel tests showed that many of them might come off in flight. This inspired a curious arrangement wherein some Rockwell employees continued to affix tiles while others tried to pull them off with vacuum pumps. Many tiles had to be removed and "densified"—treated with extra silica to permit them to bond more firmly to the adhesive—before they could be certified for flight.

As costs mounted and delays lengthened, NASA found itself neglecting other important areas of effort, particularly the planetary program. A decade earlier this agency had hoped to send astronauts to Mars and had built the Viking spacecraft that carried out automated landings. But the budget contained funds for only the two Vikings of 1976, even though additional missions would have been of great importance to scientists. At the Jet Propulsion Laboratory, there was considerable interest in a Venus orbiter that would pierce through the planet's clouds with radar to map its unseen surface. This proposal could not win support, and for want of money, the Venus mapper would not fly until 1989. Nor would NASA send a probe to Halley's Comet, even though its appearance in 1986 represented a once-in-a-lifetime opportunity.

Then there was Galileo, which was to follow up Voyager by orbiting Jupiter rather than merely flying past this planet. Jet Propulsion Laboratory won approval for this project in 1977 and could have flown it on a Titan III. But in the same year, NASA decided to buy no more such rockets. This was part of the policy of putting all its eggs into the shuttle's basket, but at first it seemed the shuttle would be ready in plenty of time. John Casani, the project manager, stated ruefully two years later: "We were originally scheduled to be taken up on the 26th launch, and then schedule slippages moved us up to the seventh. We sure thought we had enough padding." The shuttle program passed its delays on to Galileo, which could not fly without it, and Galileo's cost soared from $450 million to $850 million. In 1981 the Office of Management and Budget ordered its cancellation. This would have shut down the planetary program, leaving only the Voyager 2 encounters at Uranus and Neptune. Supporters of Galileo managed to save it, but it was a near thing.

The first flight of the space shuttle took place in April 1981, twenty years from the day when Yuri Gagarin became the first man in space. The shuttle's pilots, John Young and Robert Crippen, faced a very high degree of success-oriented management. The SSME had operated only on test stands—never in an actual launch. No preliminary flight had demonstrated that the tiles would stay in place rather than falling off. No astronauts had ever ridden atop solid-fuel boosters. Here was all-up testing at an even higher level than in Apollo, which had given the Saturn V two unmanned missions before committing a flight crew. The Roman poet Horace had written of men such as these:

> Oak and brass of triple fold
> Surrounded sure the heart that first made bold
> To the raging sea to trust a fragile bark.

But the flight lasted two days and generally went well. The cargo bay doors opened and then failed to close properly, but this caused no difficulty. The zero-gravity toilet broke down. A few tiles came off during re-entry, but the design allowed for this and their loss did not burn a hole in the spacecraft. In sum, the mission was a success.

While NASA had been wrestling with its new technology, the Soviets had been following their own star, by pursuing long-duration manned flights in Salyut. However, the 63-day mission of Salyut 4 in 1975 had pushed the limits, for its cosmonauts, Pyotr Klimuk and Vitaly Sevastyanov, had to bring with them all the supplies they would need. To stay longer, flight crews would need to be able to receive additional food, water, and oxygen while in orbit.

Moscow therefore modified the basic Soyuz and Salyut to open the way to missions of arbitrary length. Engineers converted Soyuz into an unmanned spacecraft, called Progress, that could carry the supplies. It now had a pair of TV cameras, with views that permitted ground controllers to guide it to an automated docking with Salyut. And because an orbiting Salyut had a Soyuz docked at one end, in which its cosmonauts had arrived and with which they would depart, it was necessary to add a second docking port to the other end. Salyut then would accommodate both Soyuz and Progress simultaneously, like a locomotive with a railroad car at each end.

A three-month flight was the initial goal, for which Salyut 6 took to the sky atop its Proton rocket in September 1977. Two cosmonauts went up but failed in their docking attempt. Two others followed: Yuri Romanenko, the commander, and Georgi Grechko, an engineer who had intimate familiarity with the docking mechanism. They docked at the port meant for Progress; Grechko did a spacewalk and verified that the other port was okay, while Romanenko watched from a hatch. Evidently the problem had been with the earlier Soyuz that had failed to dock, and not with Salyut.

They settled down for their lengthy stay, with plenty to keep them busy. Medical tests took much time, as did housekeeping; after all, one purpose of having cosmonauts in space was to have them look after the life-support equipment. To stay fit, their onboard equipment included a treadmill, an Exercycle, and a set of bungee cords. They didn't like these items; exercise left them sweaty and smelly, and they dodged it when they could. At times they even operated the onboard instruments. These included a multispectral camera for earth observations, a mapping camera, and a telescope that worked at far-infrared wavelengths.

Two of their friends arrived for a visit, delivering a fresh Soyuz craft and returning to earth in the old one after a five-day stay. They brought mail and newspapers. Then came the Progress supply vessel, with a treasure trove that included fresh fruit, a cassette recorder with tapes, clean clothes and linen, as well as medicines and spices. Grechko and Romanenko loaded this spacecraft with their trash, then sent it down from orbit

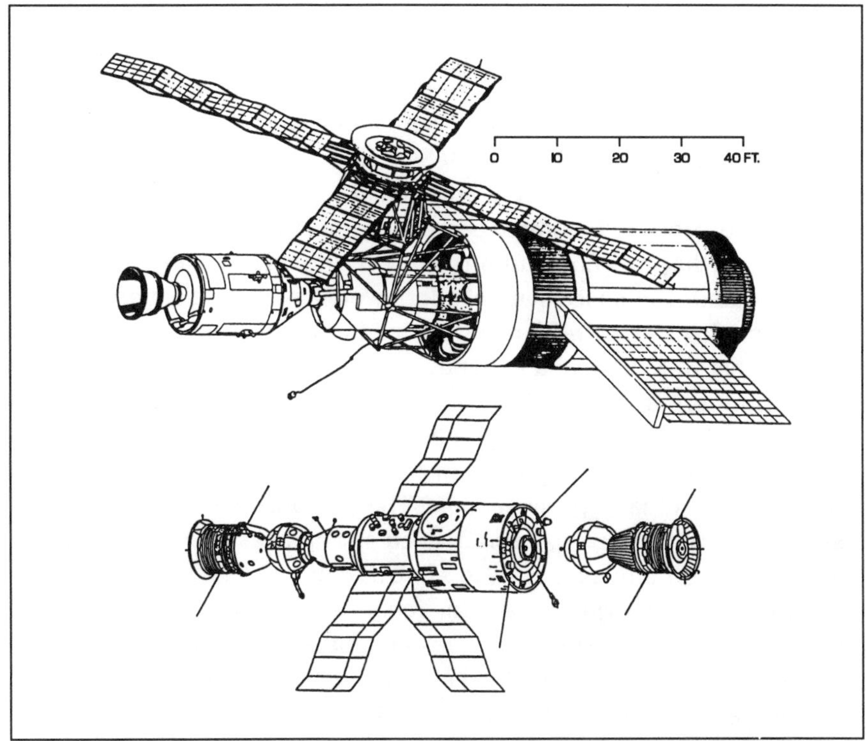

Early space stations. ***Top,*** **Skylab. Apollo spacecraft is docked at left; windmill-like attachment is Apollo Telescope Mount, used for solar observations.** ***Bottom,*** **Salyut 6. Soyuz is docked at left; Progress supply craft approaches at right. (Dan Gauthier)**

to burn up in the atmosphere, garbage and all. A second visit by a Soyuz followed, with a man from Czechoslovakia, Vladimir Remek, the first man in space who was neither Soviet nor American. This inaugurated a new program, wherein cosmonauts from other nations would fly as guests of Soviet crews.

Grechko and Romanenko came down in March 1978, having set a record of 96 days in orbit. Even longer missions would follow: 140, 175, and then 185 days. These compared with 84 days for the longest Skylab mission and provided a useful perspective on medical effects of weightlessness. The most serious effect was the loss of calcium from bones, which raised concern that spaceflight might turn a bold astronaut's spine to jelly. Soviet doctors had no useful measurements of the actual amount of calcium loss. Still, the cosmonauts' good health, following return, eased the concern.

Salyut 6 was a space station. After the initial success of the shuttle, in April 1981, NASA was ready to declare openly that it wanted one, too.

After all, a space station had been high on the agency's agenda for over a decade. The necessary equipment was already bought and paid for: a functional second Skylab, along with two Saturn V launch vehicles. Unfortunately, the Skylab was in the National Air and Space Museum, while the Saturns were on display at NASA centers in Houston and at Cape Canaveral, exposed to the wind and rain. The agency might have placed them all in protected storage, as it had done with the Saturn I-B and Apollo craft that had flown to link up with the Russians in 1975. But as matters stood, the space-station effort would seek to accomplish with new technology what NASA had disdained to do with rockets and an orbital lab that were built and at hand.

In Houston, a small group of designers had recently rekindled this flame by proposing a Space Operations Center. Their concept called for an array of modules brought up in the shuttle and assembled in orbit, with room for nine to twelve people. These people would occupy themselves assembling large spacecraft, such as antennas for use in radio astronomy and orbiting platforms to serve as communication satellites. The SOC would also provide a base for repair and maintenance of existing spacecraft. To ferry the satellites back and forth to their proper orbits, the SOC was to have its own rocket vehicle, housed in a hangar and fueled with propellants from tanks at the SOC itself.

The SOC concept was very much in line with the shuttle's background. The shuttle had been designed to launch the nation's spacecraft, while having enough lifting power to build a space station as the next step. The SOC introduced its own purpose: satellite assembly and maintenance. Still, you didn't have to be a rocket scientist to understand that its space ferry, ranging as far as geosynchronous orbit, could give way to a larger version that would carry astronauts back to the moon, as a prelude to a mission to Mars.

President Reagan took office early in 1981 and named new people to run NASA. Its administrator would be James Beggs, who had recently managed the F-16 fighter program at General Dynamics. For his deputy, Reagan detached Hans Mark from the Air Force and sent him over. Beggs and Mark both had a strong interest in a space station, but the SOC wouldn't do. It was too specific as a design concept, and would attract critics in droves. These administrators remembered that the detailed shuttle designs of 1971 had aroused strong opposition from the OMB, which had forced NASA to scale back its plans. To avoid such a fate this time around, they would have to lie low until they could get Reagan's approval.

Accordingly, Beggs set up a Space Station Task Force in May 1982. It would not present a design concept as a fait accompli; rather, it would carry out studies and seek the views of potential users. This approach played to a view that the best station would be crafted to fit those users' needs. It also avoided early commitment to a firm design that would become a magnet for opponents. "We absolutely refused to draw a picture of a space station," said Robert Freitag, one of the task force leaders.

Still, it would take more than in-house studies to win a presidential endorsement. A space-station project would represent a major step, and would touch the concerns of a host of other departments: the Pentagon, the CIA, Commerce with its weather satellites, and State with its interest in international cooperation in space. Beggs might try to present a space-station plan in a one-on-one meeting with Reagan, but Reagan would not bite; he would insist that the views of all these other players should have a fair hearing. And Beggs had a problem, for there was virtually no interest in a space station outside of NASA itself. The "users," of whose views he was so solicitous, were mostly NASA's own clients, ready to recommend whatever he wanted.

At this point a vigorous young Air Force colonel, Gilbert Rye, entered Beggs's life. Tall and lean, with a shock of curly black hair, Rye was the sort of fellow who looked good in a blue uniform. He had worked with General Alexander Haig at NATO headquarters in Europe, and Haig remembered him when he became secretary of state. Rye had also spent three years in the Pentagon, helping to set up the Air Force Space Command. That service had long played a major role in space, and it was increasing; under Reagan, military space programs would become larger than those of NASA. Rye now held the post of Director of Space Programs, a staff position in the National Security Council. He had the support of the NSC's chairman, William Clark, who was an old friend of Reagan. And Rye was a strong proponent of a space station.

Reagan was to give a major speech on the space program in July. Beggs hoped to bypass the process of interagency review by getting the president to make a personal commitment in the style of JFK, announcing this new program in the way that JFK had introduced Apollo. His minions stirred up speculation that the president would do this, as they lobbied vigorously and orchestrated a letter-writing campaign that demanded a space station. White House staffers counted seventeen newspaper and magazine articles, predicting that Reagan would announce his approval. Beggs met with Rye, who was helping to write the president's speech, and together they agreed on the words that would make it official: "We must look aggressively to the future by demonstrating the potential of the shuttle and establishing a permanent manned presence in space."

Reagan was to give this speech at Edwards Air Force Base on July 4, as part of a space extravaganza. The *Columbia* was to land there following another flight to orbit. Also another shuttle orbiter, *Challenger,* was to take off for Cape Canaveral atop the back of a Boeing 747, and Reagan would personally give the takeoff command to the 747's pilot. A band was playing at the landing site when the day arrived; flags were flying, and fifty thousand people were on hand. In addition, NASA flight controllers had thoughtfully delayed the landing of *Columbia* for an extra orbit, so Reagan could get more sleep.

The president stood with his party in the hot desert sun, as Beggs introduced him to a group of astronauts in blue flight suits. Then, as Reagan later recalled, "They hurried us up on the platform, because they said it was time to get up there, the shuttle was coming in. They said it was on its approach." The president asked, "Where is it now?" Someone replied, "Just over Honolulu." Hawaii was as far away as Washington, and in Reagan's words, "The whole miracle was brought home to me right then."

He proceeded with his speech: "We must look aggressively to the future by demonstrating the potential of the shuttle and establishing a more permanent presence in space." This meant virtually nothing; it could even refer to more unmanned spacecraft. Rye's draft of the speech had circulated in the White House, and David Stockman, the budget director, had smelled a rat. Stockman wanted no costly new initiative in space, and had arranged to take the offending word out.

Now there was no dodging the interagency review. A White House panel would conduct it, and as usual, whoever named its members could determine its recommendations. George Keyworth, Reagan's science adviser, was at the forefront, and he was very skeptical of NASA. "The main reason people would like to build a space station is to put men in space," he declared. "Why do we want to do that when we are just entering the era of robots and automation?" He argued that the Soviets were relying on the manned Salyut and Soyuz because Moscow lacked advanced electronics. "Don't emulate an inferior technology," he warned a House subcommittee. Emphasis on man in space would "take a step backward." It would be "a most unfortunate step."[2]

Keyworth had proposed to place the issue of the space station with a new Senior Interagency Group for Space, a cabinet-level committee with himself as chairman. Gil Rye intercepted the plan before it reached Reagan and talked with his boss William Clark, who supported NASA. "I wanted to ensure that we controlled the space agenda and not the science adviser," Rye later said. Clark decided that he should be chairman instead, and deprived Keyworth of the right even to join the panel as a voting member. He also took away the vote of Stockman, who headed OMB. The SIG that resulted was more to Beggs's liking.

Meanwhile, NASA's Space Station Task Force was broadening the program's justification. Satellite repair, the purpose of the SOC, might well be a fool's errand, for modern electronics gave spacecraft long life. When one of them failed, it was not hard to launch a replacement. If the station was to serve as a maintenance depot, it would have to demonstrate cost savings, which would lead the agency once again into the swamp of cost-benefit analysis. But NASA could supplement this purpose with another one: research in zero gravity, with developments of potentially large significance to medicine and perhaps to electronics as well. Significantly, several industrial firms were already becoming active in this area.

Crystal growth was of particular interest, not as a New Age therapy but as a basic exercise in molecular biology. Scientists knew that in this field, function follows form; the specific intricacies of a molecule's shape determine how it works within a cell. The way to learn its form was to grow a good crystal of the substance under study and then examine it by using a powerful beam of X-rays. This would produce data that a computer could crunch, yielding the molecular shape. Molecules of interest included nucleic acids, enzymes, and other proteins, none of which formed crystals readily. The process of crystal growth was delicate, easily upset by effects due to gravity, and there was reason to believe that zero gravity would help.

Computers and automation had taken many tasks away from space-station concepts such as the one in *Collier's*, but crystal growth offered work for a space station in which the most powerful computers would provide the best advantage. They would address the demanding task of determining a detailed molecular structure. The crystallographer Max Perutz had help from early computers but still labored from 1937 to 1967 to solve the structure of hemoglobin, locating each one of its ten thousand atoms. Perutz took fifteen years merely to learn how to approach the problem, a discovery that in itself won him a Nobel Prize; he then took another fifteen years to carry through the work. Yet this structure, once determined, showed clearly how hemoglobin in red blood cells takes up oxygen and releases it. Modern computers could solve such problems in weeks—if the crystals were good enough.

For NASA, another venture gave promise of commercial advantage. It involved a partnership between McDonnell Douglas and Ortho Pharmaceuticals, dating back to 1980, that sought to produce a protein called erythropoietin. Erythropoietin stimulates the growth of red blood cells; the kidneys make it, and lack of it causes a type of anemia. The best available production method was also very sensitive to gravity, and this partnership set out to demonstrate that zero gravity made for large improvements. An initial test aboard the fourth shuttle mission, the one that Reagan greeted when it returned, yielded four hundred times the output and up to five times the purity. Three subsequent flights during the following year confirmed these advantages, with the best test showing a 700-fold increase in yield. Clinical trials lay ahead, as did a truly dazzling prospect: that space might serve not merely for research but for routine production of commercially valuable products.

Other companies were also interested. Two start-ups, Microgravity Research Associates and Microgravity Technologies, were pursuing plans for space processing of semiconductors such as gallium arsenide, used in microelectronics. The 3M Corporation was developing a ten-year program to prepare organic crystals and thin films using a succession of shuttle flights. There was even activity at the low-tech firms of John Deere and Bethlehem Steel, where researchers hoped to study the solidification of iron in zero gravity.

A White House aide, Craig Fuller, brought these matters to Reagan's attention during 1983. The president had high hopes for NASA; he saw the space program as an expression of his morning-in-America optimism. He also made a strong commitment on the military side that March, by announcing the Strategic Defense Initiative with its goal of protecting the nation against Soviet missiles. Early in August, Fuller arranged for him to host a luncheon for a group of business executives who hoped to pursue commercial activities in space. They told him that a space station would stimulate such activities, and Reagan replied, "I want a space station, too. I have wanted one for a long time." He made no commitment, but he came away fascinated and enthralled with what his guests had told him.

But within the SIG, the interagency review group, the station was in serious trouble. Gil Rye had done what he could to stack the deck in NASA's favor, but still the military and intelligence communities were heavily represented in the SIG. And William Casey, who headed the CIA, was opposed to the station; he saw no need for it. There also was resistance from the Joint Chiefs and from Caspar Weinberger, the secretary of defense, for the same reason. This was doubly worrisome. Weinberger was the most powerful member of the cabinet; his close association with Reagan went back to the latter's governorship of California. And whereas Air Force support had allowed NASA to start the space shuttle back in 1972, Beggs now would have to try to proceed with the station over the Pentagon's outright opposition.

Beggs now was in danger of finding himself out on a limb, and critics were present to saw it off. High among them was Senator William Proxmire, a scourge of NASA, who had responded bluntly to space-station advocates: "I am concerned that it will proceed regardless of the real need for such a program because your agency needs it more than the country needs it. I have long believed that your agency has a strong bias toward huge, very expensive projects because they keep your centers open and your people employed."[3]

It was clear that the SIG would not be able to come to a consensus in favor of the station. Still, Beggs was not at a loss. He now would pitch the station as a purely civil effort, without any military ties. In addition, another interagency panel provided a forum for review: the Cabinet Council on Commerce and Trade. Its chairman was Malcolm Baldridge, the secretary of commerce. And Baldridge was willing to support NASA's plans, which might promote the commercial uses of space.

The matter came to a decision early in December, during two meetings with Reagan. Colonel Rye presented policy options; as he put it, "It was easy talking with Reagan about something he's enthused about, and he's enthused about space. He asked a lot of questions and wanted to know more about the various areas." Beggs presented the space-station concept. David Stockman declared that the deficit would never go down if such projects were to go forward. William French Smith, the attorney general and another close friend of Reagan, responded: "I suspect the comptroller to King Ferdinand and Queen Isabella made the same pitch when Christopher Columbus came to court."

Reagan made no decision—not just then—but he met again with the principals a few days later. The question now involved a specific NASA request of $150 million with which to launch the space station as a project. This time Stockman conceded that NASA might indeed receive small budget increases during future years, and Reagan said, "Done!" He added, "I do not wish to be remembered only for El Salvador."

The president announced it openly in his State of the Union speech of January 1984. It is worth remembering this address, as a moment when hype could still substitute for actual accomplishment:

> There is renewed energy and optimism throughout the land. America is back, standing tall.
>
> A sparkling economy spurs initiative and ingenuity. Nowhere is this more true than our next frontier: space. Nowhere do we so effectively demonstrate our technological leadership and ability to make life better on earth.
>
> We can reach for greatness again. We can follow our dreams to distant stars, living and working in space for peaceful, economic, and scientific gain. Tonight, I am directing NASA to develop a permanently manned space station, and to do it within a decade.
>
> A space station will permit quantum leaps in our research in science, communications, and in metals and lifesaving medicines which can be manufactured only in space.[4]

NASA could share the sunshine of this hope, for the shuttle was now beginning to operate on a schedule, carrying out missions that pointed clearly toward the promise of the station. During 1984 and 1985, three missions focused on the repair and recovery of failed satellites. The first, the Solar Maximum Satellite, had reached orbit in 1980 to study the sun while it was very active, producing many sunspots and solar flares. *Challenger* rendezvoused with this craft; crew members replaced its attitude-control system and repaired a faulty instrument by installing new electronics. This work restored Solar Max to full operational status.

Subsequent flights succeeded with similar exercises. Early in 1984, *Challenger* had lifted two communication satellites. A malfunctioning rocket stage left them stranded uselessly in low orbit. In November a sister ship, *Discovery,* went up and recovered these valuable spacecraft, bringing them back to earth for eventual relaunch. Much the same happened the following year, as a rocket glitch left a military communication satellite similarly stranded. Again *Discovery* came to the rescue, fixing the problem. The rocket stage later ignited normally and boosted the satellite to geosynchronous orbit.

Other missions carried Spacelab, the European research module that fitted into the shuttle's payload bay. The first, in November 1983, carried seventy-seven experiments that formed a hodgepodge: astronomical telescopes, solar telescopes, and instruments to examine the earth and the ionosphere. *Science* noted that ordinarily, "this would be the worst way conceivable to run a mission. Many of the experiments are utterly incompatible: *Columbia* will constantly be twisting down to point toward the earth, up toward the stars, and out toward the sun. No one experiment will be able to make full use of the time. But then Spacelab 1 is not a normal mission. It is an exercise in engineering exuberance."[5]

It also brought Europe strongly into NASA's manned space program, with a German payload specialist, Ulf Merbold, flying as one of the crew members. Spacelab flew three more times during 1985. McDonnell Douglas repeatedly flew its biomedical equipment, with Charles Walker, a scientist of this firm, accompanying this unit into space on three flights during 1984 and 1985.

Crystal growth also made headway. Everyone in the field understood that this was more an art than a science, and that there were very few artists. German investigators thus stirred much interest when they described their work with proteins aboard Spacelab 1. Their crystals of beta-galactosidase were up to twenty-seven times larger than those grown on earth; crystals of lysozyme were a thousand times larger.

At the University of Alabama, the investigators' colleague Charles Bugg decided to try his luck with a small experiment that could fit behind the flight deck and would try to crystallize thirty-four samples. All but five were lost as Bugg's shuttle, *Discovery,* shook strongly during powered flight and then accelerated while maneuvering in orbit. The rest were lost at landing, when this orbiter blew a tire. But Bugg found a welcome surprise: a crystal 1.6 millimeters across. "I don't know of any case where a lysozyme crystal has grown so big in just five days," he said. "It confirms what we thought possible."

In addition, NASA was approaching the day when the shuttle would monopolize U.S. launches. The agency was moving sharply away from expendable boosters, even though these rockets had been keeping up with the times. The Air Force had hiked the power of its Titan III by mounting larger solid-fueled strap-on rockets to its sides. The Atlas-Centaur had become a standard, replacing the less capable Atlas-Agena, which flew its last mission in 1978. In addition, McDonnell Douglas had taken advantage of Apollo-era efforts that had raised the thrust of the Saturn I-B's engines to 206,000 pounds. Similar engines powered the Thor, which originally had flown with as little as 150,000 pounds. Its builders had lengthened it by fifteen feet, to carry more propellant, and added improved upper stages. The launch vehicle that resulted was the latest version of an ongoing series of Thor-class boosters called Delta. These were NASA's most widely used satellite launchers.

As early as 1977, NASA had decided to purchase no more Titan IIIs. Subsequently, it similarly rejected the Atlas-Centaur and Delta; in May 1983 Reagan directed this agency to transfer the rockets to operators within the private sector. On its face, NASA seemed to be moving government launch vehicles into private industry. But their builders found themselves squeezed between a subsidized shuttle and a subsidized Ariane of France, both of which could undersell them by quoting lower launch costs.

In 1972, NASA had projected a cost per flight of $10.4 million for the shuttle, or $24 million in the inflated dollars of 1982. To achieve full cost recovery, the agency would have had to charge $155 million, because its flights would be infrequent and yet would have to cover the expense of its large supporting facilities. For communication satellites and other commercial users, the actual charge was $71 million for a full cargo bay, prorated to much less when a payload filled only part of this bay. Nor could NASA raise this fee or it would lose business to Delta and Atlas-Centaur, as well as to Ariane.

In 1980 the backers of this French rocket had formed a commercial firm, Arianespace, to market its launch services. It offered attractive terms, for while NASA customers had to pay in full up to three years before a shuttle flight, clients of Ariane could make a deposit and pay the balance over several years. In addition, these Europeans could also cut prices. "Arianespace has a good pipeline," said William Rector of General Dynamics, who was responsible for the Atlas-Centaur. "They know what our bid is and—wham! They underbid us every time." With Atlas and Delta squeezed in this fashion between NASA and Europe, it was clear that both would soon vanish from service. Nor would NASA officials boast of acting like robber barons. They would smile and say that expendable boosters were losing out in competition with the shuttle, just as they had expected back in 1972.

Yet the shuttle was proving to be a most fragile basket in which to place all the nation's eggs. The engines had been designed with the hope of making fifty-five flights before requiring major refurbishment. Instead, they had to be disassembled every three flights, to have their turbine blades replaced within the turbopumps. Each engine cost $36 million, and a combination of tight budgets and success-oriented management meant that there weren't many spares. As a result, minor problems with particular engines could disrupt the entire flight schedule.

In December 1981, a small accident at Rocketdyne tore loose an eight-inch pipe from the combustion chamber. Workers welded it back into place and tested the weld to be sure it would hold. It didn't. A year later, with this engine now mounted on the *Challenger* at Cape Canaveral, the weld cracked and leaked hydrogen during a test firing. This leak could have destroyed the shuttle by explosion; yet it took weeks to locate because it was in an inaccessible spot. NASA shipped a spare engine from Mississippi; it too had a welding crack. NASA shipped a second spare, and that did it, permitting a good flight.

However, those two spares were the only ones in the inventory. The second one had been intended for *Columbia*, which had already made five flights with the same three engines. *Columbia* flew once more, on the Spacelab 1 mission late in 1983, but that was it. This orbiter did not fly again until January 1986. And when *Challenger* again needed an engine, in February 1984, it proved necessary to cannibalize one from *Discovery*.

Discovery subsequently regained a full set of engines, but in June a launch attempt aborted, owing to a computer failure and a faulty valve. Three weeks of intensive investigation failed to pin down the problem with the valve, and the program director canceled the flight outright. Its payloads flew on the next flight, in late August; the original August payloads went onto two subsequent missions during autumn. And so it went.

In addition, *Challenger* came close to disaster in August 1983, because of a problem with one of its solid-rocket boosters. The lining of its exhaust nozzle eroded excessively during the rocket's two-minute burn. If it had fired for another eight seconds, the lining would have burned through. Six seconds later, the booster's hot flame would have breached the nozzle itself, as a prelude to a massive explosion.

The Pentagon had planned to become the shuttle's principal user, expecting to rely on it exclusively once it became fully operational. But in February 1984, Defense Secretary Weinberger approved a policy document stating that total reliance on the shuttle "represents an unacceptable national security risk." In June the Air Force gave NASA a sharp vote of no confidence. It declared that it would remove ten payloads from the shuttle beginning in 1988 and fly them on expendables.

James Beggs was well aware that the Air Force might do this with many more payloads during subsequent years, and he was outraged. "The shuttle is the most reliable space transportation system ever built," he spluttered. The Air Force did not agree. It still would use the shuttle occasionally, but its prime launch vehicle would be another upgrade of Martin Marietta's Titan III. This firm now rebuilt it with new liquid-fueled engines that provided a 30 percent increase in thrust. They improved the performance so much that this new rocket gained the name Titan IV. It could lift up to 39,000 pounds to orbit. This made it very competitive with the shuttle, which had come in overweight and could only boost 47,000. To add insult to injury, the Titan IV could accommodate payloads that were physically larger than those the shuttle could carry.

Another vote of no confidence came in September 1985, as Ortho Pharmaceuticals dropped out of its partnership with McDonnell Douglas. Both firms remained very interested in producing erythropoietin to control anemia, and McDonnell's zero-gravity processing unit had been flying regularly on shuttle missions. But a California start-up company, Amgen, now had a new process based on genetic engineering, wherein its scientists had spliced appropriate genes into fast-growing bacteria. A company director said: "The difference between what McDonnell is doing and what we're doing is like reproducing a book by hand in the Middle Ages and using a Xerox machine." Amgen's protein, called Epogen, went on to become the most successful product in the biotechnology industry.

Ortho abandoned McDonnell to team up with Amgen, illustrating vividly how tiny start-ups could seize opportunities. McDonnell tried to carry on, but in time it had to admit that Amgen's process was superior, particularly because it worked well in gravity. This was a serious blow to NASA, which had touted the McDonnell-Ortho effort as a harbinger of processing in space. NASA had pushed aggressively during 1984 and 1985, seeking to win over corporations as customers for commercial research aboard the shuttle and even offering free space on the flights. But the two-year effort produced no takers. In its wake, the 3M Corporation was the only major firm still committed to work in orbit. In turn, this development seriously undercut the rationale for the space station, whose backers had touted it as a boon to industry.

Then came January 28, 1986: the day of *Challenger.*

The image remains seared in the nation's memory: the shuttle rising on its pillar of rocket exhaust, then an orange fireball, two solid boosters flying off crazily, and a shower of fragments that trailed smoke against the blue sky. One of those fragments was the cabin with its flight crew, who apparently remained conscious following the explosion. It took them more than two minutes to fall to the ocean from their initial altitude of nine miles. No parachutes were on board; when they struck the water, their deaths were swift.

The crew included a teacher, Christa McAuliffe, whom NASA had picked from among 11,400 applicants. This was part of a public-relations campaign to stir interest by demonstrating that the shuttle was safe enough for ordinary people. Senator Jake Garn, a strong supporter of NASA, had flown in this manner in April 1985. Congressman Bill Nelson had gone into orbit in similar fashion earlier in January.

The shuttle was indeed approaching routine operation, having made five flights in 1984 and nine in 1985. The latter year brought about a milestone: For the first time, the nation flew more shuttles than expendables. Fifteen flights were on the schedule for 1986, and NASA was pushing toward a goal of twenty-four by the end of the decade.

But the demand for an increasing flight rate put everyone under considerable pressure. Two important missions lay directly ahead. On May 15, *Challenger* was to launch Ulysses, a European probe that would swing around Jupiter and use its gravity to fly over the poles of the sun. Five days later, the Galileo spacecraft was to head for Jupiter itself. Both were to ride atop the Centaur upper stage, which would make its first flights within the shuttle. If either Ulysses or Galileo missed its launch date, it would have to wait thirteen months until the earth and Jupiter were once again in proper alignment.

The cause of the disaster was in the solid-rocket boosters that flanked the main propellant tank. They were too large to fill with fuel or to transport in one piece; hence their builder, Morton Thiokol, had fabricated them in sections, to be assembled at the Cape. There was a gap between any two such segments, which had to be sealed to prevent hot gases at high pressure from leaking. The seal had two O-rings—thick rubber bands that encircled the twelve-foot diameter of the booster. When the booster's solid fuel ignited, a sudden pressure rise caused its casing to flex. The O-rings then needed enough resiliency to respond by flexing in concert, to maintain the seal's integrity. Each booster had three such sealed joints.

Prior to the disaster, O-rings had failed to hold on several earlier flights, because of erosion from the hot gases. A severe instance had occurred during a launch in April 1985, as rocket exhaust blew past the first O-ring and eroded up to 80 percent of the second one, within a limited area. The seal nevertheless held and the booster operated normally. But at Thiokol, engineers responsible for the seals became very concerned.

In July Roger Boisjoly, the company's senior seal specialist, wrote a memo to his vice president "to insure that management is fully aware of the seriousness of the current O-ring problem. It is my honest and very real fear that if we do not take immediate action to dedicate a team to solve the problem, we stand in jeopardy of losing a flight along with all the launch pad facilities."[6] Thiokol set up such a team, whose boss wrote a memo in October that began with the word "HELP!" and ended, "This is a red flag." However, there was no real urgency. Better seals were just one more improvement that would come along in good time, as the shuttle design continued to mature.

No clear criterion determined just when they might fail, but experience had shown that they worked best in warm weather, which made their rubber flexible and resilient. By contrast, chilly weather left them stiff, unable to flex quickly, and this was when they tended to show erosion and charring from the booster's flame. As the date for the *Challenger* flight approached, Florida was in the grip of an unusual cold wave. At Thiokol, Boisjoly and thirteen other engineers held a meeting and recommended unanimously that the launch be postponed.

This did not sit well with senior managers. They knew of the O-ring problem, but the experience of two dozen successful shuttle flights had bred complacency. The seals had worked before, however imperfectly; why not again? NASA had recently invited other firms to compete for Thiokol's solid-booster contract, placing the company under great pressure to show that it had a proper can-do spirit. Its vice president for space boosters, Joseph Kilminster, overruled Boisjoly's group and recommended launching *Challenger,* calling the available evidence "not conclusive."

Lawrence Mulloy, who managed Thiokol's contract at NASA's Marshall Space Flight Center, also wanted to press ahead. "My God, Thiokol, when do you want me to launch, next April?" he asked during the discussions. George Hardy, a deputy director at Marshall, added, "I'm appalled at your recommendations." Neither Hardy nor Mulloy would make the final decision, but they refrained from sharing the concerns of Thiokol's engineers with the Cape Canaveral officials who had the authority to delay the launch. Still, despite the opposition from NASA-Marshall, Thiokol's engineering managers held their ground. Presented with a formal statement approving the launch, Allan McDonald, the company's senior man at the Cape, refused to sign.

Three-foot icicles were forming on the launch tower during the launch preparations, hanging downward like daggers. But late that morning the sun came out and melted the ice, relieving last-minute concerns. As the engines fired and the shuttle started to lift, a puff of black smoke appeared alongside the bottom joint in one of the solid boosters. The seal had momentarily failed to flex. It quickly did so, and no more smoke appeared, but the harm had been done. This seal now was a damaged gasket that soon would blow.

It gave way completely in a small area some fifty-nine seconds into the flight, as a small flame appeared along the side of this rocket. The flame quickly spread and burned like a blowtorch, both at the propellant tank and at a strut that secured this booster. Within seconds, the liquid hydrogen began to ignite. The tank ruptured; the solid booster broke free of its strut and slammed into it, breaking it open like an egg full of explosives. The tank and the shuttle orbiter disintegrated. "I do not know how many seconds it took for the sound of the blast to travel down," wrote a reporter from the Los Angeles *Times*. "When it arrived, it did so like a thunderclap, rattling the metal grandstands. And then it abruptly ceased, replaced by a strange and terrible quiet."

This stillness at Canaveral led to the swift appointment of a presidential commission, which found flaws not only in O-rings but in NASA itself. The commission's chairman was William Rogers, a former secretary of state. Noting that Thiokol had faced the possible loss of its contract, he told Mulloy that its officials "were under a lot of pressure to give you the answer you wanted. And they construed what you and Mr. Hardy said to mean that you wanted them to change their minds."

John Young, the chief of NASA's astronauts, wrote bitterly of the seals, noting: "There is only one driving reason that such a potentially dangerous system would ever be allowed to fly—schedule pressure." The commission's report stated that as problems with the seals "grew in number and severity, NASA minimized them in briefings and reports; as tests and then flights confirmed damage to the sealing rings, the reaction by both NASA and Thiokol was to increase the amount of damage considered 'acceptable.' " Richard Feynman, a Nobel Prize–winning physicist from Caltech who was known for speaking his mind, warned that "NASA exaggerates the reliability of its product to the point of fantasy." He then pronounced the *Challenger*'s epitaph: "For a successful technology, reality must take precedence over public relations, for Nature cannot be fooled."[7]

Three orbiters remained: *Discovery, Columbia,* and *Atlantis,* which had joined the fleet in 1985. In addition, Gil Rye had arranged for Rockwell International to build major portions of a new orbiter. It later took shape as a replacement named *Endeavour*. But in other respects, NASA and the Air Force took several large steps away from the shuttle. It still would play its role, but the goal now was a mixed fleet, in which expendable launch vehicles would receive due attention.

In NASA's hour of crisis, Reagan turned to James Fletcher, prevailing upon him to return to this agency as its head. In June, NASA scrubbed plans to use Centaur as a shuttle upper stage, leaving Galileo and Ulysses in the lurch. This decision reflected new concern for danger; Centaur burned liquid hydrogen and oxygen, and there was no way to abort a shuttle flight and vent the propellants for a safe landing. Late that summer, NASA also canceled most of its planned Spacelab flights. It had flown four to date; the next one would not fly until December 1990.

The Air Force did more, and it had the clout to do it. Prior to the Reagan administration, the Pentagon's space budget had been only half the size of NASA's. In 1986, with the Strategic Defense Initiative setting the pace amid a host of highly classified new programs, this budget reached $17 billion, more than twice that of NASA.

The Air Force had built a launch complex for the shuttle at Vandenberg that cost $3 billion. It now went into mothballs. Air Force Secretary Edward Aldridge expanded the earlier purchase of ten heavy-lift Titan IV launchers, ordering another thirteen. He also had nearly seventy Titan II ICBMs, most of them deployed operationally in their silos; he now stepped up efforts to refurbish thirteen of them, for use as launch vehicles. He was well aware that both versions could fly with the Centaur, whereas the shuttle couldn't.

Still, for the moment the nation's space program rested on its limited stockpile of Titans, Atlases, and Deltas that were left over from earlier purchases. The Air Force had seven Titan IIIs, and on April 18, one of them lifted off from Vandenberg, carrying a Hexagon reconnaissance satellite. This mission was important; another Titan III had splashed into the Pacific the previous August. This one did not even get that far. Eight seconds after liftoff, one of its solid boosters blew up, touching off an explosion of the entire vehicle. A blazing pyrotechnic shower caused damage to the launch pad that took seven months to repair, while the Titan III went in for the same scrutiny as the shuttle. It did not fly again for a year and a half.

On May 3 it was the turn of NASA, as it launched a GOES weather satellite aboard a Delta. This rocket was known for its reliability, having flown successfully on its previous forty-three launches, but this time its number came up. Its engine shut down some seventy seconds into the flight. This was serious, though less so than in the case of the *Challenger;* a military Delta would fly again, with success, as early as September. But NASA had exactly two Deltas left in its inventory.

Just then, in the spring of 1986, the only NASA boosters certified for flight were three Atlas-Centaurs, all committed in advance to specific payloads. The Air Force had its own supply of Atlases, but these lacked significant upper stages and could only lift spacecraft of modest size. During 1986, NASA and the Air Force together carried out only four successful launches via Titan, Atlas, and Delta, though twenty would have been more like it if they could have used the shuttle. This was the smallest number since 1959. It rose to seven in 1987 and fell back to six in 1988; not until 1989 would the national space program begin to recover. "We have an unqualified disaster on our hands," said Albert Wheelon, the man behind Hexagon, shortly after the Titan III explosion of April 1986. "We are essentially out of business."

Here was a situation made to order for Arianespace, which had been vigorously promoting its services. But late in May of the same year, an

Ariane failed on a flight out of Kourou when its third stage did not ignite. This too was nothing to sneeze at, for it was the fourth failure of the rocket in eighteen launches. In addition, three of the failures had involved malfunctions of the third-stage ignition. Frédéric d'Allest, chairman of Arianespace, received a report from a board of inquiry and concluded that this stage had a "fundamental and generic problem." It shut down Europe's launch program as well, for Ariane did not fly again until September 1987.

In the entire world, the Soviet program was the only one still moving ahead at full throttle. And it was advancing toward greater strength, both in manned flight and in the development of new and powerful launch vehicles. In particular, while Americans might talk of space stations, Moscow already had them.

Salyut 6 had given way to the similar Salyut 7, which flew to orbit in April 1982. Between 1977 and 1985 these stations hosted a total of ten long-term visits by cosmonauts, which lasted up to 237 days. Together, these visits kept the stations inhabited for nearly half of those eight years. They provided a solid base of experience with long-duration flight, as life in space, resupplied by Progress craft, became routine. Automated rendezvous and docking emerged as a standard practice, as did on-orbit refueling.

The cosmonauts performed research, too. They were restricted by the limited capacity of their spacecraft, but one mission had a radio telescope that unfolded to a diameter of thirty-five feet. In the course of a 175-day flight, the cosmonauts Vladimir Lyakhov and Valery Ryumin used it for tests that included coordinated operation with another radio telescope in the Crimea, using a technique that allows such instruments to form particularly sharp images. Then, when this dish antenna became snagged and could not be jettisoned, Ryumin took a spacewalk and cut it loose.

The next space station, Mir, went up in February 1986. The word *mir* means both "peace" and "world," which was appropriate: It would host additional visitors from other nations, including the United States, and at the same time it was complete enough to operate as a world in itself. It represented a major upgrade of Salyut, with more electric power and with automatic systems to take care of the life support. Eight onboard computers provided power of a different sort, making it possible to operate very sophisticated equipment. Better yet, Mir was built to grow. It had a docking adapter with five ports, allowing it to take on major additions virtually at will.

The first one, Kvant (quantum), carried a major set of instruments for use in astrophysics. It went up in April 1987, when the world's astronomers were dazzled by an exploding star, the closest such supernova in four hundred years. Kvant carried an observatory that could study X-rays from this stellar explosion, which could not penetrate the atmosphere. But when Kvant arrived, it could not dock firmly. The cosmonauts aboard Mir took a spacewalk and found that a trash bag was in the way. They cleared it, so that Kvant, now solidly in place, could contribute uniquely to worldwide studies of the rare celestial event.

Other research stations reached orbit and also attached themselves to Mir: a second Kvant in December 1989 and then Kristall in June 1990, for studies in zero gravity of metallurgy and crystal growth. With a Soyuz docked to the adapter and a Progress craft up for a visit, Mir formed a cluster of six substantial sections, several of which had far-spreading solar panels. Nor did Mir's residents neglect their long-duration records. In December 1988, Vladimir Titov and Musa Manarov returned safely after spending an entire year in space—the first people to do so.

In addition, Valentin Glushko, the country's longtime leader in rocket-engine design, had pursued to the full his interest in energetic propellants. He had prevailed over his rival Vasily Mishin, Korolev's successor, and with the N-1 moon rocket scrapped, Glushko had developed a heavy-lift vehicle that combined many of the best features of the Saturn V and the space shuttle. Indeed, it could serve in either role.

It closely resembled the shuttle in general appearance, but there were important differences. The shuttle relied on solid boosters, which had brought *Challenger* to grief. In addition, the shuttle mounted its main engines at the rear of the airplane-like orbiter. Glushko used no solid boosters; instead he relied on liquid-fueled units of proven reliability. In addition, he mounted his main engines at the base of the propellant tank, and not on the orbiter. This meant that the rocket could fly without the orbiter, as a major launch vehicle in its own right.

This was Energiya. It flew from Tyuratam in May 1987, while NASA was still in its doldrums, and set a world record with a payload of some 220,000 pounds, a hundred metric tons. This was more than even the Saturn V had ever carried, and liquid hydrogen was the fuel that made it possible. Its spacecraft, called Polyus (pole), was a prototype of an orbiting battle station equipped with a powerful laser. Unfortunately, its control system failed to achieve the proper orientation, and it fired its rocket engine in the wrong direction. Polyus failed to reach orbit, burning up in the atmosphere over the Pacific. But Energiya would fly again.

It did this in November 1988, when it carried the Soviet space-shuttle orbiter on an unmanned test flight. This orbiter was Buran (snowstorm). It closely resembled its American counterpart, both in size and in the shape of its wings and cargo bay, though it had no major engines of its own and reached orbit through those of Energiya. Buran made two orbits and then executed an automated landing despite high winds, by using an advanced onboard computer.

Yet like the brilliant colors of sunset that foretell the dark of night, this burst of Soviet achievement proved to be the prelude to a major retreat from space. It is clear that Buran was to serve Mir; the Kristall module had a docking port intended specifically for use by that shuttle. But Mir, Kristall, Energiya, Buran, and much else found themselves caught up in the economic disaster that followed the fall of Communism in August 1991.

One should not say that this second Russian revolution replaced Communism with capitalism. It is more appropriate to describe it as the overturn of an authoritarian system that maintained public order, in favor of a chaotic arrangement in which long-established practices could no longer hold. Hyperinflation reduced the ruble to a thousandth of its former value. Violent crime flourished, in a society that had achieved a high level of public safety. Former Party officials did well by using their connections, while workers struggled on the equivalent of a few dollars per month. Industrial production dropped precipitously, in a manner rarely seen since the Great Depression.

The space program continued for a while at its accustomed pace, largely by using up a reserve of launch vehicles and spacecraft. But major budget cuts began almost immediately. In 1992 the military space program received only half the funding of 1990 and 1991, while the civilian program took a cut to two-thirds of its previous level. Two new modules had been slated as additions to Mir: Spektr and Priroda ("spectrum" and "nature"), both intended for use in studies of the earth's surface and environment. Their launches now were delayed indefinitely; they would not go up until 1995 and 1996.

Russian officials nevertheless reserved funds to continue the Buran project. Yuri Semenov, the head of Korolev's organization, predicted a new launch as early as 1993. But during that year, with budgets continuing to drop, both Buran and Energiya were canceled. A full-size version of Buran, used for structural and engineering tests, wound up in Moscow's Gorky Park for public display, not far from a Ferris wheel.

In 1995 the military space program was down to 10 percent of its 1989 funding. The civilian program was at 20 percent—some $400 million at the current exchange rate. By contrast, NASA received some $14 billion in 1995. During that year, Russia carried out only thirty-two space launches, the smallest number since 1964. The total dropped again in 1996, to only twenty-three. As recently as 1990, the tally had numbered seventy-five. The heads of Russia's civil and military programs warned that if funding did not increase, the entire space program might end in two to three years.

The decline was particularly dramatic at Tyuratam and the adjacent city of Leninsk, where a major riot took place in February 1992. Military

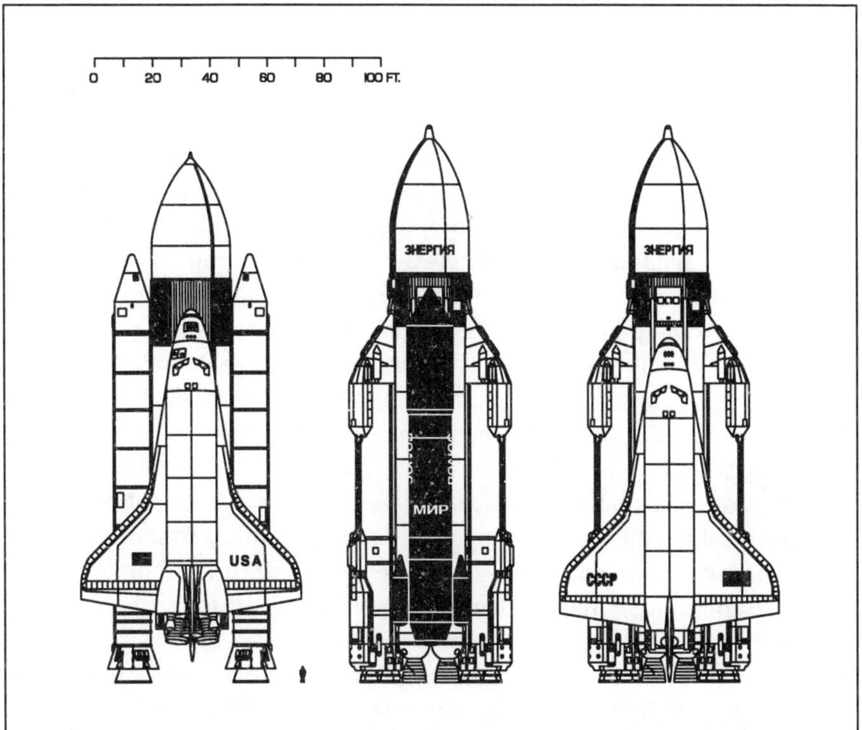

Shuttle-class launch vehicles. ***Left,*** **space shuttle.** ***Middle,*** **Energiya; booster rockets flank the central propellant tank, while a heavy payload is mounted along the length of this tank.** ***Right,*** **Energiya with Buran space shuttle. (Dan Gauthier)**

draftees had been sent there to do maintenance work, but they ate very poor food and went for months without a bath. As one of them put it, "I came here whole, healthy, unharmed, and now I have been serving here two years and I am going home sick, a cripple."

The riot began when a military worker was arrested and sent to the commandant's office. Other troops tried to rescue him by seizing a vehicle and ramming the gates, whereupon a sentry opened fire, wounding an attacker. Several hundred other soldiers then joined the uprising, commandeering some fifteen additional army cars and setting fire to buildings. Three barracks burned; rescuers later found the bodies of three men inside.

Authorities restored order not only by arresting the leaders but by granting furloughs to soldiers who were enduring difficulties such as poor health. Still, the spread of decay was evident. A writer noted in 1992 that "the main thing is to restore the launch complex, where the necessary routine inspection and maintenance haven't been done for a year and a half. I visited the launch pad for Energiya/Buran recently. It was a dismal sight. You could have shot science-fiction films there about a dying space civilization!"

A few months later, a reporter observed that in Leninsk, "it can be seen at first glance how catastrophically the situation in the city has deteriorated. Whereas during recent years there has been almost no food, now there are also no quarters. Equipment and materials are being plundered. Mothballed launching complexes are deteriorating."[8]

There were fires amid the mothballs. In 1994, a Moscow correspondent wrote: "The garrison officers' center burned down this past winter. But 'burned down' is not quite accurate. To be precise, it was set on fire, after all the windows were broken out: 'for fun.' Afterwards, it smoked for three days, like Fujiyama in Japan." Another fire broke out in the Soyuz assembly building. Months later, visitors could still smell the smoke. Some launch pads were no longer usable, and another correspondent noted that "people steal anything, even copper cable or sheet metal from the roofs of buildings." When a supply rocket reached Mir, part of its complement of food was missing. It had apparently been looted on the ground by the launch crew.

In March 1995 the American analyst Jim Oberg came to look and presented his own view of the situation in the journal *IEEE Spectrum:*

> There was plenty of decay and abandonment to see. By the early 1990s, there was often no heat or running water in workers' homes, no social services from schools to medical care, and only the drabbest items of food in the stores. Security collapsed and squatters moved in while looters lurked in the city's outskirts. Public health declined rapidly and diseases spread, especially among the children. Many civilian workers, unpaid for months, abandoned their posts.
>
> The abandoned buildings, broken fences, and thickly-strewn junk piles reminded me of the worst extremes of U.S. urban decay. Abandoned apartments, sometimes in entire blocks, stare windowless at the dusty sun. The dust, blown from the pesticide-laden salt flats of the Aral Sea, is gradually poisoning the city's inhabitants, the weakest first.
>
> Somehow the hard-core space workers struggle on, enduring, improvising, cannibalizing, and making do. Members of this fanatic cadre, on average well over 50, have been through so much that no future challenge frightens them.[9]

Conditions were much the same in Mirnyy, a military city serving the major launch center of Plesetsk. *Izvestiya* reported that "the imprint of abject poverty of the soldiers can be seen everywhere." People were eating gruel three times a day. A cook said, "I can no longer remember when there was fresh meat. We have not tasted fresh cabbage this year." A general added that "a total of 255 families of officers have no quarters."

The Soviets had relied on tracking ships: *Komarov, Korolev*, and the flagship, *Gagarin*. These were rusting in port at Odessa, and the latter two were up for sale. Mir remained occupied, but Oberg reported a curious incident in the mission control center: "The cosmonauts displayed several pieces of hardware over their television downlink, described the items, and asked what they were for. After a determined research effort, ground controllers had to tell the cosmonauts that nobody on earth recognized the gear or knew its purpose. There were no written records and no surviving experts to consult."

In the United States the shuttle program had resumed, with *Discovery* making the first new flight in September 1988. The following year was the first really good one since 1985, for the shuttle flew five more missions. Thirteen others went up aboard a reviving fleet of expendables, along with seven on Ariane. Still the effects of *Challenger* lingered, particularly in the planetary program.

The loss of Centaur as a shuttle upper stage had forced the Galileo mission to fall back on a less capable stage that used solid propellant. Its managers coped with this by relying anew on planetary swingbys to provide gravity assist. The mission plan now called for the probe to fly past Venus, execute two flybys of the earth to add more energy, and only then to go on to Jupiter. Venus lay sunward of the earth, in the opposite direction of Galileo's destination. Hence the spacecraft would head outward by first proceeding inward.

The spacecraft, built at Jet Propulsion Laboratory, had ridden to Canaveral by truck prior to the loss of *Challenger*. It went back to California the same way, then returned to the Cape aboard another truck as its rescheduled launch date approached. In the course of these peregrinations, lubricant rubbed off some of its antenna's ribs. Then, once in space, unlubricated metal surfaces cold-welded, bonding tightly through nothing more than contact in vacuum. When mission controllers tried to unfurl the antenna, they found it was stuck. This should have been as simple as opening an umbrella, which the antenna closely resembled. Instead JPL was left with what the chief scientist, Torrence Johnson, called "a useless, twisted sack of metal mesh." Galileo was to have transmitted photos at a rate of 134,000 bits per second, but it now had only a low-gain antenna that was over three thousand times slower.

Fortunately, JPL had a great deal of experience in nursing sick spacecraft. Its scientists could reprogram the onboard computer from earth, and they instructed it to apply techniques of data compression, to increase tenfold the information content in each bit. The computer also would edit the data—for example, by deleting strings of bits that portrayed only the blackness of space. In addition, JPL increased the sensitivity of the Deep Space Network used for tracking. These changes boosted the effective capacity of the communications link to a thousand bits per second—enough to ensure a productive mission. Still, despite these heroics, much data would be lost.

Approved as a project in 1977, scheduled for flight as early as 1982 and later for 1986, and launched at last in October 1989, Galileo finally reached Jupiter in December 1995. Yet this was a mission with relatively high priority. Other scientists were enduring similar waits, for while the shuttle was back in service, it would fly only a few times a year. And even when Spacelab missions resumed, late in 1990, they rarely offered the prospect of breakthroughs. This stemmed from the nature of science itself.

The world's scientists depend on their instruments and equipment in every field of study, and it is very unusual to achieve sudden advances in any area that lead to fundamental discoveries. This does sometimes happen; when the researchers James Watson and Francis Crick found the structure of DNA in 1953 and launched the modern era of molecular biology, they relied on a few good X-ray images and a great deal of superb insight. For the most part, however, even the best scientists work year by year using standard lab apparatus and hoping to solve specific problems. They rarely seek more than modest advances in their art, for the difficulties are too formidable to hope for more.

The accomplishments of the Spacelab flights followed this pattern. Crystal growth attracted a good share of attention, and some substances indeed grew good crystals. The 1988 flight of *Discovery*, for instance, included eleven experiments, each of which sought to grow crystals of a different protein. Three of them yielded crystals that were markedly superior to the best grown on earth; they were larger and yielded better data on molecular structure. Later work aboard subsequent shuttle flights raised the number to twenty-five proteins, out of a hundred investigated, whose crystals were better than any on earth. The studies did not result in cures for diseases. However, they suggested two new but modest directions for future research.

The first would involve systematic experimental investigations seeking to learn the methods and molecules for which zero gravity could indeed offer advantage. This work would seek to improve the batting average to more than 25 percent, with the recognition that for the seventy-five other proteins studied, spaceflight had not been of benefit. In particular, there were subtle effects that could disrupt the delicate process of crystal growth even in the total absence of gravity.

The second direction reflected the fact that shuttle flights are rare and encounter long delays, cramped facilities, and limited time in orbit. Hence, rather than seeking to create perfect crystals as a matter of routine, the emphasis would call for growing them in a limited number of substances, and then using these crystals to set a standard. Specialists in their earthside labs, knowing now what was possible, could work to improve their techniques so as to match the standard without having to fly into space.

These strategies would reflect science as it is actually done. Yet they carried danger for space-station advocates, who certainly had not sold their project by pitching it as a better way to do protein crystallography. On the contrary, they had come close to asserting that crystals grown in space, like thrice-blessed waters of Lourdes, would take on magical powers that would allow the lame to walk and the blind to see.

Reagan, echoing such advocates, had spoken of "lifesaving medicines which can be manufactured only in space." "We could find a beautiful cure for cancer in space," said Krafft Ehricke, a very capable rocket man who had developed the Centaur but who had no background or specialized knowledge in medical research. The respected journal *Astronautics & Aeronautics* had set the lead of an article, "Manufacturing in Space," in large type: "It is not simply that through space processes we can create new materials and structures, but that in doing this we inaugurate a new age of technical civilization." Amid such hype, the cost of the space station emerged as a matter of considerable concern.

In Washington, standard procedures exist for turning dubious proposals into permanent programs. One starts by quoting an unrealistically low cost and then spending money at a measured pace for several years. This builds a constituency of people who depend on the program, and who are prepared to lobby with vigor against cuts in it. It also builds support for the view that any cutback, let alone outright abandonment, would waste the sums appropriated to date. Advocates then can argue that having gone this far, there can be no turning away from this national commitment.

The shuttle had followed this approach. Through much of the 1970s, NASA continued to tout its low cost and high utility, using success-oriented management to make headway despite its limited budget. Then, as costs and delays mounted, NASA's Air Force allies asserted that the nation was too deeply committed to go back. NASA now played the same game with the space station, and with similar results.

At the outset, around 1980, backers of the Space Operations Center had cited a development cost of $8 billion. Prior to Reagan's approval in 1984, and for some time thereafter, NASA continued to speak of $8 billion. However, this sum did not reflect the type of analysis that building contractors rely on when they quote you a price for a new house. Instead, it reflected close attention to what the political traffic would bear.

It drew in part on a view that one could design a station to meet a specific cost goal. It also drew on what a NASA official called "the scream level": People outside the space program started screaming at about $9 billion. The analyst Howard McCurdy notes that James Beggs "made a political decision to set the cost of the initially occupied station at $8 billion. Supporters of the initiative believed that any price estimate above the $8 to $9 billion range would provide opponents of the program with the 'show stopper' they needed to slay the proposal."

Still, some people saw what was coming. They included Douglas Pewitt, who had recently been responsible for NASA's budget within the Office of Management and Budget. "They'll spend $30 billion before they have it completed," he predicted in December 1983, before Reagan approved the project. "The $8 billion is the buy-in cost. It's not what NASA really wants, which is a manned space platform with a crew of some size."[10]

The design that emerged was the "dual keel," featuring a big rectangular truss crossed by a long arm. Instruments, crew quarters, research labs, and solar panels all would attach to these structures. Then in January 1987, the agency discovered—surprise!—that it would cost $14.5 billion. This brought on a high-level review that stretched out the project, delaying major expenses until the next president would be in the White House. The new plan also chopped the design in two. The long crossarm would go up first, for $12.2 billion. Later, maybe, the big rectangle could go on as well, for another $3.8 billion.

Reagan christened the new concept Freedom, but as Janis Joplin had sung back in 1970, freedom's just another word for nothing left to lose. By 1991 the cost was up to $38 billion, and the space station was losing its friends. "The work that's been done to date is crap," a senior manager told *Science*. "It's not even good engineering." Thomas Paine, a former head of NASA, wrote that it "is delaying, not advancing, the President's goals." Other critics noted that Freedom would approach the size of a football field and would require at least twenty-eight shuttle flights, placing new pressure on NASA to meet a launch schedule. Nor would it grow by stages, with opportunities to learn along the way; it would be at full size from the start, with no means to test its full assembly on the ground. A former Air Force Secretary put it bluntly: "What can't be tested, can't be trusted."

Technologists had already voted with their feet by abandoning thoughts of using the shuttle for commercially oriented research. Even the 3M Corporation dropped out, with this firm's coordinator of space research, Earl Cook, warning that the commercial potential "was grossly oversold." Now the scientific community weighed in. Robert Park, a director of the American Physical Society, held a news conference and declared that the station was a waste of money. He added an op-ed piece in the *Washington Post* that mocked the station as an "orbiting pork barrel." The presidents of fourteen societies joined him in a written statement from the APS, criticizing the station, with ten more such organizations signing on with Park during subsequent months. This was unprecedented;

the nation's scientists depended on federal funds for their research grants, which meant they were biting the hand that fed them. And despite all the talk about growing crystals in space, one group was notable for its support of Park: the American Crystallographic Association.

These societies lacked the clout to prevail, but no one could deny that $38 billion was a bit much. Nor did it help when the comptroller general noted that in addition to its construction costs, Freedom would incur heavy continuing expenses during a planned thirty-year life. "When these costs are added together, what we actually have is at least a $118 billion program," he stated in congressional testimony.

This meant it was time for another redesign, with the goal now of *cutting* the construction cost to $30 billion. The new plan won White House approval, and it had its advantages: It featured smaller modules that now could be loaded with equipment and tested on the ground. Even so, with the realization sinking in that Freedom wasn't free, the new plan ended up as only a stopgap. In mid-1993, the space station survived an attempt to kill it in the House by a margin of one vote: 216 to 215. This margin increased during a subsequent vote, but only because President Clinton and Vice President Gore lobbied heavily to persuade key opponents to stay away.

Yet this would prove to be the dark before the dawn. The unmanned space program had already come back strongly, with Delta, Atlas, and Titan again at the forefront. The space station was also about to make a comeback, gaining new purpose and strength. The world's major space programs were entering a time of renewal, rising again in a surge of achievement. And amid the accomplishment, two important themes prevailed: international cooperation and commercial leadership.

12

Renewal and Outlook

Commerce and Cooperation in Space

ARE TODAY'S SPACE PROGRAMS of America and Russia merging into a single enterprise? Recent years have certainly seen these nations take cooperation to new heights. With both powers' space-station efforts having become increasingly shaky, they paradoxically strengthened the projects by combining them. Amid shortages of rubles, the firm of Lockheed Martin has stepped in with dollars, and is marketing Moscow's Proton as a commercial launch vehicle. A new Atlas-Centaur is under development—with Russian engines. NASA's head since 1992, Daniel Goldin, is vigorously pushing for more such cooperation.

He came to NASA from the world of classified reconnaissance satellites at TRW, where he had kept a very low profile. Jeffrey Richelson, a close observer of the programs and an author of several books about them, confessed, "I've never heard of him." Yet since 1987 Goldin had been a vice president and general manager. His background in unmanned spacecraft put him in a position to shake things up, for it contrasted sharply with a powerful emphasis on manned flight that characterized NASA's previous leaders.

The administrator whom Goldin replaced, Richard Truly, had begun his career as a Manned Orbiting Laboratory astronaut. He became a shuttle astronaut, managed the shuttle program during the post-*Challenger* recovery, and then took on NASA's top job. His deputy, James R. Thompson, had been director of the Marshall Space Flight Center, which was responsible for the design and testing of the shuttle's engines. William Lenoir, head of the office of spaceflight, was another former astronaut, as was Robert Crippen, the boss of the shuttle program.

They shared a commitment to big, costly programs such as the shuttle and space station. The writer Gregg Easterbrook, of *Newsweek* and the *Atlantic Monthly*, describes them as believing that "even turning on a light bulb in space required a half-dozen astronauts in a billion-dollar space chariot." By contrast, Goldin preferred automated approaches that were simple and inexpensive. In particular, he expected to rely on expendable boosters rather than the shuttle.

A turn toward expendables had already begun in August 1986, when Reagan gave the go-ahead to build *Endeavour* to replace *Challenger*. At the same time, he issued a decision: NASA was not to use the shuttle to carry commercial payloads that could fly aboard Delta, Atlas, or Titan. As early as 1983, the president had established a new policy of placing these expendables in the hands of commercial operators, but this effort had come to nothing amid competition from the highly subsidized shuttle. But after 1986, the lid was off.

Not only could the expendables win customers on their merits; they also had a made-to-order market in payloads marked for the shuttle that now needed rides to orbit. As early as September 1986, NASA issued a post-*Challenger* shuttle manifest that left twenty-five communication satellites looking for alternatives. The Pentagon had a similar problem, with a backlog of over twenty satellites and more coming through the pipeline.

Air Force action in 1984 had given Martin Marietta a head start in renewing production of its Titan III and in getting on with its more powerful Titan IV. McDonnell Douglas with its Delta and General Dynamics, builder of the Atlas-Centaur, had no such advantage; both firms had to restart production lines that had gone stone-cold. But in January 1987, Air Force Secretary Edward Aldridge awarded McDonnell a $316-million contract for seven newly developed Delta II launchers, with an option on an additional thirteen. Three months later, this firm announced that nine paying customers had booked flights of communication satellites on this same rocket.

The Delta II had enough power to encroach on the domain of the Atlas, while these orders gave McDonnell a leg up on General Dynamics. But General Dynamics could upgrade its Atlas-Centaurs by fitting them with the improved Rocketdyne engines that its rival was purchasing. In 1988 the Air Force ordered eleven such launch vehicles, thereby ensuring that the Atlas would enter this new era as well. General Dynamics went on to offer a family of Atlas-Centaur models in four versions that its managers declared could cover ninety percent of the commercial satellite market. These executives also invested $300 million in facilities and spare parts, while announcing plans to build sixty-two of the rockets through 1997.

In addition to the eleven Atlas-Centaurs, Air Force orders also included twenty Delta IIs, thirteen Titan IIs, and twenty-three Titan IVs. Commercial orders accounted for nine more Titans. Ariane was also in the picture; as of January 1989 it had thirty-eight launches on order. Yet there was business for all, for as of that month, launch vehicles for as many as fifty-eight commercial payloads had yet to be selected.

The payloads' builders definitely faced a buyers' market, for by 1990 there was launch capacity galore. A Paris research firm, Euroconsult, estimated that the market for communication satellites, the principal commercial category, would saturate at about fifteen launches per year. By contrast, Atlas, Titan, Delta, and Ariane together offered as many as forty annual launches, and China hoped to compete, too. Everyone thus faced continuing pressure to quote the lowest possible prices, and China, a newcomer, succeeded for a time in winning by underselling.

That nation had developed a family of launchers called Long March, which first flew to orbit as early as 1970. Export restrictions might have barred China from launching American payloads, by treating communication-satellite electronics as a strategic technology. But to promote better relations with Beijing, the U.S. government eased the restrictions, while allowing China to offer initial launches at a low, low price.

Joseph Allen, a former astronaut, put it this way: "If you're chairman of an American satellite company and your choice is $80 million at Cape Canaveral or $15 million with Great Wall [Beijing's national launching company], you put up the $15 million and travel to China. It's a no-brainer." The first such launch took place in April 1990. Its payload was a communication satellite built by Hughes, Asiasat 1, that in an earlier life had been rescued by the space shuttle *Discovery* in November 1984. It represented China's entry into the commercial market.

The Soviets also had their eyes on this business. They were ready for opportunity after the fall of *Challenger,* for in June 1985 they had set up an office, Glavkosmos, through which they intended to sell their launch services. Their experience was unmatched, for whereas the United States had specialized in modest numbers of long-lived spacecraft, the Soviets lacked reliable electronics and so had launched large numbers of short-lived craft. Since the 1960s they had maintained a launch rate that at times approached a hundred per year. They also had a versatile fleet of long-proven rockets that included the powerful Proton and the old R-7 with several choices of upper stages, as well as newer designs.

They got nowhere at first, for they ran afoul of American export restrictions, which barred most advanced technology from crossing the Soviet border. An event in 1987 set the pattern, as Hughes Aircraft failed to obtain an export license and so could not allow Glavkosmos to launch one of its satellites. Still, Soviet officials continued to try, though old habits died hard. In 1990 *Scientific American* reported a conversation with Yuri Glazkov, a director of training. He boasted: "We've got Americans, French, Canadians, and they all want to train their cosmonauts here."

Q. How much does it cost to train a cosmonaut?
A. That's a secret.
Q. How can it be a secret if you're trying to sell your services?
A. Oh. More than $10 million. My secrets are leaking out![1]

But the fall of Communism, in August 1991, permitted the easing of the export restrictions and the granting to Moscow of the favorable terms that Washington had previously extended to Beijing. This led to a massive upsurge in joint ventures, which came close to merging major American and Russian activities into a coordinated international program.

The two nations achieved a milestone in June 1992, when President George Bush met with his Russian counterpart, Boris Yeltsin. A new arms-control agreement was at the top of their agenda, as they pledged formally to dismantle many of the bombers, ICBMs, and nuclear warheads that they had aimed at each other. In addition, they renewed and expanded a 1987 agreement on cooperation in space. The new pact provided for Russian spacefarers to fly aboard American spacecraft, and vice versa. It revived the spirit of the Apollo-Soyuz mission of 1975, but the agreement demonstrated a clear intent that this new mutuality would involve much more than a one-shot mission.

The pact lent strong encouragement to executives at Lockheed, who were in hot pursuit of Russia's Proton launch vehicle. The Proton had fallen on hard times, for its engine manufacturer, Motorostroitel, had received no recent orders and was not being paid for its work. The firm nevertheless was continuing to build its engines, for as a company director put it, "They need them, and how! The problem is that they don't have any money to pay for them." But Motorostroitel was sustaining huge losses, and faced a shutdown.

The Protons' builder, the Khrunichev factory near Moscow, had operated as the manufacturing arm of Vladimir Chelomei's rocket empire. Lockheed officials came to visit in 1992 and proceeded to set up a joint venture, International Launch Services, that went on to market the Proton internationally as a commercial launch vehicle. A 1993 intergovernmental agreement set terms for Moscow's entry into this commercial market, as Russia agreed to carry out up to eight launches to geosynchronous orbit prior to 2001. Within months, the backers of Proton snared their first order, when Space Systems/Loral, a builder of communication and weather satellites, arranged to purchase as many as five such launches. In April 1996, a Proton launched its first Western satellite as a flight of ILS.

America's rocket-engine companies were also active. The demise of the N-1 moon rocket twenty years earlier had left a trove of engines in storage, featuring designs by Nikolai Kuznetsov, a protégé of Korolev. Their manufacturer, a firm called Trud, now wanted to sell them in the United

States. In California, Aerojet was particularly interested in the NK-33, a production version of the first-stage engine of the N-1. Similarly, the abandonment of the big Energiya rocket had left its engines in need of buyers. Energiya had four strap-on boosters, each powered by the RD-170, with 1.63 million pounds of thrust. It was the world's mightiest rocket engine, matching the power of the Apollo-era F-1. Pratt & Whitney approached its builder, NPO Energomash, and agreed to act as its U.S. representative.

A half-century earlier, the fall of Nazi Germany had unlocked a treasurehouse of advanced technology in rockets, missiles, and jet aircraft. The fall of Communism now created a similar opportunity, with Lockheed and Martin taking the lead in tapping the Soviet treasure as they set out to challenge Europe's Ariane. Arianespace had cultivated its customers with care; in the words of its managing director, Charles Bigot, "NASA treated its clients like a lord accepting peasants on his land. Arianespace treated them like a retailer."

Arianespace inspired customer loyalty by offering attractive terms, thereby undercutting Delta, Atlas, and Titan. NASA had used the shuttle to undercut these expendables, too. And even after 1986, when their builders won the freedom to compete, they faced long delays in restarting production lines. Air Force orders helped considerably, but in the commercial realm, no one could touch the Europeans. Their Ariane 4 series was setting a standard. They also were readying Ariane 5, with a pair of big solid-fueled boosters for the pep to shoot two three-ton satellites to geosynchronous orbit on a single launch. Arianespace had as much as 60 percent of the commercial market, twice the share that belonged to the Yankees.

But Martin Marietta was preparing an answer to Ariane 5, in the form of an advanced Titan IV. Like the other rockets of this type, it represented an upgrade of the Titan III, which first flew in 1965. However, this definitely was not your father's Titan, and one could appreciate its merits in the light of a design for a nuclear spaceship that the rocket engineer Max Hunter had presented in the mid-1960s. It had featured a cargo bay larger than the propellant tank, and Hunter had written that this was how a spaceship should be.

The ancestral Titan II had stood a bit over a hundred feet tall. The latest Titan IV doubled this height to 204 feet, and much of this was a payload fairing, which measured eighty-six feet long by nearly seventeen feet across. It was wider than the payload bay of the shuttle and twenty-six feet longer. Twin solid boosters ran alongside most of the length of the liquid-fueled core, while a Centaur upper stage, which the shuttle could not accommodate, added further power. With increased thrust from the solids, this Titan IV could lift 49,000 pounds to low orbit—more than twice the load of the early Titan III and six times that of the Titan II.

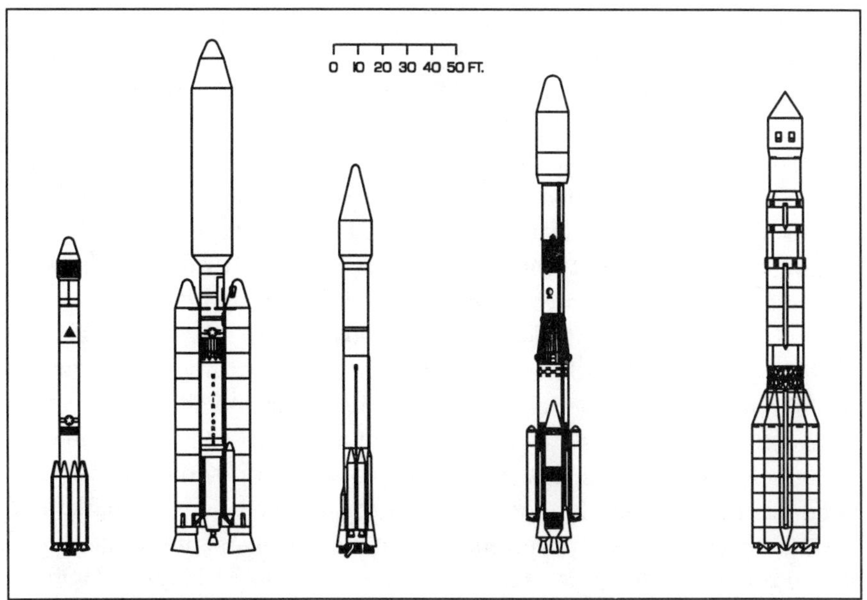

Expendable boosters of 1995: U.S. Delta II, Titan IV with 86-foot payload shroud, Atlas 2AS; Europe's Ariane 4; Soviet Proton. (Dan Gauthier)

In addition, a flurry of corporate activity brought Proton, Atlas, and Titan under a single roof. In 1994 Martin purchased the division of General Dynamics that built the Atlas-Centaur. Then in 1995, this firm merged with Lockheed to form a new aerospace giant, Lockheed Martin. This was part of a consolidation of major aerospace firms that stemmed from a falloff of military contracts in the wake of the Cold War. Officials of Lockheed Martin and Khrunichev declared that they would combine Atlas and Proton into a mutually supporting family of launch vehicles, and Charles Bigot, now the chairman of Arianespace, was outraged. Warning of "a systematic dumping of Protons onto the market," he added: "This is a declaration of war."

If such a war indeed were to break out, it would see Washington and Moscow standing together as allies against the French. An immediate topic of concern involved the Atlas 2AS, the top-of-the-line version of the Atlas-Centaur. Like the models dating to 1957, Atlas had two liquid-fueled booster engines along with a sustainer engine to power it to orbit. It also had acquired four small solid strap-on boosters, for a total of seven rocket motors. This was too much of a good thing, and Norman Augustine, the president of Lockheed Martin, decided that the time was right for simplification. He wanted a new version of Atlas with only one or two engines to replace the seven.

The resulting competition pitted Rocketdyne, which had been building engines for Atlas before that missile program even existed, against two Russian rivals. Trud offered its NK-33, with Aerojet, its partner, running this engine on a test stand and verifying its performance. Energomash, in turn, proposed to chop its RD-170 in half. This engine had four thrust chambers fed by a single set of turbopumps, a layout that Valentin Glushko had pioneered in the long ago. The new proposal called for a variant, the RD-180, that would use only two thrust chambers along with a smaller set of turbopumps suited to its reduced thrust.

In November 1995, Rocketdyne dropped out of the three-way race. Its president, Paul Smith, determined that his rocket engine could not be ready in time to meet the schedule of Lockheed Martin and remarked that its executives "were basically committed to a Russian engine." Somewhat plaintively, he set his sights on a new Pentagon effort, the Evolved Expendable Launch Vehicle: "We hope that the Air Force sees the benefit of maintaining an all-U.S. space vehicle." But the EELV was somewhere off in the future, while the new Atlas was close at hand. In January 1996 its builders picked the RD-180 as the winner. A Lockheed Martin official declared that this Atlas would hit the "sweet spot" of the commercial market, by launching single communication satellites of the largest size.

The choice of the RD-180 represented a posthumous victory for Valentin Glushko, who had died in 1989. Its builder, NPO Energomash, had grown out of the rocket-engine organization that Glushko founded in the Moscow suburb of Khimki following World War II. And while this engine had competed on an equal footing with Kuznetsov's NK-33, the outcome of the competition mirrored that of the battle over Korolev's N-1 moon rocket. The NK-33 represented a legacy of the N-1, which Glushko had scrapped. But the RD-180 had a close association with Glushko's own Energiya. The old rivalry between Korolev and Glushko, played out again on American soil, had ended anew in the same way. In addition, Pratt & Whitney would build the RD-180 in West Palm Beach, Florida, and would pay royalties to Moscow—in hard currency.

Meanwhile, Dan Goldin was making his presence felt in NASA. President Bush had dismissed his predecessor, Richard Truly, early in 1992, and named Goldin as the new administrator. Goldin had strong support from Vice President Dan Quayle, chairman of a space council that wanted to break with traditional practices, and Goldin came in with a mandate to shake things up. He and Quayle agreed that a basic problem was a tendency to pursue a handful of large, glamorous projects that were prone to long delays and big cost overruns. The manned space program had certainly seen its share of such efforts, but they plagued the unmanned side of NASA as well. Just then, a case in point was the Hubble Space Telescope.

It was a true astronomical telescope in orbit, fitted with instruments, solar panels for power, a very accurate system to point it at a particular star or galaxy, and a data system. It drew on the same base of technology that had made possible the KH-11 reconnaissance satellite, including large mirrors of exquisitely accurate shape and charge-coupled devices to form its images. However, it had not been possible to purchase a KH-11 and simply point it at the stars. The very existence of the KH-11 had been under wraps for a time, and the details of its design were matters of strict security. Only the intelligence community could work with this spacecraft, behind closed doors. By contrast, the world's astronomers worked in the open, without security clearances. The CIA wouldn't talk to them.

The Hubble Telescope thus came into being as a project in its own right. But whereas the Air Force and CIA had drawn on their extensive experience to build reconnaissance satellites, NASA pursued the Hubble by resorting to the success-oriented management of the shuttle program. The agency did this because, once again, it was trying to do more than its budget would allow. NASA had never built a spacecraft like Hubble, and this lack of experience caused delays and cost overruns. As a result, project managers faced such intense pressure that they let things slide. They pooh-poohed proposals for additional tests, for these would cost money, and the project had none to spare.

This led to a major error in the shaping of Hubble's main mirror. The contractor, the firm Perkin-Elmer, had fabricated many such mirrors for the CIA. In working on the mirror for Hubble, its specialists relied on a highly sensitive optical instrument, a "reflective null," that could determine very slight deviations from proper curvature. Unfortunately, no one checked the accuracy of the reflective null, which was off by 1.3 millimeters, more than enough to spoil the final result. Other test instruments were available, which indeed pointed to a misshapen mirror, but people ignored those results. After all, they conflicted with the reflective null, which was the one people trusted. Nor did NASA take advantage of an Air Force offer to test the complete telescope in its own optical lab.

The flaw remained undetected as the mirror went into the complete Hubble Telescope, a $1.5-billion craft that astronomers hailed as the world's premier observatory. The shuttle *Discovery* took it to orbit in April 1990. Only then, in space, did it undergo its optical tests as an integrated system. These promptly disclosed the faulty shape, which had left Hubble slightly nearsighted. NASA was crestfallen, for once again it had made a billion-dollar blunder.

The telescope was not useless, for computer processing of its images made it possible to sharpen the focus in a number of instances. But NASA still lived by public relations, and this latest mistake brought about such unwelcome publicity as an opening scene in the comedy film *Naked Gun*

2 1/2. It depicted a dark and depressing lounge with pictures of disasters: the *Hindenburg*, the *Titanic*, the Hubble Space Telescope.

The planetary program constituted another example of what was wrong with NASA. Following the successes of the 1960s and 1970s, it had languished while the shuttle and space station were given higher priority. Across a span of more than ten years, from 1978 to 1989, the nation did not launch a single new mission to any of the planets.

That decade saw plenty of work on the ground, aimed at building three approved probes: Galileo for Jupiter, Mars Observer, and Magellan, to map Venus using radar. Yet these spacecraft displayed anew the agency's penchant to put too many eggs into too few baskets. Magellan performed brilliantly, producing maps to rival those of the moon and Mars. But Galileo was hobbled by its undeployed main antenna, which sharply cut its data rate and its scientific return. The $1-billion Mars Observer did even worse. It arrived at Mars in August 1993, received commands to fire a retro-rocket and enter orbit—and vanished. A review panel blamed an onboard rupture that could have left the spacecraft spinning wildly, unable to point its antenna toward the earth.

This eggs-in-one-basket approach contrasted markedly with that of the 1960s. The style of the sixties had emphasized frequent flights, often in duplicate to guard against a single launch failure. The Mariner spacecraft of the day had come in at five hundred to a thousand pounds. By contrast, Galileo and Magellan approached four tons, and were vastly more capable. But with so few missions in prospect, strong trends pushed each one toward greater cost and lengthier delays.

As one scientist put it, any particular mission "looked like the only bus out of town, so you wanted to pile everything on it." Everyone wanted to raise the scientific return by adding new or better instruments, even though NASA couldn't always handle what resulted. Mars Observer had gotten out of hand in this fashion, with the head of its review panel pointing to sloppy management. He stated bluntly that "if you reflew Mars Observer and did nothing different, there's a high probability you would lose it again."

When Goldin came to NASA, he found one more such dinosaur on the agenda: Cassini, a Galileo-like orbiter of Saturn, planned for launch in 1997. Goldin described it as an overburdened "battleship Galactica." It was too far along to cancel, but it represented an institutional mind-set that he wanted to change. He declared that "the time was ripe to begin a cultural revolution and develop a new NASA." Calling for "smaller, cheaper, faster, better" missions, he vowed to "change the culture" within the agency, turning it from a stodgy bureaucracy into a risk-taking group of innovators.

An existing program, the Earth Observing System, gave him an early opportunity to pursue this approach. It had grown out of a NASA ploy to win support for the space station in the scientific community, by building large unmanned satellites to study the earth's atmosphere and environment—and paying for them out of the space-station budget. When EOS received approval, in 1989, its managers envisioned two thirteen-ton spacecraft and a budget of $17 billion. This plan did not last; in 1991 the Senate chopped the funding to $11 billion.

Goldin now went further. In mid-1992 he approved a plan that would cut the program even more, to $8 billion, by emphasizing smaller spacecraft that would focus more closely on specific scientific issues. "There is always a tendency to want to get 100 percent of desired results," he told *Aerospace America*. "The difference between getting 85 percent and 100 percent can be a very significant difference in costs. Someone asked me, are you going to get the last two percent of data from the Magellan mission? My answer was no, because the last two percent costs a lot of money."[2]

Goldin wanted to return to the 1960s, by reviving an era of frequent launches and modest costs. Such an approach would permit NASA to experience mission failures and to take them in stride. He had fond memories of the Surveyor program, which had carried out automated moon landings in preparation for Apollo: "We had two failures out of seven launches. It was a wonderfully successful program."

He had come into his office as an appointee of President Bush, but he survived the change in administrations and retained his post after President Clinton came to the White House. A year later, a Defense Department program called Clementine showed that advances in instrumentation meant that unmanned missions might be small, inexpensive—and highly productive. Like that miner's daughter, Clementine would identify different types of rock, on the moon. It then would proceed onward to look at an asteroid, to be "lost and gone forever" in deep space.

Clementine grew out of the Strategic Defense Initiative. At Lawrence Livermore National Laboratory near San Francisco, the physicist Lowell Wood had pushed a concept called Brilliant Pebbles, wherein small but highly capable spacecraft would orbit in large numbers to destroy enemy missiles by colliding with them in flight. This effort cut the weight of a spaceborne camera from fifty pounds to two pounds or less. The Clementine that resulted weighed less than a thousand pounds, nearly half of which was onboard propellant. It carried six cameras able to make images at ultraviolet, visible, and near- and far-infrared wavelengths. These included a laser ranging device. It might once have been intended to find the distance to missile warheads, but now it would make accurate measurements of the height of lunar mountains and the depth of valleys and craters, complementing the photos with topographic studies.

Previous photographic missions included the five Lunar Orbiters from 1966 to 1967, as well as two Soviet orbiters in 1971 and 1974. They had compiled a superb lunar photo atlas, but had lacked the multispectral capability that subsequently demonstrated great value in observations of the earth. Clementine now brought this capability to the moon, orbiting it for two months and returning nearly a million images. It mapped the moon anew, not in black-and-white but in eleven visible and near-infrared colors, while charting its topography as well. The results proved highly interesting to geophysicists, for they shed light on the moon's early evolution. They also made it possible to prepare geological maps by showing the moon's various types of rock. Other data even indicated the presence of ice near the lunar south pole, which might provide water and rocket propellants for a future manned base. What was more, the spacecraft cost only $55 million.

Though it was a Pentagon project, Goldin embraced Clementine as his darling. In April 1994 he spoke at a conference on low-cost planetary missions, declaring that in prior years "we all became accustomed to a way of doing business where we didn't have to pay attention to cost. Getting the mission done was more important than cost. That is gone and forgotten, finished." Instead he endorsed a program called Discovery, featuring projects with a cost cap of $150 million. These would not demand the resources of Jet Propulsion Laboratory; instead, they would result from proposals prepared by individual scientists, known as principal investigators. "We're asking for PIs to come in with a whole mission," a NASA associate administrator told the conference. "If we like it—if we like your science, if we like the way you're going to manage it, if we like the cost—we'll buy it, pay you, and you do it."[3]

While Goldin was cutting back EOS and rejuvenating the planetary program, the Air Force was broadening the scope of the nation's unmanned space efforts by introducing major new services in navigation and air traffic control. These stemmed from a military project, the Global Positioning System, which had come about as a navigation system that could improve on the Navy's Transit satellites. Transit required costly signal-processing equipment and usually needed two separate passes of a satellite, an hour and a half apart. This worked for Polaris submarines, the main clients, whose skippers could wait all day for a navigational fix. But the Air Force wanted real-time position determination, to enable aircraft to drop bombs with better accuracy. The system that resulted was GPS.

The system is a constellation of twenty-four satellites, in twelve-hour orbits at an altitude of 12,543 miles. Each of them carries an atomic clock for precise determination of time, while ground-based tracking permits each one to know its position with similar accuracy. A ground receiver then accepts signals from the spacecraft in view, learning their positions as well as the exact times when the signals were transmitted. The receiver has its own internal clock, which is not very accurate, but the data from space allows it to synchronize this clock with those of the satellites. The receiver then calculates the length of time each signal has been in transit, traveling at the speed of light. This translates into an accurate determination of distance to each satellite. Through triangulation, the receiver then determines its own location.

The first such satellite reached orbit in 1977; by 1983 enough were available for preliminary tests. In that year, a Boeing 747 of Korean Airlines strayed into Soviet airspace over Sakhalin Island, a sensitive military area, and was shot down. Better navigation might have prevented this, and President Reagan ordered the Air Force to make GPS signals available for use internationally. The complete 24-satellite array entered service in March 1994; a Russian counterpart, Glonass, became operational some two years later.

Receivers for the GPS are indeed simple; the firm Magellan Systems is selling them for $199. This puts them within reach of boaters finding their way back to fogbound harbors and trucking companies that want to keep track of the locations of their vehicles. A variant allows emergency 911 calls to report position via satellite. It will find buyers among motorists who fear a breakdown in a remote area. More elaborate versions include computerized maps. Installed in rental cars, they not only show location, but actively direct people to their destinations in unfamiliar cities.

Commercial aviation also stands to benefit. It relies on extensive arrays of ground-based radio transmitters used in navigation, as well as the Instrument Landing System, which guides airliners to bad-weather landings at over a thousand airports. The systems have shown their value during decades of routine use, and one gets a sense of this longevity by rereading Ernest K. Gann's bestselling novel of 1953, *The High and the Mighty*. The airplane that figured in the novel was a DC-4, with all of twenty people on board, struggling to get from Honolulu to San Francisco following loss of a piston engine. The plane's navigator took star sightings, as in the days of Captain Cook. The pilot approached the airport with help from an ancient radio system that no doubt had been used during the era of the biplanes. But during final approach, the flight crew made an instrument landing by relying on the ILS.

Still, these systems have their limitations. Aircraft do not follow the shortest routes along their courses, but fly instead from one navigational station to another, in a connect-the-dots fashion. On long over-ocean routes, navigational accuracy often deteriorates, and airliners avoid collisions by being spaced as much as a hundred miles apart. And when the weather goes down, most airports go down as well, reducing their landing operations to a single runway equipped with ILS. Moreover, ILS permits only straight-in approaches and does not allow aircraft to turn or maneuver while landing. This severely restricts the permitted traffic patterns by allowing aircraft to fly only in straight lines, and sharply cuts the capacity at major cities with their multiple airports. New York, for one, has three world-class airports. But when the weather is poor, La Guardia can handle only twenty-four landings or takeoffs per hour, compared to eighty on fair days.

Today, GPS offers the prospect of a unified system called Free Flight, which will transcend these limitations. Instead of relying on ground-based transmitters, pilots will turn to GPS receivers and onboard computers. This will allow them to fly preferred routes that save fuel. In addition, during the next fifteen years the FAA expects to phase out many of its traditional radio systems. It has also decided that GPS-based systems in time will replace ILS, and has already issued formal rules whereby flight crews can use GPS satellites to make instrument landings.

With GPS and Goldin's reforms pointing to new directions for unmanned space activities, the space shuttle was also gaining fresh success, as it finally made a first-class contribution in the world of science. It did this by conducting a mission that repaired the Hubble Telescope as it flew in orbit. The telescope had been designed for service and maintenance in space. In December 1993, *Endeavour* rendezvoused and captured it with a long manipulator arm. Members of the flight crew replaced a wobbly set of gyros, installed new solar panels, and put in new internal equipment that featured small mirrors, shaped to precise perfection. The mirrors served as eyeglasses, correcting Hubble's faulty vision.

The telescope responded with a burst of new images that raised disturbing questions. A prime task involved making observations that could determine the age of the universe. Indeed, the telescope was named after Edwin Hubble, the first astronomer to attempt this determination. The astronomer Wendy Freedman, working near Caltech, had clear new images from this telescope and discovered a disturbing paradox: The universe appeared younger than the stars it contained.

This was science at its best, raising deep issues at the foundations of astrophysics. It was not a simple matter of routine experiments and modestly improved findings. Rather, the Hubble Telescope was calling into question the basic methods of astronomy, casting doubt on what scientists truly know and how they can claim to know it. Yet this telescope would stand at the forefront of work aimed at resolving this paradox, by sharpening the estimated age of the cosmos.

With the shuttle winning new plaudits by rescuing Hubble, the space station was settling in comfortably to a spending level of close to $2 billion per year. In 1984, Reagan had called for its completion "within a decade." Nine years later, as President Clinton took office, the program had received over $9.8 billion in appropriations; yet amid its frequent redesigns, it had produced little more than concepts on paper. The space station was well on its way to becoming the ultimate government project, spending billions annually with no end in sight while yielding little more than subsidized jobs.

Goldin didn't like this; he referred to the station as a "thirty-year entitlement program." Clinton disliked it, too. Early in 1993 he directed Goldin to set aside the existing $30-billion plan and to come up with yet another redesign, chopping the cost by up to 50 percent. The resulting paper space station looked rather like a lead balloon; it was during its fifteen minutes of fame, in June, that the House came within a single vote of recommending cancellation of the program. Nevertheless, Goldin saw that he still could get the big station that NASA's clients all wanted, via cooperation with Russia.

He had followed up the 1992 accord with Moscow by proceeding with plans that would fly cosmonauts on the shuttle while sending astronauts to do research aboard the Mir space station. In addition, Clinton had authorized Goldin to seek support from Russia in building NASA's station. Russia had pursued plans for a second Mir, but the plans were on hold, amid budget pressures even more severe than those of NASA. But in mid-March of 1993, Goldin wrote a letter to his Russian counterparts, proposing to build a joint space station that would merge Mir 2 with the NASA design. At the same time, a NASA delegation went to Moscow and invited NPO Energiya, the builder of Salyut and Mir, to take part in the latest redesign of America's station.

After this, events moved swiftly. Within weeks President Boris Yeltsin gave his blessing to the joint effort. In late June, NASA and the Russian space agency struck a formal agreement, opening the way to expanded cooperation. The head of NPO Energiya, Yuri Semenov, then led a group that visited Washington to brief NASA and its contractors on Russia's potential contributions. The group also took part in discussions held in Crystal City, close to Washington, where a NASA team was continuing to look at space-station designs.

Late that August, Prime Minister Viktor Chernomyrdin flew to Washington and met with Vice President Albert Gore. They signed a new agreement that transformed the space station into a true mutual effort, calling for the strongest cooperation between the two nations in the history of space exploration. The United States agreed to rescue the Russian manned space program by paying $400 million in four annual installments. In Washington such a sum would fail to cover the cost of a single shuttle flight, but in Moscow it would go far. In that country one might purchase the services of a well-staffed research center for as little as $100,000 per year.

The payments were aimed primarily at supporting Mir, by reviving its Spektr and Priroda modules and funding their launch into space. In turn, the focus on Mir was part of the joint program of manned flights envisioned in the Bush-Yeltsin agreement of 1992. The first such flight took place in February 1994, as *Discovery* flew an eight-day mission whose crew included the cosmonaut Sergei Krikalev. A year later, the same shuttle carried Vladimir Titov, a veteran of a year in space, on a similar mission. It rendezvoused with Mir, but did not dock; that delicate operation would await the next flight.

Meanwhile, Russia had been readying Spektr. It had initially been intended for environmental studies using multispectral cameras, but it changed this mission due to the law that whoever has the gold, rules. NASA had the gold; part of its $400 million paid for a new set of instruments for use in biomedical studies. The astronaut Norman Thagard would use them during a planned stay aboard Mir, for with Russian help, he would break the 84-day American record aboard Skylab.

Thagard flew to Mir aboard a Soyuz, in March 1995. Unfortunately, Spektr was still on the ground. The Yankees had run into delays in shipping their equipment; they then encountered further delays in Moscow, as customs officials detained some instruments because proper fees had not been paid. Spektr finally flew from Tyuratam in May and linked up with Mir. That gave Thagard about a month in which he could use it; then the shuttle *Atlantis* came calling, late in June. It carried two cosmonauts up to Mir and brought two others home, along with Thagard. He had spent 115 days in orbit.

As *Atlantis* maneuvered near Mir, the two returning cosmonauts watched from their Soyuz and took photos. This was not *2001: A Space Odyssey*, with a Pan Am spaceliner docking at a station shaped like an enormous rotating wheel. Still it was impressive enough: a shuttle orbiter as large as a DC-9 airliner, close to a sprawling array of fat cylinders and broad solar panels weighing a total of 130 tons. Then, in a similar joint mission during 1996, the astronaut Shannon Lucid set her own record by living in space for 188 days.

In addition, engineers from the two nations had a design for the international station that would replace Mir. The firm of Khrunichev would build its core, called the Functional Cargo Block (FGB, in Russian). It would fly atop the Proton, which Moscow would provide as a heavy-lift vehicle, complementing the shuttle. Russia promised to build other major sections, too, and would also sell to NASA several of the systems used for automated rendezvous and docking.

The overall plan was enough to take your breath away. In addition to the United States and Russia, Canada, Japan, and the European Space Agency all would actively participate. Construction would begin in November 1997 with the launch of the FGB core and would continue through June 2002. Those five years would see seventy-three flights to orbit, three-fifths of them Russian. The completed station would span a length of 290 feet and a width of 361 feet; its weight of 443 tons would match that of a Boeing 747. It would support a crew of six, with enormous solar arrays providing 110 kilowatts, enough for twenty single-family homes. It would cost $94 billion over thirty years.

Here was more than ambition; here was empire-building on a scale seldom seen, and if the plan was far-reaching, at least it left ample scope for downsizing and program stretch-out. Nor was it lacking in cheerful optimism, particularly as reflected in the schedule for spacewalks during assembly. NASA, which had traditionally seen them as hazardous, was now doing an about-face. The final version of the design for space station Freedom, prior to 1993, had called for 365 hours of spacewalks during assembly, and Goldin had viewed this as excessive. As of March 1994, the new plan called for 434 hours. A year later, amid better understanding of project requirements, it was up to 648 hours. The policy analyst Marcia Smith pointed to "NASA's sudden change in philosophy about the risk involved in spacewalks. No longer does NASA consider spacewalks a 'last resort'; now they are an 'opportunity.' " The agency was showing its can-do spirit, but it was also finding what indeed might be an opportunity, for foolhardiness.

Today, other pitfalls stem from the very nature of the partnership, for NASA is staking much of its future on collaboration with a shaky government and an uncertain economy. A nationalist regime may challenge the West with newly aggressive policies, while denouncing the accord with NASA as a sellout. A friendly and supportive government might nevertheless plead poverty and renege on its financial commitments. Should this happen, the joint space-station effort could end up as nothing more than a candle of hope amid the hurricanes of human folly.

Further weakness derives from the focus on manned spaceflight, which is still a sport of governments, having no life of its own. Meteorologists and communications managers, among others, have long relied routinely on unmanned spacecraft. No such community exists to support manned missions, and the reason is simple: They are far too costly for the limited benefit they can provide to users.

One can see this in the basic matter of flight to orbit. The shuttle requires 1.2 million separate procedures to make it ready for launch. To carry them out, NASA relied on 2,800 in-house employees, along with

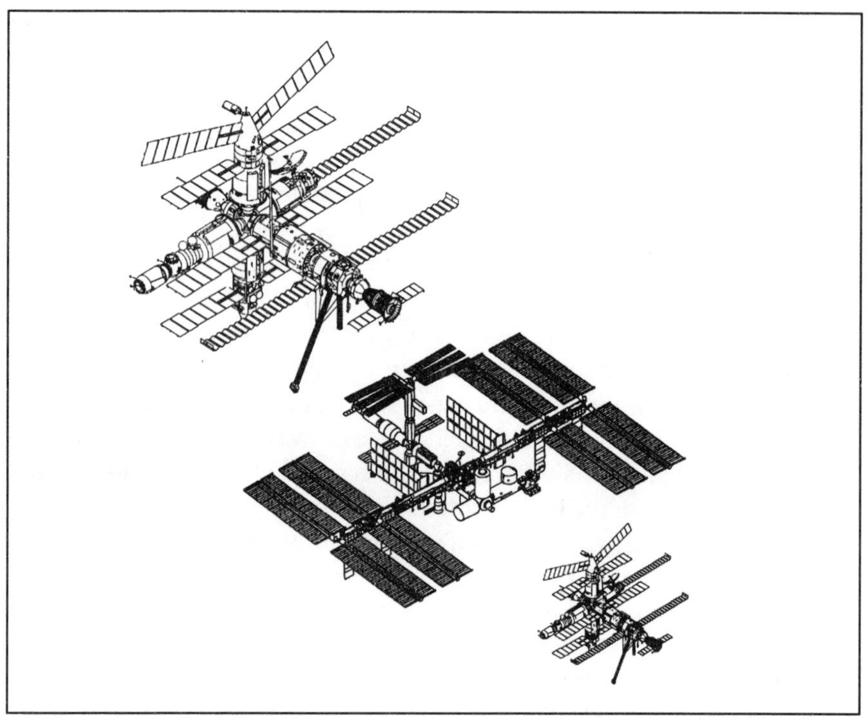

Space stations of the 1990s. ***Top,*** **Mir in 1996.** ***Middle,*** **proposed International Space Station.** ***Lower right,*** **Mir to same scale as ISS. (Dan Gauthier)**

19,700 more from contractors, as of 1995. This staff was on a par with the 20,000 people it took to launch an Apollo lunar mission, a quarter-century earlier. Early in 1996, *Aviation Week and Space Technology* quoted the cost of a shuttle launch at $550 million. This contrasted with the quoted cost for expendable boosters of similar capacity: $110 million for Ariane 5, and $60 million for Proton.

This disparity of as much as ten to one is likely to persist. The reason is that while a shuttle mission may launch spacecraft, its true goal is not to carry that payload to orbit. Rather, it is to bring its crew back safely. After all, the experience of *Challenger* shows that the loss of spacefarers can lead to national mourning, while shaking the agency to its foundations.

In spite of all this, the space station today represents a milestone. For the first time since Apollo, NASA proposes to build its future on something more than selling the nation a bill of goods. In this sense, too, the agency is returning to the 1960s, as it pursues manned flight primarily for its political value. The joint effort promises to show vividly and dramatically that Russia's best prospects may indeed lie in partnerships with the West, that Europe and the United States value and admire the work of its industries and engineers, and that an era of confrontation indeed can give way to one of cooperation.

Looking back upon this sorry century, one notes a striking contrast between the utopian hopes of Lenin's revolutionaries and the limited achievements that they and their descendants in time could claim. They defeated the Nazis, bearing the brunt of battle. They established an admirable system of public education, which in important respects continues to surpass that of the United States. They led the way into space. But these three accomplishments come close to exhausting the list. Their failures range from Stalin's terror through suppression of liberty and onward to general poverty and low standards of public health. They left Russia as a Third World nation with First World weapons. Yet their space program continues to shine in the eyes of their people. If Russia in the end will go her own way regardless of what America does, still for now we may cherish the hope that her partnership with NASA will indeed herald a new era, pointing toward a new century that will be vastly less bloody and more peaceful than the previous one.

How, then, can we understand this basic paradox of manned spaceflight? What can we say of the contrast between its high cost and almost nonexistent utility, and its potentially great political value? One can begin by noting that aviation, closely akin to space flight, has long carried similar political significance. Nations of Africa, Asia, and Latin America, flying the Boeing 747, reap the benefit of effective air transportation. They also enjoy powerful symbols of modernity, and of their governments' ability to purchase the newest and best.

The 747 offers genuine value and usefulness along with its symbolism, and thus avoids this contrast. But the contrast appears in full force in a different commercial aircraft: the Concorde. Its cachet is undeniable; a photo of this airliner flying past the Eiffel Tower summarizes a century of French technical leadership. Yet only a handful of Concordes are in service, for a simple reason: their high cost. The one-way fare on British Airways from New York to London is $4,509. The round trip at a discount costs $7,574. As a result, this airliner serves an elite clientele that is nearly as remote as the astronauts.

The history of the Concorde also recalls that of manned spaceflight, for it also evolved as a project of governments, heavily weighted with concern for national prestige. As noted before, its sponsor, Charles de Gaulle, had fretted over what he called "America's colonization of the skies," a commercial dominance that had left Europe's planebuilders unable to compete. Nor did the Concorde face the standard economic tests that firms such as Boeing applied when considering new ventures. De Gaulle and his British counterpart, Prime Minister Harold Macmillan, operated their aircraft industries as arms of the state, and covered their losses with subsidies.

High cost also characterizes manned spaceflight, and may well be its basic feature. To achieve a low level of risk, the shuttle cannot pursue the frequent, routine launch schedules that supporters had promised in 1972. Instead, NASA today treats the shuttle as a rare and valuable commodity, separating its operations from the ongoing world of space applications and commercial opportunities. This approach shows itself today in the shuttle launch schedule, which calls for up to two-thirds of its flights to serve the space station.

This at least has the merit of ensuring that if the shuttle experiences another failure, the effects will remain confined largely to the manned-mission ghetto. But it hammers home the point that manned flight has become an enterprise unto itself, offering political value and the drama of symbolism, but little real utility. In turn, this suggests that the enterprise may fade in time into history, for successful technologies do not survive merely as government-funded monuments. They serve communities of users, including the general public.

One gains further insight by placing manned flight in the context of what we commonly view as technical progress. Classic examples include the transition at sea from sail to steam, from steam locomotives to diesel, from piston-powered aircraft to jets, from vacuum-tube electronics to integrated circuits, and from ocean liners to commercial airliners. The difference is that in all these instances, the technologies that became obsolete were entirely adequate to their work.

No basic flaw characterized the sailing vessels that plied the seas for centuries. Indeed, even when challenged by steamships, they proved so effective that the transition from sail to steam took a hundred years to run its course. Piston-powered aircraft helped win World War II and spurred the rise of major airlines; they simply proved to be less capable than jets. Steam locomotives hauled freight and passengers for a century.

Vacuum tubes provided an entirely adequate basis for radio, radar, television, and long-distance telephone service; they also brought a good start in computers. Indeed, these tubes and their circuits continued to show new strength and capability even as the transistor appeared. And ocean liners such as the *Normandie* were much loved, with many people mourning them as they passed from the scene.

Sail, steam locomotives, piston aircraft, vacuum tubes, and ocean liners were all very capable and widely used. Yet the relentless press of the market, and of better alternatives, swept them all away and relegated them to museums. It is useful to view space stations in a similar light, with an important caveat: Because of its intrinsically high costs, the space station never got the opportunity to offer its potential services.

If the development of good microelectronics had been long delayed, then the scenario of *Collier's* might have come true. The space station might have emerged as the initial method for achieving such benefits as military reconnaissance, weather observation, and transoceanic telephone service. The eventual rise of microelectronics, decades later, would then have worked against the space station like jet engines against piston motors, or diesel locomotives against the steam variety. In fact, the rapid advent of advanced electronics made the space station obsolete before it could be built, by spurring the development of unmanned spacecraft. In the words of the playwright Nigel Kneale, the space-station concept was "almost worn out before anything arose to claim it."

Hence, in proposing an outlook for space, one finds sharply different prospects for the two realms of astronautics. The unmanned program will continue to find its future in modest numbers of highly capable spacecraft, whose uses will continue to expand. Today these include the worldwide Internet, which relies routinely on satellite communications. Tomorrow the GPS system, along with similar navigation satellites, will find major use in air traffic control.

The launches of these craft will continue as vivid, exciting events. NASA has fired its Delta rockets by the hundred; yet each flight remains unforgettable to people who see it for the first time. However, the launches rarely appear on TV; the public has long since stopped paying attention to them. People also take for granted the services of the unmanned craft. Thus, despite the importance of these services, the unmanned program is virtually invisible. This testifies to its success, for truly useful technologies fade into the background and do not intrude upon our awareness as we use them.

Manned flight faces a different outlook. The experience of three decades has failed to lead to serious proposals whereby it might in itself become useful. Yet it has continued to fascinate the public. Being useless but visible, it thereby shows a twofold contrast with the usefulness and near-invisibility of the unmanned program. This fascination also provides the manned program with its best opportunity and prospect.

The focus of the manned program today is the space station, and there is some chance that it will actually be completed. Should this happen, it will leave a legacy in the form of industries and organizations in a number of major nations that are now cooperating to build it. These institutions will not go away; on the contrary, political leaders will press for further and similar efforts. These could include a return to the moon, or a flight of astronauts and cosmonauts to Mars. These leaders will point anew to the high symbolism of spaceflight, and to the hope that future international ventures can dramatize their era as one of peace.

Thus, the ultimate purpose of the manned space program might amount to an exercise in theater. Yet theater has long been an essential element of the most serious and productive politics. Flourishes marked the British Empire, with its colorful hussars on parade and its bewigged judges, its naval reviews and bands playing "God Save the Queen." Performance played an essential role in the work of Winston Churchill and John Kennedy, masters of the rhetoric of challenge and hope. By dramatizing the ideals of internationalism in the new century, and by giving them life, the visions of von Braun and Korolev may yet come true.

Notes

Chapter 1 *Wonder-Weapons and Prison Camps*

1. Stuhlinger. Interview, Jan. 18, 1995.
2. Dornberger, *V-2*, pp. 12, 17.
3. Winter, *Prelude*, p. 42.
4. Ley, *Rockets* (1957), p. 148.
5. McDougall, *Heavens*, p. 38. See also Kerber, *Stalin's*, p. 151.
6. Dornberger, *V-2*, pp. 108–110.
7. Emme, *History*, p. 281.

Chapter 2 *Ingenious Yankees*

1. Truax. Interview, Jan. 20, 1989.
2. AAS History Series, vol. 6, p. 148.
3. *Astronautics & Aeronautics*, Oct. 1972, p. 41.
4. Emme, *History*, p. 68.
5. Jeanne Bollay. Interview, Jan. 24, 1989.
6. Atwood. Interview, July 18, 1988.
7. Hoffman. Interview, July 28, 1988.

Chapter 3 *Racing to Armageddon*

1. Rhodes, *Dark Sun*, pp. 28–29.
2. Zaloga, *Target*, pp. 69–70.
3. Zaloga, *Target*, p. 128.
4. Siddiqi, draft of manuscript, NASA SP-4409.
5. Hall. Interviews, Jan. 25, 1989, and Dec. 1, 1994.
6. Tea Pot report. J. Neufeld, *Development*, pp. 254–261.
7. Koch, *Estimates*, p. 159; McDougall, *Heavens*, p. 104.
8. Ramo, *Business*, pp. 113–114.
9. *Time*, Jan. 30, 1956, p. 53.

Chapter 4 *The Mid-1950s*

1. Ley, *Rockets* (1957), pp. 366–368.
2. Emme, *History*, p. 75.
3. Powers, *The Man*, p. 94.
4. Killian, *Sputnik*, pp. 72–79.

5. McDougall, *Heavens,* pp. 123–124.
6. JPRS-USP-88-001, Feb. 26, 1988, pp. 48–52.
7. JPRS 82970 (Space No. 20), Feb. 28, 1983, pp. 143–146.
8. *Astronautics & Aeronautics,* Oct. 1972, p. 58.
9. Agnew. Interview, Aug. 19, 1985.
10. Hartt, *Mighty Thor,* p. 79.
11. Castenholz. Interview, Aug. 18, 1988.
12. Ezell. Interview, July 22, 1988.
13. *Time,* Jan. 30, 1956, p. 54.

Chapter 5 *"The Russians Are Ahead of Us!"*

1. Lieutenant Colonel Lee James. Interview, Jan. 19, 1995; Castenholz. Interview, Aug. 18, 1988; Hoffman. Interview, July 28, 1988.
2. Johnson, Dorrenbacher, Thiel. Interviews, respectively, Feb. 10, 1995; Jan. 13, 1995; Dec. 4, 1994; Ezell. Interview, July 22, 1988.
3. JPRS-USP-88-001, Feb. 26, 1988, pp. 48–52. Also Golovanov, *Korolev* (1994), pp. 505–506.
4. *Khrushchev Remembers,* p. 46.
5. Golovanov, *Sergei Korolev* (1975), pp. 7–11.
6. Medaris, *Countdown,* pp. 155, 157; McDougall, *Heavens,* p. 141; *Newsweek,* Oct. 14, 1957, p. 37.
7. Oberg, *Red Star,* p. 33; McDougall, *Heavens,* p. 137.
8. Kearns, *Lyndon Johnson,* p. 145.
9. Halberstam, *Fifties,* p. 627.
10. *Time,* Dec. 16, 1957, p. 12.
11. Gunther, *Russia,* pp. 257–258.
12. *Time,* Dec. 29, 1958, p. 7.
13. McKee. Interview, Mar. 29, 1989.
14. Barr and Howard, *Polaris!,* p. 36; Spinardi, *Polaris,* p. 25. Also *Astronautics & Aeronautics,* Oct. 1972, p. 64.
15. Ruffner, *Corona,* pp. 8–9.
16. Mosley, *Dulles,* p. 432.
17. Ruffner, *Corona,* p. 130.

Chapter 6 *A Promise of Moonglow*

1. *Time,* Oct. 20, 1958, p. 18.
2. Ishlinsky, *Academician,* pp. 214–215.
3. Sloop, *Hydrogen,* p. 183.
4. Logsdon, *Decision,* p. 35.

Chapter 7 *Afternoon in May*

1. JPRS-USP-93-005, Oct. 5, 1993, p. 19.
2. Dwiggins, *SST,* p. 4.
3. *New York Times,* Sept. 13, 1962, p. 16.
4. McDougall, *Heavens,* pp. 221–222; Logsdon, *Decision,* p. 65.
5. JPRS-USP-91-004, Sept. 20, 1991, pp. 71–77.
6. Logsdon, *Decision,* p. 106; AAS History Series, vol. 3, p. 54.
7. Reston. Quoted in Manchester, *Glory,* p. 910.
8. Logsdon, *Decision,* pp. 109–110; AAS History Series, vol. 3, pp. 56–59.
9. *New York Times,* May 26, 1961, p. 12.
10. Von Braun. Quoted in *U.S. News & World Report,* June 1, 1964, p. 54; Weisner. Quoted in Logsdon, *Decision,* p. 111.

11. Nedelin. Quoted in Sakharov, *Memoirs*, p. 194.
12. JPRS-UMA-89-015, June 15, 1989, pp. 34–45.
13. Radio Moscow. Quoted in Abel, *Missile Crisis*, p. 181.

Chapter 8 *High-Water Mark*

1. Ezell. Interview, July 22, 1988.
2. Mansfield, *Vision*, pp. 294–295.
3. Houbolt. Quoted in Hacker and Grimwood, *Shoulders*, p. 15.
4. JPRS-USP-91-002, Apr. 16, 1991, p. 68.
5. Castenholz. Interview, Aug. 18, 1988.
6. CBS Television Network coverage, Dec. 21, 1968. Also *Air & Space*, June/July 1989, p. 60.

Chapter 9 *Lunar Aftermath*

1. Wolfe, *Right Stuff*, p. 52.
2. *Time*, Jan. 3, 1969, p. 12; *Newsweek*, Jan. 6, 1969, pp. 17–18; Benson and Faherty, *Moonport*, p. 457; Archibald MacLeish. Quoted in AAS History Series, vol. 3, p. 100.
3. Benson and Faherty, *Moonport*, p. 476; Brooks et al., *Chariots*, pp. 344–347; Armstrong et al., *First*, pp. 245, 247, 267–268; *Newsweek*, July 28, 1969, p. 18.
4. Mondale. Quoted in Chaikin, *Man On the Moon*, p. 336.
5. Mondale. Quoted in *National Journal*, Aug. 19, 1972, p. 1329.
6. Heiss. Quoted in *National Journal*, Aug. 12, 1972, p. 1299.
7. Simpson, *Space Station*, pp. 269–271.
8. JPRS-USP-89-010, Nov. 22, 1989, p. 44.
9. JPRS-USP-89-010, Nov. 22, 1989, p. 39.

Chapter 10 *Electrons in the Void*

1. Richelson, *Secret Eyes*, pp. 93, 265.
2. Koppes, *JPL*, p. 164.

Chapter 11 *Space in the Eighties*

1. Hoffman. Interview, July 28, 1988.
2. Reagan and Keyworth. Quoted in McCurdy, *Space Station*, pp. 56, 61, 129.
3. Proxmire. Quoted in McCurdy, *Space Station*, p. 128.
4. *New York Times*, Jan. 26, 1984, p. B8.
5. *Science*, Oct. 28, 1983, p. 405.
6. Boisjoly. Quoted in *Scientific American*, Aug. 1986, pp. 62–63.
7. *Science*, Mar. 14, 1986, p. 1238; Mar. 28, 1986, p. 1495; June 20, 1986, p. 1488; Feynman, *What Do You Care*, pp. 223, 237; Los Angeles *Times*, Jan. 28, 1996, p. A3.
8. JPRS-USP-92-006, Nov. 17, 1992, p. 24; 93-002, May 18, 1993, p. 32.
9. *IEEE Spectrum*, Dec. 1995, pp. 22, 26–27.
10. Space station cost. Discussed in Collins and Fries, *Spacefaring*, pp. 14–15; Pewitt. Interview, Dec. 14, 1983.

Chapter 12 *Renewal and Outlook*

1. *Scientific American*, July 1990, p. 85.
2. *Aerospace America*, Nov. 1992, p. 6.
3. *Science*, May 27, 1994, pp. 1244–1245.

Bibliography

Books

Abel, Elie. *The Missile Crisis.* Philadelphia: Lippincott, 1966.

Anderson, Frank W. *Orders of Magnitude: A History of NACA and NASA, 1915–1980.* NASA SP-4403. Washington, D.C.: U.S. Government Printing Office, 1981.

Armstrong, Neil, Michael Collins, and Edwin E. Aldrin. *First on the Moon.* Boston: Little, Brown, 1970.

Baar, James, and William E. Howard. *Polaris!* New York: Harcourt, Brace, 1960.

Barnett, Lincoln. *The World We Live In.* New York: Time, 1955.

Beatty, J. Kelly, Andrew Chaikin, and Brian T. O'Leary, eds. *The New Solar System.* New York: Cambridge University Press, vol. 1, 1981; vol. 3, 1990.

Benson, Charles D., and William Barnaby Faherty. *Moonport: A History of Apollo Launch Facilities and Operations.* NASA SP-4204. Washington, D.C.: U.S. Government Printing Office, 1978.

Bilstein, Roger E. *Stages to Saturn: A Technological History of the Apollo/Saturn Launch Vehicles.* NASA SP-4206. Washington, D.C.: U.S. Government Printing Office, 1980.

Bonnet, Roger M., and Vittorio Manno. *International Cooperation in Space.* Cambridge, Mass.: Harvard University Press, 1994.

Boyne, Walter. *Messerschmitt 262: Arrow to the Future.* Washington, D.C.: Smithsonian Institution Press, 1980.

———. *Clash of Wings: Air Power in World War II.* New York: Simon & Schuster, 1994.

Brooks, Courtney G., James M. Grimwood, and Loyd S. Swenson. *Chariots for Apollo: A History of Manned Lunar Spacecraft.* NASA SP-4205. Washington, D.C.: U.S. Government Printing Office, 1979.

Burroughs, William E. *Deep Black: Space Espionage and National Security.* New York: Random House, 1986.

Carpenter, M. Scott, et al. *We Seven.* New York: Simon and Schuster, 1962.

Chaikin, Andrew. *A Man on the Moon.* New York: Viking, 1994.

Chaisson, Eric J. *The Hubble Wars.* New York: HarperCollins, 1994.

Chapman, John L. *Atlas: The Story of a Missile.* New York: Harper & Brothers, 1960.

Clark, Phillip. *The Soviet Manned Space Program.* London: Salamander Books, 1988.

Clarke, Arthur C. *The Exploration of Space.* London: Temple Press, 1951.

———. *Profiles of the Future.* New York: Harper & Row, 1963.

Collins, Martin J., and Sylvia Fries, eds. *A Spacefaring Nation: Perspectives on American Space History and Policy.* Washington, D.C.: Smithsonian Institution Press, 1991.

Compton, W. David, and Charles D. Benson. *Living and Working in Space: A History of Skylab.* NASA SP-4208. Washington, D.C.: U.S. Government Printing Office, 1983.

Conquest, Robert. *Kolyma: The Arctic Death Camps.* London: Macmillan, 1978.

———. *The Great Terror: A Reassessment.* New York: Oxford University Press, 1990.

Cooper, Henry S. F. *A House in Space.* New York: Holt, Rinehart and Winston, 1978.
Cortright, Edgar M. *Exploring Space with a Camera.* NASA SP-168. Washington, D.C.: U.S. Government Printing Office, 1968.
Dixon, Don. *Universe.* Boston: Houghton Mifflin, 1981.
Dornberger, Walter. *V-2.* New York: Viking Press, 1954.
Dunne, James A., and Eric Burgess. *The Voyage of Mariner 10.* NASA SP-424. Washington, D.C.: U.S. Government Printing Office, 1978.
Dwiggins, Don. *The SST: Here It Comes Ready or Not.* Garden City, N.Y.: Doubleday, 1968.
Emme, Eugene M., ed. *The History of Rocket Technology.* Detroit: Wayne State University Press, 1964.
Ezell, Edward Clinton, and Linda Neuman Ezell. *The Partnership: A History of the Apollo-Soyuz Test Project.* NASA SP-4209. Washington, D.C.: U.S. Government Printing Office, 1978.
———. *On Mars: Exploration of the Red Planet, 1958–1978.* NASA SP-4212. Washington, D.C.: U.S. Government Printing Office, 1984.
Ezell, Linda Neuman. *NASA Historical Data Book.* Vol. II, *Programs and Projects 1958–1968.* NASA SP-4012. Washington, D.C.: U.S. Government Printing Office, 1988.
Feynman, Richard P. *What Do* You *Care What Other People Think?* New York: W. W. Norton, 1988.
Fimmel, Richard O., James Van Allen, and Eric Burgess. *Pioneer: First to Jupiter, Saturn and Beyond.* NASA SP-446. Washington, D.C.: U.S. Government Printing Office, 1980.
Fimmel, Richard O., Lawrence Colin, and Eric Burgess. *Pioneer Venus.* NASA SP-461. Washington, D.C.: U.S. Government Printing Office, 1983.
Frisbee, John L., ed. *Makers of the United States Air Force.* Washington, D.C.: Office of Air Force History, 1987.
Galloway, Jonathan F. *The Politics and Technology of Satellite Communications.* Lexington, Mass.: Lexington Books, 1972.
Gann, Ernest K. *The High and the Mighty.* New York: William Morrow, 1953.
George, David Lloyd. *War Memoirs.* Boston: Little, Brown, 1933.
Glushko, Valentin P. *Rocket Engines GDL-OKB.* Moscow: Novosti Press Agency Publishing House, 1975.
Golovanov, Yaroslav. *Sergei Korolev: The Apprenticeship of a Space Pioneer.* Moscow: Mir Publishers, 1975.
———. *Korolev: Facts and Myths.* Moscow: Nauka, 1994.
Gordon, Theodore J., and Julian Scheer. *First into Outer Space.* New York: St. Martin's Press, 1959.
Graham, Loren. *The Ghost of the Executed Engineer.* Cambridge, Mass.: Harvard University Press, 1993.
Gray, Mike. *Angle of Attack: Harrison Storms and the Race to the Moon.* New York: W. W. Norton, 1992.
Green, Constance, and Milton Lomask. *Vanguard: A History.* Washington, D.C.: Smithsonian Institution Press, 1971.
Gump, David P. *Space Enterprise.* New York: Praeger, 1990.
Gunston, Bill. *Fighters of the Fifties.* Osceola, Wis.: Specialty Press, 1981.
Gunther, John. *Inside Russia Today.* New York: Harper, 1958.
Hacker, Barton C., and James M. Grimwood. *On the Shoulders of Titans: A History of Project Gemini.* NASA SP-4203. Washington, D.C.: U.S. Government Printing Office, 1977.
Halberstam, David. *The Fifties.* New York: Villard, 1993.
Hall, R. Cargill. *Lunar Impact: A History of Project Ranger.* NASA SP-4210. Washington, D.C.: U.S. Government Printing Office, 1977.
Hallion, Richard, ed. *The Hypersonic Revolution.* Wright-Patterson Air Force Base, Ohio: Aeronautical Systems Div., 1987.
Harford, James. *Korolev.* New York: Wiley, 1997.
Hartt, Julian. *The Mighty Thor.* New York: Duell, Sloan and Pearce, 1961.
Heppenheimer, T. A. *Toward Distant Suns.* Harrisburg, Pa.: Stackpole, 1979.
———. *Turbulent Skies: The History of Commercial Aviation.* New York: Wiley, 1995.

———. *Air Traffic Control.* Arlington, Va.: Pasha, 1995.

Herken, Gregg. *Cardinal Choices: Presidential Science Advising from the Atomic Bomb to SDI.* New York: Oxford University Press, 1992.

Hewlett, Richard G., and Francis Duncan. *Nuclear Navy.* Chicago: University of Chicago Press, 1974.

Holloway, David. *Stalin and the Bomb: The Soviet Union and Atomic Energy 1939–1956.* New Haven, Conn.: Yale University Press, 1994.

Hooper, Gordon R. *The Soviet Cosmonaut Team.* Woodbridge, England: GRH Publications, 1986.

Howarth, David. *The Dreadnoughts.* Alexandria, Va.: Time-Life Books, 1979.

Hoyle, Fred. *Frontiers of Astronomy.* New York: Harper & Brothers, 1955.

Hudson, Heather E. *Communication Satellites.* New York: Free Press, 1990.

Hughes, Patrick. *A Century of Weather Service.* New York: Gordon and Breach, 1970.

Ishlinsky, Aleksandr, ed. *Academician S. P. Korolev: Scientist, Engineer, Person.* Moscow: Nauka, 1986.

Jackson, Robert. *Airships.* Garden City, N.Y.: Doubleday, 1973.

Johnson, Clarence "Kelly," and Maggie Smith. *Kelly: More Than My Share of It All.* Washington, D.C.: Smithsonian Institution Press, 1989.

Johnson, Nicholas L. *Handbook of Soviet Lunar and Planetary Exploration.* American Astronautical Society, Science and Technology Series, vol. 47. San Diego: Univelt, 1979.

———. *The Soviet Reach for the Moon.* Huntsville, Ala.: Cosmos Books, 1994.

Johnson, Paul. *Modern Times: The World from the Twenties to the Eighties.* New York: Harper & Row, 1983.

Judson, Horace Freeland. *The Eighth Day of Creation: Makers of the Revolution in Biology.* New York: Simon & Schuster, 1979.

Kagan, Donald. *On the Origins of War and the Preservation of Peace.* New York: Doubleday, 1995.

Kearns, Doris. *Lyndon Johnson and the American Dream.* New York: Harper & Row, 1976.

Keldysh, Mstislav, ed. *Creative Legacy of Academician Sergei Pavlovich Korolev.* Moscow: Nauka, 1986.

Kennedy, Gregory. *Vengeance Weapon 2: The V-2 Guided Missile.* Washington, D.C.: Smithsonian Institution Press, 1983.

Kerber, L. L.; edited by Von Hardesty. *Stalin's Aviation Gulag.* Washington, D.C.: Smithsonian Institution Press, 1996.

Khrushchev, Nikita. *Khrushchev Remembers: The Last Testament.* Boston: Little, Brown, 1970.

Killian, James R. *Sputnik, Scientists and Eisenhower.* Cambridge, Mass.: MIT Press, 1977.

Kissinger, Henry A. *Nuclear Weapons and Foreign Policy.* New York: Harper & Brothers, 1957.

Kistiakowsky, George B. *A Scientist at the White House.* Cambridge, Mass.: Harvard University Press, 1976.

Klass, Philip J. *Secret Sentries in Space.* New York: Random House, 1971.

Koch, Scott A., ed. *Selected Estimates on the Soviet Union.* Washington, D.C.: Central Intelligence Agency, 1993.

Kohn, Richard H., and Joseph P. Harahan, eds. *Strategic Air Warfare.* Washington, D.C.: Office of Air Force History, 1988.

Koppes, Clayton R. *JPL and the American Space Program.* New Haven, Conn.: Yale University Press, 1982.

Lasby, Clarence G. *Project Paperclip: German Scientists and the Cold War.* New York: Atheneum, 1971.

Lay, Beirne. *Earthbound Astronauts: The Builders of Apollo-Saturn.* Englewood Cliffs, N.J.: Prentice-Hall, 1971.

Lehman, Milton. *This High Man: The Life of Robert Goddard.* New York: Farrar, Straus, 1963.

Lehmann, Ernst. *Zeppelin.* New York: Longmans, Green, 1937.

Ley, Willy. *Rockets, Missiles and Space Travel.* New York: Viking Press, 1957.

———. *Rockets, Missiles, and Men in Space.* New York: Viking Press, 1968.

Logsdon, John M. *The Decision to Go to the Moon.* Cambridge, Mass.: MIT Press, 1970.

Lovell, Sir Bernard. *The Story of Jodrell Bank.* New York: Harper & Row, 1968.

Lovell, Jim, and Jeffrey Kluger. *Lost Moon: The Perilous Voyage of Apollo 13.* New York: Houghton Mifflin, 1994.

MacKenzie, Donald. *Inventing Accuracy: A Historical Sociology of Nuclear Missile Guidance.* Cambridge, Mass.: MIT Press, 1990.

Malia, Martin. *The Soviet Tragedy: A History of Socialism in Russia, 1917–1991.* New York: Free Press, 1994.

Manchester, William. *The Arms of Krupp.* Boston: Little, Brown, 1968.

———. *The Glory and the Dream: A Narrative History of America, 1932–1972.* Boston: Little, Brown, 1974.

———. *American Caesar: Douglas MacArthur 1880–1964.* Boston: Little, Brown, 1978.

Mansfield, Harold. *Vision: The Story of Boeing.* New York: Popular Library, 1966.

McCurdy, Howard E. *The Space Station Decision.* Baltimore: Johns Hopkins University Press, 1990.

McDougall, Walter. *. . . The Heavens and the Earth: A Political History of the Space Age.* New York: Basic Books, 1985.

McPhee, John. *The Curve of Binding Energy.* New York: Farrar, Straus and Giroux, 1973.

Medaris, Major General John B. *Countdown for Decision.* New York: G. P. Putnam's Sons, 1960.

Miller, Jay. *The X-Planes, X-1 to X-29.* Marine on St. Croix, Minn.: Specialty Press, 1983.

Miller, Ron, and Fred Durant. *Worlds Beyond: The Art of Chesley Bonestell.* Norfolk, Va.: Donning Co., 1983.

Morrison, David. *Voyages to Saturn.* NASA SP-451. Washington, D.C.: U.S. Government Printing Office, 1982.

Morrison, David, and Jane Samz. *Voyage to Jupiter.* NASA SP-439. Washington, D.C.: U.S. Government Printing Office, 1980.

Mosley, Leonard. *Dulles: A Biography of Eleanor, Allen and John Foster Dulles and Their Family Network.* New York: Dial, 1978.

Neufeld, Jacob. *The Development of Ballistic Missiles in the United States Air Force 1945–1960.* Washington, D.C.: Office of Air Force History, 1990.

Neufeld, Michael. *The Rocket and the Reich: Peenemunde and the Coming of the Ballistic Missile Era.* New York: Free Press, 1995.

Oberg, James E. *Red Star in Orbit.* New York: Random House, 1981.

———. *The New Race for Space.* Harrisburg, Pa.: Stackpole, 1984.

———. *Uncovering Soviet Disasters.* New York: Random House, 1988.

Oberg, James E., and Alcestis R. Oberg. *Pioneering Space.* New York: McGraw-Hill, 1986.

Ordway, Frederick, and Mitchell Sharpe. *The Rocket Team.* New York: Thomas Y. Crowell, 1979.

Pedigree of Champions: Boeing Since 1916. Seattle: Boeing, 1985.

Pierce, J. R. *The Beginnings of Satellite Communications.* San Francisco: San Francisco Press, 1968.

Ploman, Edward W. *Space, Earth and Communication.* Westport, Conn.: Quorum Books, 1984.

Powers, Robert M. *Shuttle: The World's First Spaceship.* Harrisburg, Pa.: Stackpole, 1979.

Powers, Thomas. *The Man Who Kept the Secrets: Richard Helms & the CIA.* New York: Alfred A. Knopf, 1979.

———. *Heisenberg's War: The Secret History of the German Bomb.* Boston: Little, Brown, 1993.

Prados, John. *Presidents' Secret Wars: CIA and Pentagon Covert Operations Since World War II.* New York: Morrow, 1986.

———. *The Soviet Estimate: U.S. Intelligence Analysis and Soviet Strategic Forces.* Princeton, N.J.: Princeton University Press, 1986.

Ramo, Simon. *The Business of Science.* New York: Hill and Wang, 1988.

Ranelagh, John. *The Agency: The Rise and Decline of the CIA.* London: Weidenfeld and Nicolson, 1986.

Reeves, Robert. *The Superpower Space Race.* New York: Plenum, 1994.

Rhea, John, ed. *Roads to Space: An Oral History of the Soviet Space Program.* New York: McGraw-Hill, 1995.

Rhodes, Richard. *Dark Sun: The Making of the Hydrogen Bomb.* New York: Simon & Schuster, 1995.
Richelson, Jeffrey. *American Espionage and the Soviet Target.* New York: William Morrow, 1987.
———. *America's Secret Eyes in Space: The U.S. Keyhole Spy Satellite Program.* New York: Harper & Row, 1990.
———. *The U.S. Intelligence Community.* Boulder, Colo.: Westview Press, 1995.
Robinson, Douglas H. *Giants in the Sky.* Seattle: University of Washington Press, 1973.
Rockwell, Theodore. *The Rickover Effect.* Annapolis, Md.: Naval Institute Press, 1992.
Roland, Alex. *A Spacefaring People: Perspectives on Early Spaceflight.* NASA SP-4405. Washington, D.C.: U.S. Government Printing Office, 1985.
Romanov, Aleksandr. *Spacecraft Designer: The Story of Sergei Korolev.* Moscow: Novosti Press Agency Publishing House, 1976.
———. *Korolev.* Moscow: Molodaya Gvardiya, 1990.
Rosen, Milton. *The Viking Rocket Story.* New York: Harper, 1955.
Roth, Ladislav E., and Stephen D. Wall. *The Face of Venus: The Magellan Radar-Mapping Mission.* NASA SP-520. Washington, D.C.: NASA, 1995.
Ruffner, Kevin C., ed. *Corona: America's First Satellite Program.* Washington, D.C.: Central Intelligence Agency, 1995.
Ryan, Cornelius. *The Last Battle.* New York: Simon & Schuster, 1966.
Sagan, Carl, et al. *Murmurs of Earth: The Voyager Interstellar Record.* New York: Random House, 1978.
Sagdeev, Roald Z. *The Making of a Soviet Scientist.* New York: Wiley, 1994.
Sakharov, Andrei. *Memoirs.* New York: Alfred A. Knopf, 1990.
Sapolsky, Harvey M. *The Polaris System Development.* Cambridge, Mass.: Harvard University Press, 1972.
———. *Science and the Navy: The History of the Office of Naval Research.* Princeton, N.J.: Princeton University Press, 1990.
Scarboro, C. W. *Pictorial History of Cape Kennedy 1950–1965.* Indialantic, Fla.: South Brevard Beaches Chamber of Commerce, 1965.
Schwiebert, Ernst G. *A History of the U.S. Air Force Ballistic Missiles.* New York: Praeger, 1965.
Serling, Robert J. *Legend and Legacy: The Story of Boeing and Its People.* New York: St. Martin's Press, 1992.
Shirer, William. *The Rise and Fall of the Third Reich.* New York: Simon & Schuster, 1960.
Siddiqi, Asif. *The Soviet Human Space Program: A History, 1945–1974.* NASA SP-4409. Washington, D.C.: U.S. Government Printing Office, 1998.
Simpson, Theodore R., ed. *The Space Station: An Idea Whose Time Has Come.* New York: IEEE Press, 1985.
Sloop, John L. *Liquid Hydrogen as a Propulsion Fuel, 1945–1959.* NASA SP-4404. Washington, D.C.: U.S. Government Printing Office, 1978.
Smith, Melvyn. *Space Shuttle.* Newbury Park, Calif.: Haynes Publications, 1985.
Smith, Robert W. *The Space Telescope.* New York: Cambridge University Press, 1993.
Solzhenitsyn, Aleksandr. *The Gulag Archipelago.* New York: Harper & Row, 1973.
Spinardi, Graham. *From Polaris to Trident.* Cambridge, England: Cambridge University Press, 1994.
Spitzer, Cary R., ed. *Viking Orbiter Views of Mars.* NASA SP-441. Washington, D.C.: U.S. Government Printing Office, 1980.
Stares, Paul B. *Space Weapons and US Strategy.* London: Croom Helm, 1985.
———. *Space and National Security.* Washington, D.C.: Brookings Institution, 1987.
Steury, Donald P., ed. *Estimates on Soviet Military Power 1954 to 1984: A Selection.* Washington, D.C.: Central Intelligence Agency, 1994.
———. *Sherman Kent and the Board of National Estimates.* Washington, D.C.: Central Intelligence Agency, 1994.
Stockton, William, and John Noble Wilford. *Spaceliner.* New York: Times Books, 1981.

Sullivan, Walter. *Assault on the Unknown: The International Geophysical Year.* New York: McGraw-Hill, 1961.

Swenson, Loyd S., James M. Grimwood, and Charles C. Alexander. *This New Ocean: A History of Project Mercury.* NASA SP-4201. Washington, D.C.: U.S. Government Printing Office, 1966.

Thirty Years of Rocketdyne. Canoga Park, Calif.: Rocketdyne Division, Rockwell International Corp., 1985.

Thompson, Tina D., ed. *TRW Space Log.* Redondo Beach, Calif.: TRW Inc., vol. 27, 1992; vol. 28, 1993; vol. 29, 1994; vol. 30, 1995; vol. 31, 1996.

Toland, John. *Ships in the Sky.* New York: Henry Holt, 1957.

Tuchman, Barbara. *The Guns of August.* New York: Macmillan, 1962.

Vaeth, J. Gordon. *Weather Eyes in the Sky: America's Meteorological Satellites.* New York: Ronald Press, 1965.

Von Karman, Theodore, and Lee Edson. *The Wind and Beyond: Theodore von Karman, Pioneer in Aviation and Pathfinder in Space.* Boston: Little, Brown, 1967.

Walker, Martin. *The Cold War: A History.* New York: Henry Holt, 1993.

Wells, Helen T., Susan H. Whitely, and Carrie E. Karegeannes. *Origin of NASA Names.* NASA SP-4402. Washington, D.C.: U.S. Government Printing Office, 1976.

Whitnah, Donald R. *A History of the United States Weather Bureau.* Urbana, Ill.: University of Illinois Press, 1965.

Winter, Frank. *Prelude to the Space Age: The Rocket Societies 1924–1940.* Washington, D.C.: Smithsonian Institution Press, 1983.

Winterberg, Friedwardt. *The Physical Principles of Thermonuclear Explosive Devices.* New York: Fusion Energy Foundation, 1981.

Wolfe, Tom. *The Right Stuff.* New York: Farrar, Straus & Giroux, 1979.

Yeager, General Chuck, and Leo Janos. *Yeager: An Autobiography.* New York: Bantam, 1985.

York, Herbert F. *Making Weapons, Talking Peace.* New York: Basic Books, 1987.

Zaloga, Steven. *Target America: The Soviet Union and the Strategic Arms Race, 1945–1964.* Novato, Calif.: Presidio, 1993.

Zinn, Walter H., Frank K. Pittman, and John F. Hogerton. *Nuclear Power, U.S.A.* New York: McGraw-Hill, 1964.

Reports and Monographs

Arms, W. M. *Thor: The Workhorse of Space—A Narrative History.* MDC G3770. Huntington Beach, Calif.: McDonnell Douglas, July 31, 1972.

Baker, Michael E., Kaylene Hughes, and James D. Bowne. *Redstone Arsenal Complex Chronology.* Part II, *Nerve Center of Army Missilery, 1950–62.* Section A, "The Redstone Arsenal Era (1950–55)"; Section B, "The ABMA/AOMC Era (1956–62)." Redstone Arsenal, Ala.: Historical Division, 1994.

Development of a Strategic Missile and Associated Projects, AL-1347. Los Angeles: Aerophysics Laboratory, North American Aviation, Oct. 1951.

Grimwood, James M., and Frances Strowd. *History of the Jupiter Missile System.* Redstone Arsenal, Ala.: U.S. Army Ordnance Missile Command, 1962.

McConnell, James M. *The Soviet Shift in Emphasis from Nuclear to Conventional: The Mid-Term Perspective.* CRC-490-Vol. 2. Alexandria, Va.: Center for Naval Analyses, June 1983.

National Aeronautics and Space Administration. *Historical Origins of the George C. Marshall Space Flight Center.* Huntsville, Ala.: NASA, Dec. 1960.

National Aeronautics and Space Administration. *Missile and Space Project Information Manual.* MTP-MS-IS-61-4. Huntsville, Ala.: NASA, vol. 2, *Space Carrier Vehicle Firing Histories,* Dec. 1962; vol. 4, *Spacecraft,* Jan. 1963.

National Park Service. *Minuteman Missile Sites.* Denver: National Park Service, 1995.

Newkirk, Dennis, and Asif A. Siddiqi. *The FGB Core Module of the International Space Station Alpha: A Historical Overview of Its Lineage and Organizational Origins.* Charlottesville, Va.: Society for the History of Technology, Oct. 21, 1995.

Probing the Lunar Area. Redstone Arsenal, Ala.: Army Ballistic Missile Agency, May 15, 1959.

Special Compilations

Hall, R. Cargill, series ed. *AAS History Series,* 18 vols. American Astronautical Society. San Diego: Univelt, 1977 to 1995. In the following list, the volume number precedes the page number.

Viking (1946–1954): 7-II, p. 429.
Von Karman, Theodore: 15, p. 427.
V-2: 8, p. 215; 9, p. 121; 11, p. 39; 15, pp. 335, 377.

JPRS Report: Science & Technology. Joint Publications Research Service, U.S. Department of Commerce. Springfield, Va.: National Technical Information Service. In the following list, the publication number precedes the page number. All are in the series JPRS-USP.

Accidents and fatalities: JPRS-USP-91-001, p. 66; 91-002, pp. 74, 76; 92-002, p. 48.
Babakin, Georgi: 85-003, pp. 122, 127.
Baikonur. *See* Launch centers, Tyuratam.
Barmin, Vladimir P.: 89-005, p. 67; 93-006, p. 38.
Chelomei, Vladimir N.: 85-001, p. 81; 93-005, p. 32.
design bureau of: 89-010, pp. 45, 48.
Gagarin, Yuri: 85-001, p. 85; 89-001, p. 41; 91-002, p. 65; 91-004, p. 71. *See also* Spaceflight, manned.
Glushko, Valentin: 89-005, p. 58; 90-001, p. 54.
Keldysh, Mstislav: 86-004, p. 128; 87-003, p. 146.
Korolev, Sergei: 88-001, p. 40; 90-002, p. 56; 91-004, p. 79; 93-003, p. 35.
design bureau of: 93-002, p. 28; 94-007, pp. 11, 13.
Kuznetsov, Viktor I.: 89-006, p. 73.
Launch centers: 93-002, pp. 29, 32; 93-006, p. 35.
Plesetsk: 93-004, p. 32; 94-008, p. 43.
Tyuratam: 90-001, p. 49; 91-004, p. 85; 92-002, pp. 44, 51; 92-006, p. 24; 93-004, p. 31; 93-006, p. 37; 94-006, p. 31; 94-007, p. 16.
Launch vehicles and missiles: 87-001, p. 220; 88-001, pp. 38, 44; 90-003, p. 51; 90-005, p. 85; 93-002, p. 27; 93-003, p. 36.
Buran and Energiya: 89-005, pp. 32–48; 92-006, p. 24; 94-006, p. 21.
Proton: 92-004, p. 96; 92-006, p. 22.
Moscow Aviation Institute: 94-006, p. 30.
Mozzhorin, Yuri: 91-003, p. 29.
Nedelin's disaster. *See* Accidents and fatalities.
Pilyugin, Nikolai A.: 89-007, p. 87.
Rocket engines: 90-003, p. 63; 91-001, p. 62; 93-001, p. 69.
Spaceflight, manned: 89-001, p. 43; 89-005, p. 58; 89-010, p. 40; 90-002, p. 44; 91-002, p. 67. *See also* Gagarin, Yuri.
advanced projects: 92-003, special issue.
Mir space station: 90-004, p. 34; 91-004, p. 38; 93-002, p. 30.
Moon, flight to: 89-010, p.35; 90-002, p. 45; 90-005, p. 90; 91-002, pp. 71, 73; 91-006, special issue; 92-003, pp. 9–18; 92-004, p. 95; 94-006, p. 34.
Spaceflight, unmanned:
Luna missions: 91-004, p. 77.
Mars, missions to: 90-002, p. 44; 92-004, p. 94; 92-006, p. 32.
Reconnaissance satellites: 92-004, p. 97; 93-001, p. 70; 93-005, p. 15.
Sputniks: No. 20 (1983), pp. 140, 143; 88-001, p. 48.
Space plans and management: 89-005, p. 64; 90-003, p. 60; 92-006, pp. 22, 26.
cooperation with United States: 85-005, p. 133; 90-003, p. 64; 91-001, p. 64; 92-005, p. 36; 92-006, pp. 27, 29, 30, 33; 94-006, p. 34.
Ustinov, Dmitri: 85-003, p. 136.

Periodicals and Papers

Air & Space
Apollo: Special issue, June/July 1989; Apr./May 1994, p. 46.
Europe in Space: Apr./May 1996, p. 62.
Reaction Motors, Inc.: Dec. 1988/Jan. 1989, p. 76.

American Heritage
Consequences of World War I: July-Aug. 1992, p. 80.

American Institute of Aeronautics and Astronautics
Space shuttle: AIAA Paper 71-804, 71-805, 73-60, 73-62, 73-604, 73-606.

Astronautics & Aeronautics (after 1983: *Aerospace America*)
Europe in space: Apr. 1978, p. 12; Mar. 1980, p. 16; July 1994, p. 4.
Future plans in space: Jan. 1970, pp. 30, 62.
Global Positioning System: Aug. 1995, pp. 4, 18.
Goldin, Daniel: Nov. 1992, p. 6.
History of rockets and missiles: Oct. 1972, p. 38.
Landsat: Apr. 1971, pp. 24, 41; Sept. 1973, pp. 32–68.
Mars: Sept. 1972, p. 26.
Nerva nuclear rocket: May 1968, p. 42.
Outer planet missions: Dec. 1966, p. 26; Oct. 1968, p. 42; Sept. 1970, pp. 36–94.
Skylab: June 1971, pp. 20–77; Feb. 1974, p. 36.
Space manufacturing: Sept. 1972, p. 42.
Space shuttle: Mar. 1969, p. 24; Aug. 1969, p. 50; Jan. 1970, p. 52; Jan. 1981, p. 24.
Space station: Sept. 1994, p. 8; Jan. 1995, p. 22.
Titan IV: July 1991, p. 34.

Aviation Week & Space Technology
Air traffic control: July 31, 1995, pp. 38–53.
Commercial space processing: Special issue, June 25, 1984; Feb. 25, 1985, p. 18; Apr. 22, 1985, pp. 22, 85; Sept. 16, 1985, p. 21; Sept. 30, 1985, p. 50; Jan. 12, 1987, p. 102.
Global Positioning System: Oct. 9, 1995, pp. 50–59.
International cooperation: June 22, 1992, p. 24; May 24, 1993, p. 26; Sept. 6, 1993, p. 22; Nov 27, 1995, p. 42.
Launch vehicles: Dec. 19, 1988, p. 73; Aug. 14, 1995, p. 51; Nov. 6, 1995, p. 66; Jan. 8, 1996, p. 111.
Ariane: Nov. 20, 1995, p. 92.
Atlas: Mar. 23, 1987, p. 25; May 9, 1988, p. 17; Apr. 10, 1989, p. 24; Sept. 12, 1994, p. 55; Nov. 6, 1995, pp. 60, 65; Nov. 20, 1995, p. 35; Jan. 22, 1996, p. 59.
Long March: Feb. 6, 1995, p. 62.
Proton: Jan. 4, 1993, p. 24; Sept. 20, 1993, p. 90; Apr. 24, 1995, p. 40; June 19, 1995, p. 27.
Space shuttle: Aug. 28, 1995, p. 64.
Titan IV: Sept. 12, 1994, p. 54; Nov. 6, 1995, p. 61.
Missile programs: Dec. 1, 1958, p. 81.
NASA budget: Sept. 7, 1970, p. 18.
Ramo-Wooldridge: 1954: Oct. 11, p. 48. 1956: Feb. 20, p. 28. 1958: July 7, p. 28; Aug. 25, p. 24; Dec. 1, p. 77.
Rocket engines: Mar. 30, 1992, p. 21; Nov. 2, 1992, p. 25; Oct. 20, 1993, p. 29; Oct. 24, 1994, p. 25; Apr. 24, 1995, p. 46; Oct. 30, 1995, p. 26.
Russian space program: Jan. 15, 1996, p. 26.
Soviet cruise missile (1950s): Nov. 2, 1992, p. 50.
Space station: Aug. 23, 1993, p. 22; Nov. 8, 1993, p. 25.

Business Week
Commercial space processing: Apr. 28, 1986, p. 86.
Nuclear power: Sept. 15, 1956, p. 188; Sept. 15, 1962, p. 88.
Ramo-Wooldridge: Jan. 15, 1955, p. 66; Sept. 6, 1958, p. 64; Dec. 28, 1958, p. 110.

Collier's
Space flight: Mar. 22, 1952, p. 23; Oct. 18, 1952, p. 51; Oct. 25, 1952, p. 28; Feb. 28, 1953, p. 42; Mar. 7, 1953, p. 56; Mar. 14, 1953, p. 38; June 27, 1953, p. 33; Apr. 30, 1954, p. 21.

Discover
Space shuttle: Nov. 1985, p. 29.

Economist
Europe in space: Mar. 4, 1995, p. 79; June 1, 1996, p. 77.
Planetary missions: Feb. 24, 1996, p. 88.
Space station: Jan. 27, 1996, p. 73.

Forbes
Commercial space processing: Mar. 24, 1986, p. 176.
Proton launch vehicle: Oct. 25, 1993, p. 42.

Fortune
Apollo: Special issue, June 1962.
Bell Labs: Mar. 1953, p. 128; May 1954, p. 120; Nov. 1958, p. 148; Dec. 1958, p. 117.
Communication satellites: July 1961, p. 156; Dec. 1962, p. 188; Oct. 1965, p. 128.
General Dynamics: Feb. 1959, p. 87.
Missile programs: Dec. 1955, p. 138; May 1956, p. 101; Aug. 1957, p. 94; Oct. 1958, p. 119.
Nuclear power: Aug. 1955, p. 113; July 1963, p. 172.
Ramo-Wooldridge: Feb. 1954, p. 116; Feb. 1963, p. 95.
Space shuttle: Jan. 1972, p. 83.
Thiokol Chemical Corp.: June 1958, p. 106.
Thor missile: July 1958, p. 86.

IEEE Spectrum
Russian space program: Dec. 1995, p. 18.

International Astronautical Federation
Apollo fire: IAA-96-IAA.2.1.06.
Luna 3 spacecraft: IAA.2.2-93-677.
Rocketdyne engines: IAA-91-673.

Journal of Guidance, Control and Dynamics
Missile guidance: May–June 1981, p. 225; Sept.–Oct. 1981, p. 449; Mar.–Apr. 1982, p. 97.

Journal of Spacecraft and Rockets
Missile nose cones: Jan.–Feb. 1982, p. 3.
Polaris: Sept.–Oct. 1978, p. 265.

Journal of the Aero/Space Sciences
Missile nose cones: May 1960, p. 377.

Journal of the British Interplanetary Society
R-7 rocket: Oct. 1981, p. 437; Feb. 1982, p. 79.
Thor-Able launch vehicle: May 1984, p. 219.

National Journal
Space shuttle: Mar. 13, 1971, p. 539; Aug. 12, 1972, p. 1289; Aug. 19, 1972, p. 1323.

Newsweek
Apollo: 1961: May 22, p. 59; June 5, pp. 17, 84; June 19, p. 59; Dec. 11, p. 76. 1962: Mar. 19, p. 65. 1964: Dec. 21, p. 61. 1967: Feb. 6, p. 25; Mar. 13, p. 94; Apr. 24, p. 89. 1968: Dec. 30, p. 14. 1969: Jan. 6, p. 10; Jan. 13, p. 16; July 21, p. 68; July 28, p. 18; Aug. 4, p. 14; Aug. 11, p. 64; Aug. 25, pp. 20B, 73. 1970: Apr. 20, p. 71; Apr. 27, p. 21; May 4, p. 70.
Apollo-Soyuz: 1975: July 21, p. 46; July 28, p. 40.

Communication satellites: 1961: May 8, p. 84. 1962: June 25, p. 59; July 2, p. 50; July 23, pp. 13, 62. 1963: Feb. 18, p. 57; Feb. 25, p. 56; Aug. 12, p. 75. 1964: Mar. 16, p. 85; Aug. 31, p. 48. 1965: Apr. 19, p. 74; May 24, p. 70. 1969: June 16, p. 79.
Europe in space: Aug. 7, 1961, p. 47; Dec. 12, 1965, p. 66.
Gemini: 1962: Mar. 30, p. 68. 1965: June 11, p. 24; June 18, p. 20; Sept. 10, p. 64; Dec. 17, p. 62; Dec. 24, p. 32. 1965: Mar. 25, p. 38; June 17, p. 48; July 29, p. 28; Sept. 23, p. 98; Sept. 30, p. 114; Nov. 25, p. 71.
Mariner: 1962: July 30, p. 46; Sept. 10, p. 93; Sept. 17, p. 54; Dec. 10, p. 53; Dec. 24, p. 48. 1964: Dec. 7, p. 67. 1965: July 26, p. 54; Aug. 9, p. 58.
Mercury program: 1960: July 11, p. 55; Dec. 12, p. 57. 1961: Mar. 6, p. 66; May 8, p. 68; May 15, p. 27. 1962: Mar. 5, p. 19.
Missile programs: Feb. 10, 1958, p. 91.
Pioneer: 1958: Mar. 31, p. 67; Apr. 7, p. 50; Aug. 18, p. 57; Aug. 25, p. 56.
Ranger: 1962: Feb. 5, p. 18; May 7, p. 61; Oct. 29, p. 59. 1964: Feb. 10, p. 72; Feb. 17, p. 60; Aug. 10, p. 15; Aug. 17, p. 51. 1965: Mar. 1, p. 64; Mar. 8, p. 56.
Skylab: 1973: May 14, p. 64; May 28, p. 67; June 4, p. 60; June 11, p. 113; June 18, p. 109; July 30, p. 48; Aug. 6, p. 58; Sept. 24, p. 111; Nov. 12, p. 75. 1974: Feb. 18, p. 78.
Soviet nuclear weapons: 1961: Oct. 30, p. 54; Nov. 13, p. 64.
Space race: Jan. 2, 1961, p. 42.
Sputnik: 1957: Oct. 14, p. 37; Oct. 21, p. 29.
Test pilots: Aug. 4, 1958, p. 45.
Venus missions: 1961: Feb. 20, p. 62; Feb. 27, p. 64; Mar. 13, p. 60; Mar. 27, p. 65.
Vietnam War: Feb. 12, 1968, p. 23.
Vostok: 1961: Mar. 20, p. 68; Apr. 3, p. 55; Apr. 24, p. 22.
Weather satellites: Sept. 26, 1960, p. 77; June 26, 1961, p. 84; July 2, 1962, p. 50; July 8, 1963, p. 62; Feb. 24, 1964, p. 86; Sept. 7, 1964, p. 57; Feb. 1, 1965, p. 52.
Zond: 1968: Oct. 7, p. 66; Dec. 9, p. 59.

New Yorker
Pierce, John: Sept. 21, 1963, p. 49.

Quest (first issue titled *Liftoff*)
Adam project: Summer/Fall 1994, p. 46.
Buran: Winter 1993, p. 49.
Manned Orbiting Laboratory: Fall 1995, pp. 4, 18.
Nedelin's disaster: Winter 1994, p. 38.
Reconnaissance: Summer 1995, pp. 4, 22; Fall 1995, p. 28.
Salyut: Spring 1992, p. 17.
Solid-fueled rockets: Spring 1993, p. 26.
Soviet lunar program: Winter 1992, p. 16; Summer 1993, p. 32.
X-20: Special issue, Winter 1994.

Science
Aerobee: Dec. 31, 1948, p. 746.
Air Force space program: June 29, 1984, p. 1407; Mar. 22, 1985, p. 1445; May 9, 1986, p. 702; July 4, 1986, p. 15; Aug. 15, 1986, p. 717.
Challenger disaster: 1986: Feb. 14, pp. 661–667; Feb. 28, p. 909; Mar. 7, p. 1064; Mar. 14, p. 1237; Mar. 28, p. 1495; June 20, p. 1488; June 27, p. 1596; July 4, p. 21.
Clementine: 1994: Feb. 4, p. 622; June 17, p. 1666; Dec. 16, pp. 1835–1862; 1996: Nov. 29, p. 1495.
Commercial space processing: Sept. 30, 1983, p. 1353; Aug. 24, 1984, p. 812.
Communication satellites: Mar. 18, 1977, p. 1125; Feb. 10, 1984, p. 553.
Crystal growth in space: Feb. 14, 1975, p. 527; July 13, 1984, p. 203; July 26, 1985, p. 370; Nov. 3, 1989, pp. 580, 651; Aug. 14, 1992, p. 882; Apr. 28, 1995, p. 498; Dec. 22, 1995, p. 1921.

Earth Observing System: Sept. 27, 1991, p. 1481; Feb. 12, 1993, p. 912.
Europe in space: Mar. 9, 1973, p. 984; Nov. 9, 1973, p. 562; Sept. 10, 1982, p. 1010; June 13, 1986, p. 1340; July 25, 1986, p. 411; Sept. 25, 1987, p. 1561.
Expendable boosters: May 16, 1986, p. 822; May 15, 1987, p. 766.
Heliopause: June 11, 1993, p. 1591; Oct. 8, 1993, p. 199.
Hubble Telescope: Aug. 17, 1990, p. 735; Feb. 12, 1993, p. 887.
Jupiter and Saturn: Special issues, Jan. 25, 1974; June 1, 1979; Nov. 23, 1979; Jan. 25, 1980; Apr. 10, 1981; Jan. 29, 1982. *See also* Mar. 2, 1979, p. 892.
Landsat: June 18, 1971, p. 1222; Apr. 6, 1973, p. 49; Apr. 13, 1973, p. 171; Nov. 29, 1974, p. 808; May 2, 1975, p. 434; Apr. 29, 1977, p. 511; Feb. 20, 1981, p. 781; Mar. 26, 1982, p. 1600.
Mars: Special issues, Jan. 21, 1972; Aug. 27, 1976; Oct. 1, 1976; Dec. 17, 1976. *See also* Aug. 16, 1996, pp. 864, 924; Dec. 20, 1996, p. 2119.
Mars Observer: Sept. 3, 1993, p. 1264; Jan. 14, 1994, p. 167.
Mercury and Venus. Special issues, Mar. 29, 1974; July 12, 1974; Feb. 23, 1979; July 6, 1979.
NASA administrators: Oct. 26, 1990, p. 499; Feb. 21, 1992, p. 915; Mar. 20, 1992, p. 1506; Apr. 3, 1992, p. 20; Oct. 2, 1992, p. 20.
Planetary missions: Apr. 17, 1981, p. 317; Dec. 18, 1981, p. 1322; Nov. 12, 1982, p. 665; Nov. 20, 1992, p. 1296; May 27, 1994, p. 1244.
Russia's space program: June 12, 1992, pp. 1508, 1510; June 26, 1992, p. 1756; Jan. 20, 1995, p. 323; June 9, 1995, p. 1423.
Spacelab: Oct. 28, 1983, p. 405; special issue, July 13, 1984; Sept. 19, 1986, p. 1255.
Space shuttle: Mar. 21, 1971, p. 991; Jan. 28, 1972, p. 392; Apr. 27, 1973, p. 395; Nov. 23, 1979, p. 910; May 30, 1986, p. 1099; June 13, 1986, p. 1335.
engines: Mar. 11, 1983, p. 1195; Nov. 4, 1983, p. 489; July 20, 1984, p. 292; July 27, 1984, p. 395.
flight cost: Apr. 16, 1982, p. 278; July 2, 1982, p. 35; Nov. 9, 1984, p. 674; Jan. 25, 1991, p. 357.
Solar Max mission: Dec. 21, 1984, p. 1381.
Space station: 1982: July 23, p. 331; Sept. 10, p. 1018. 1983: Oct. 7, p. 34. 1984: Feb. 24, p. 793. 1986: May 30, p. 1089. 1987: Apr. 10, p. 143; July 17, p. 242. 1990: Oct. 19, p. 364. 1991: Mar. 22, p. 1421; Mar. 29, p. 1556; May 10, p. 774; July 19, p. 256. 1993: May 28, pp. 1228, 1230; June 25, p. 1873; July 2, p. 19; Aug. 6, p. 671; Nov. 12, p. 984. 1994: July 15, p. 311. 1995: Oct. 27, p. 571.

Scientific American
Challenger disaster: Aug. 1986, p. 62.
Communication satellites: Feb. 1977, p. 58.
European strategic arms: Aug. 1986, p. 33.
Expendable launch vehicles: July 1986, p. 58; Mar. 1990, p. 34; July 1990, p. 72.
Global Positioning System: Feb. 1996, p. 44.
Hydrogen bomb: Oct. 1975, p. 106.
Jupiter and Saturn: Sept. 1975, p. 118; Jan. 1980, p. 90; Nov. 1981, p. 104; Jan. 1982, p. 100; Feb. 1982, p. 98; Dec. 1983, p. 56.
Landsat: Jan. 1968, p. 54.
Mars: May 1970, p. 26; Jan. 1973, p. 48; Sept. 1975, p. 106; Jan. 1976, p. 32; Feb. 1977, p. 30; July 1977, p. 34; Nov. 1977, p. 52; Mar. 1978, p. 76; May 1986, p. 54.
Mercury and Venus: Sept. 1975, pp. 58, 70; Aug. 1980, p. 54; July 1981, p. 66; Mar. 1985, p. 46; Feb. 1988, p. 90; Dec. 1990, p. 60.
Nuclear power: Feb. 1968, p. 21.
Planetary missions: June 1976, p. 58; Nov. 1986, p. 36; Dec. 1995, p. 44.
Reconnaissance satellites: Jan. 1991, p. 38; Mar. 1996, p. 78.
Soviet space program: Feb. 1989, p. 32; June 1994, p. 36.
Space station: Jan. 1986, p. 32.

Weather satellites: July 1961, p. 80; Jan. 1969, p. 52.

Spaceflight
CIA estimates: July 1993, p. 224; Aug. 1995, p. 256.
Glushko, Valentin: Mar. 1991, p. 88.
Korolev, Sergei: Apr. 1978, p. 144.
Luna spacecraft: Oct. 1992, p. 318.
R-7 rocket: Nov.–Dec. 1980, p. 340; Apr. 1981, p. 109; Aug. 1995, p. 260.
Soviet launch failures: Nov. 1995, p. 393.
Soviet organization, missiles and space: Aug. 1994, p. 283; Sept. 1994, p. 317.
Thor launch vehicles: June 1970, p. 240.

Space Frontiers
Pioneer spacecraft: Jan.–Feb. 1988, p. 3.

Threshold (Rocketdyne)
Rocket engines: Summer 1988, p. 34; Spring 1989, pp. 3, 35; Summer 1991, p. 14; Spring 1992, p. 34; Summer 1993, p. 18.
history of: Dec. 1987, p. 17; Summer 1991, p. 52; Spring 1992, p. 20; Summer 1993, p. 40.
nuclear: Fall 1992, p. 2.

Time
Apollo: 1961: June 2, p. 11; Dec. 8, p. 88. 1962: Aug. 10, p. 52. 1967: Feb. 3, p. 13; Feb. 10, p. 18; Apr. 14, p. 86; Nov. 17, p. 84. 1968: Nov. 22, p. 62; Dec. 6, p. 90; Dec. 27, p. 8. 1969: Jan. 3, p. 9; Jan. 10, p. 40; Jan. 17, p. 45; July 18, p. 18; July 25, p. 10; Aug. 1, pp. 11, 18; Aug. 8, p. 20; Aug. 22, p. 10. 1970: Apr. 20, p. 52; Apr. 27, p. 12; May 4, p. 82.
Apollo-Soyuz: 1975: July 21, p. 53; July 28, p. 34.
Atlas: 1957: Dec. 30, p. 12. 1958: Jan. 20, p. 76; Dec. 8, p. 15; Dec. 29, pp. 7, 31.
Cape Canaveral: Mar. 17, 1958, p. 22.
Communication satellites: 1960: Aug. 22, pp. 11, 44; Oct. 17, p. 47. 1963: Aug. 2, p. 49; Aug. 9, p. 70. 1965: Apr. 9, p. 92; Apr. 16, p. 64; May 14, p. 84.
Europe in space: Apr. 25, 1960, p. 26; Jan. 8, 1965, p. 25; Mar. 14, 1969, p. 42.
Gagarin, Yuri. *See* Vostok spacecraft.
Gemini: 1962: Mar. 30, p. 68. 1965: June 11, p. 24; June 18, p. 20; Sept. 10, p. 64; Dec. 17, p. 62; Dec. 24, p. 32. 1966: Mar. 25, p. 38; June 17, p. 48; July 29, p. 28; Sept. 23, p. 98; Sept. 30, p. 114; Nov. 25, p. 71.
Luna program: 1959: Jan. 12, pp. 19, 64; Jan. 19, pp. 20, 54; Sept. 28, p. 68; Oct. 12, pp. 22, 67; Nov. 9, p. 71.
Mariner: 1962: Sept. 7, p. 59; Dec. 21, p. 47. 1963: Jan. 4, p. 36; Mar. 8, p. 76. 1964: Dec. 11, p. 76. 1965: July 23, p. 36; Aug. 6, p. 58.
Mercury program: Apr. 20, 1959, p. 17; Feb. 10, 1961, p. 58; May 12, 1961, pp. 10, 52; Mar. 2, 1962, p. 11; Mar. 9, 1962, p. 22.
Minuteman: Mar. 10, 1958, p. 16; Jan. 25, 1960, p. 48.
Mir space station: July 10, 1995, p. 54.
Missile gap: 1959: Feb. 9, p. 12; Feb. 23, p. 22. 1960: Jan. 18, p. 12; Feb. 1, p. 9; Feb. 8, pp. 16, 18; Apr. 18, p. 22; May 9, p. 17. 1961: Feb. 10, p. 16; Feb. 17, p. 12.
Missile guidance: Apr. 29, 1957, p. 60.
Missile nose cones: June 20, 1960, p. 70.
Missile programs: May 21, 1951, p. 84; Jan. 30, 1956, p. 53; Jan. 20, 1958, p. 18.
North American Aviation: June 29, 1953, p. 82.
Pioneer program: 1958: Oct. 13, p. 61; Oct. 20, p. 17; Dec. 15, p. 41. 1959: Mar. 16, p. 60. 1960: Mar. 21, pp. 13, 71; Mar. 28, p. 44; July 18, p. 52.
Polaris: 1958: Mar. 3, p. 14. 1960: Jan. 11, p. 13; Aug. 1, p. 15; Nov. 28, p. 21.

Ramo-Wooldridge: Apr. 29, 1957, p. 84.
Ranger: 1962: Feb. 9, p. 67. 1964: Feb. 7, p. 58; Feb. 14, p. 70; Aug. 7, pp. 17, 39. 1965: Feb. 26, p. 58; Apr. 2, p. 48.
Reconnaissance: 1958: Dec. 15, pp. 15, 41. 1959: Apr. 27, pp. 16, 65; May 4, p. 15. 1960: Feb. 8, p. 48; Mar. 7, p. 80; June 6, p. 56; Aug. 22, p. 45; Aug. 29, p. 41. 1961: Feb. 10, p. 59.
Saturn rocket: Aug. 11, 1958, p. 38; Nov. 2, 1959, p. 10.
Schriever, Gen. Bernard: Apr. 1, 1957, p. 12.
Skylab: 1973: Apr. 23, p. 58; May 28, p. 94; June 4, p. 53; June 11, p. 67; June 18, p. 57; June 25, p. 59; July 2, p. 39; July 9, p. 44; July 30, p. 43; Aug. 20, p. 50; Oct. 8, p. 61. 1974: Feb. 11, p. 74; Feb. 25, p. 63.
Sputniks: 1957: Oct. 14, pp. 27, 46; Oct. 21, pp. 21, 50; Oct. 28, pp. 17, 77; Nov. 18, pp. 19, 91.
Test pilots: Apr. 18, 1949, p. 64; Apr. 27, 1953, p. 68.
Transit program: 1960: Apr. 25, p. 53; July 4, p. 52.
Van Allen, James: May 4, 1959, p. 64.
Vanguard program: Dec. 16, 1957, pp. 9, 12.
Vietnam War: Feb. 9, 1968, pp. 16, 22.
Von Braun, Wernher: Feb. 17, 1958, p. 13.
Vostok spacecraft: 1960: May 30, p. 56; Aug. 29, pp. 11, 40; Dec. 12, p. 24. 1961: Mar. 31, p. 52; Apr. 21, pp. 15, 46.
Weather satellites: 1960: Apr. 11, p. 62; Aug. 1, p. 56; Dec. 5, p. 75.
Zond: 1968: Sept. 27, p. 64; Oct. 4, p. 73.

Index